Engineering Materials

Volume 2

Engineering Materials

Volume 2

Second Edition

R. L. Timings

Pearson Education Limited
Edinburgh Gate, Harlow
Essex CM20 2JE, England
and Associated Companies throughout the world

First published 1991
Second edition 2000

British Library Cataloguing in Publication Data
A catalogue entry for this title is available from the British Library

ISBN 0-582-40466-5

Set by 32 in Times 9½/12 and Frutiger
Printed in Singapore

Contents

Preface

Engineering Materials, Volume 2, was originally written not only to satisfy the requirements of the Business and Technician Education Council (BTEC) standard units for Engineering Materials at levels three and four, but also to satisfy the main requirements of those students following 'college devised' syllabuses. At the design stage of this text, most of the colleges and new universities offering HNC/HND courses incorporating engineering materials were sent a questionnaire requesting information concerning the topic areas and treatment they would like to see incorporated. The response was a record for questionnaires and the author was swamped with information and spoilt for choice. However, after analysing the returns and asking supplementary questions of the respondents, a specification was drawn up and became the synopsis for the first edition of this book.

The time has now come to bring out a new edition of this book and the author has been assisted by Dr Kemal Ahmet, Dr Rob Hearing, and Mr Carl Westwood of the University of Luton in correcting, expanding and updating the original text. At the same time the book has been completely restructured. It is now in three closely related parts, namely:

- **Part A** — Metallic materials
- **Part B** — Non-metallic materials
- **Part C** — Materials in service

There are two further changes: firstly, the material on bearings and bearing materials has been transferred to this book from Volume 1 and, at the same time, enlarged and updated; secondly, a short, additional chapter on polymeric materials has been added to develop some of the topics introduced in Volume 1 and to provide a lead into the expanded chapter on synthetic adhesives.

Whilst still satisfying the original aims, this revised edition not only reflects changes in technology but, at the same time, the opportunity has been taken to:

- Make the language less formal and more reader friendly.
- Expand some of the explanations to improve their clarity.
- Correct some residual errors and omissions.
- Modernise the general format of the book.
- Introduce self-assessment exercises at key points in the text.
- Make more use of bulleted summaries.

Each chapter is now prefaced by a summary of the topic areas to be covered and finishes with a selection of practice exercises.

This revised text follows on naturally from the second edition of *Engineering Materials*, Volume 1. Students who have made a 'direct entry' onto their HNC/HND course will need to read Volume 1 in conjunction with Volume 2 for a full understanding of the materials and processes described.

The broad coverage of *Engineering Materials*, Volume 2, not only satisfies the requirements of Higher Technician Engineering Students with reference to the latest EDEXEL modules, but also provides essential background reading for undergraduates studying for a degree in an engineering discipline or in materials science.

R. L. Timings
2000

Acknowledgements

The author and publishers are grateful to the following for permission to reproduce copyright material.

Inco Europe Ltd for our Tables 1.3, 1.4 & 1.5; University of Luton for our Figs 4.2, 4.6, 5.15, 8.7, 11.17, 11.18, 12.8 & 12.12; Arnold Ltd for our Figs 5.4, 8.2, 8.4, 8.5, 8.11 & 8.12; Pearson Education Ltd for our Figs 5.7, 6.3, 6.4 & 10.21; Loctite Corporation for our Figs 7.1, 7.3, 7.9, 7.10, 7.11, 7.12, 7.13, 7.17 & 7.18 and Tables 7.3, 7.4, 7.5, 7.6, 7.7, 7.8, 7.10 & 7.11; Butterworth-Heinemann Ltd for our Fig. 7.19; National Design Council for our Fig. 7.20; Glacier Vandervell Ltd for our Figs 9.21, 9.22 & 9.23; Wacker Chemitronic GmbH for our Figs 12.1, 12.2, 12.3, 12.4, 12.5, 12.6, 12.7, 12.9, 12.10 & our Tables 12.1, 12.2 & 12.3;

The author would also like to thank the following individuals and organisations:

- Dr Kemal Ahmet, Dr Rob Hearing and Mr Carl Westwood of the University of Luton for all their assistance in the production of this new edition.
- Loctite Corporation for their assistance in the writing of Chapter 7 – sections 7.2, 7.6, 7.7, 7.8, 7.9, 7.10, 7.12, 7.13 and 7.16 are derived, wholly or in part, from the *Loctite World Wide Design Handbook*; the publishers are grateful to Loctite for permission to adapt and reproduce their copyright material.
- Mr Brian Campbell of Glacier Vandervell Ltd for his assistance in updating the sections of Chapter 9 concerned with plain bearing materials; the publishers are grateful to Glacier Vandervell Ltd for permission to adapt and reproduce copyright material from their data sheets and other published material.
- SKF (UK) Ltd for their assistance in updating the sections of Chapter 9 concerned with rolling bearing materials.
- Wacker Chemitronic GmbH (Burghausen, Germany) for their assistance in the writing of Chapter 12 on semiconductor materials and manufacturing processes.

How to use this book

There are many ways of using a textbook. You can read through it from cover to cover and try to remember all that you have read, but this is rarely successful unless you have a photographic memory, and few of us have. You may consider using it just as a reference book by looking up an individual topic area as and when you need that specific information. This can be useful as a reminder once you know the subject thoroughly but, until then, it can lead to misconceived ideas. So, here are a few thoughts on how to maximise the benefit that you can get from this book.

This book is divided into three main parts:

- Part A is concerned with the *Metallic materials* (Chapters 1–4).
- Part B is concerned with the *Non-metallic materials* (Chapters 5–7).
- Part C is concerned with *Materials in service* (Chapters 8–12).

Although interrelated, it is not essential that Parts A, B, and C, are read sequentially. However, the chapters within each part should be read sequentially. It is assumed that the reader will already be familiar with *Engineering Materials*, Volume 1, as many topic areas build on the foundations laid in that book.

Each part consists of a number of chapters. Each chapter covers a major syllabus area. For example, Chapter 1 covers 'Alloy steels'. It is divided up into **sections**. If you turn to the list of contents, you will find that:

- Section 1.1 deals with the need for alloying.
- Section 1.2 deals with alloying elements.
- Section 1.3 deals with the effects of alloying elements, and so on.

Sometimes it is necessary to divide these sections up further. For example, Section 1.2 subdivides into sections that deal with:

- aluminium as an alloying element
- chromium as an alloying element
- cobalt as an alloying element, and so on

As previously stated it is not essential to read each part of this book sequentially but, to obtain the maximum benefit from this book it may seem obvious but start with the first chapter of the part you wish to study and then work sequentially through that part. If you wish to work in some other order and be selective in your reading, then this should only be done in consultation with your tutor. Each chapter is prefaced with a list of the main topic areas that you are going to find in it.

As you work through the chapters, you will come across **self-assessment tasks** at key points. If you have understood what you have read previously, you should be able to complete these exercises *without looking back* at the text. If you have to look back, then you are not yet sure of your ground and there is no point in moving on. Try again and, if you are still not sure, have a chat with your tutor to clear up the difficulty.

At the end of all the chapters there are a selection of more extended exercises. Your tutor will guide you as to their use and check your responses to ensure that you have the background knowledge required for your understanding of the next topic area to be studied.

No matter whether you work through the book from beginning to end sequentially, or you choose to work only through a specialist area you require, you must complete the self-assessment tasks as you come to them, and also the end of chapter exercises, before moving. This will ensure that you will understand the next stage of your journey through your study of engineering materials.

Finally, once you have qualified, you can still obtain benefit from this book. The numbering of the sections, the comprehensive list of contents and the extended index, all assist you in using this text and its companion volumes (*Engineering Materials*, Volume 1, *Manufacturing Technology*, Volumes 1 and 2) as quick reference books in your future career in engineering.

Part A
Metallic materials

1 Alloy steels

The topic areas covered in this chapter are:

- The need for alloying.
- Alloying elements.
- The effects of alloying elements.
- The classification of steels.
- Structural steels.
- Corrosion resistant steels.
- Heat-resistant steels.
- Maraging steels.
- Tool and die steels.

1.1 The need for alloying

The limitations of plain carbon steels can be summarised as follows:

- A high critical cooling rate that leads to cracking when quench hardening.
- Poor hardenability and a corresponding low value of ruling section.
- Compared with alloy steels, carbon steels can only attain relatively low values of tensile strength even after quench hardening and tempering unless such properties as ductility and toughness are reduced in value to unacceptable levels.

Alloying elements are added to plain carbon steels to overcome these limitations and, in some instances, to improve the corrosion and heat resistance as well. However, alloy steels are more expensive and more difficult to process than plain carbon steels and should only be used where their special properties can be fully exploited.

1.2 Alloying elements

Steels containing iron and carbon with traces of phosphorus, silicon, and not more than 1.5 per cent manganese are referred to as *plain carbon steels*. The composition, properties and simple heat treatment of such steels were fully discussed in *Engineering Materials*, Volume 1.

Alloy steels are carbon steels, normally containing less than 1.0 per cent carbon, to which other metals and some non-metals (alloying elements) have been added in sufficient quantities to alter the properties of the steels to a significant extent. Let's now consider the effects of the various alloying elements commonly used in alloy steels.

Aluminium

The presence of up to 1 per cent aluminium in alloy steels enables them to be given a hard, wear-resistant skin by the heat-treatment process of *nitriding*.

Chromium

The presence of small amounts of chromium stabilises the formation of hard carbides and improves the susceptibility of steels to heat treatment. Unfortunately the presence of chromium also promotes grain growth, therefore chromium is rarely used as an alloying element on its own (Section 1.3). The presence of large amounts of chromium improves the corrosion resistance and heat resistance of steels (stainless steels).

Cobalt

The presence of cobalt induces sluggishness into the heat-treatment transformations and improves the ability of tool steels to operate at high temperatures without softening. It is an important alloying element in super high-speed steels.

Copper

The presence of up to 0.5 per cent copper improves the corrosion resistance of alloy steels.

Lead

The presence of up to 0.2 per cent lead improves the machinability of steels but, unfortunately, it also reduces the strength of the steel to which it is added.

Manganese

This element is always present in plain carbon steels, alloy steel and cast irons, as it combines with residual sulphur from the smelting process and reduces the brittleness caused by the presence of *iron sulphide*. It also stabilises the γ-phase (austenite) and helps to promote the formation of stable carbides. In large quantities (up to 12.5 per cent), manganese improves the wear resistance of steels by causing them to form a hard skin spontaneously when subjected to abrasion.

Molybdenum

The presence of molybdenum in alloy steels raises their strength and creep resistance at high temperatures. It also stabilises their carbides and improves the ability of cutting tools to retain their hardness at high temperatures. The presence of molybdenum reduces the susceptibility of nickel–chromium steels to *temper brittleness* and *weld decay*. Molybdenum is also present in chromium steel alloys to reduce grain growth (*see* Chromium).

Nickel

The presence of nickel in alloy steels results in increased strength by grain refinement. It also improves the corrosion resistance of steels. Unfortunately nickel is a powerful *graphitiser* and, by reducing the stability of any carbides present, nickel tends to reduce the hardenability of any steel in which it is present. Nickel and chromium are often used together in alloy steels where they complement each other's properties (Section 1.3).

Phosphorus

This is a residual element from the smelting process. It causes weakness in the steel and is considered as an undesirable impurity. Normally considerable care is taken to keep its presence below 0.05 per cent. However, where maximum strength and toughness are not required, increased quantities of phosphorus can improve machinability and also the fluidity of casting steels.

Silicon

The presence of up to 0.3 per cent silicon improves the fluidity of casting steels without causing the deterioration in mechanical properties associated with phosphorus. Up to 1 per cent silicon improves the heat resistance of steels. Unfortunately silicon, like nickel, is a powerful graphitiser and is never added in large amounts to high-carbon steels.

Sulphur

This, like phosphorus, is also a residual element carried over from the smelting process. It is also considered an undesirable impurity since the presence of iron sulphide reduces the strength and toughness of steels. Fortunately sulphur has a greater affinity for manganese than it has for iron, and the presence of manganese sulphide does not impair the mechanical properties of steels. However, sulphur is sometimes alloyed with low-carbon steels to improve their machinability where a reduction in component strength can be tolerated.

Titanium

Up to 1.0 per cent titanium in stainless and maraging steels helps to reduce weld decay and temper brittleness. Niobium has the same effect, and steels containing either of these alloying elements are said to have been 'proofed'.

Tungsten

The presence of tungsten in alloy steels promotes the formation of very hard carbides. It also induces sluggishness into the heat-treatment transformations. This enables steels containing tungsten to retain their hardness at high temperatures. Tungsten is mainly found in high-speed steels that are used for cutting tools and in high-duty die steels that have to operate at high temperatures.

Vanadium

Vanadium is not used on its own in alloy steels, but is used to enhance the benefits of the other alloying elements. The effects of this element in alloy steels are many and various.

- It promotes the formation of carbides.
- It stabilises martensite and thus improves hardenability.
- It reduces grain growth.
- It enhances the 'hot hardness' of tool steels and die steels.
- It improves the fatigue resistance of steels.

1.2.1 *Classifications*

It has been mentioned above that some alloying elements, such as chromium, promote the formation of carbides whilst others, such as nickel, promote the formation of free graphite. Therefore, alloying elements can be classified as:

- Carbide promoters
- Graphitisers
- Austenite stabilisers
- Stabilisers.

Carbide promoters

Some alloying elements form very stable carbides that are harder than iron carbide. The formation of such carbides increases the overall hardness of the steel and makes it suitable for tooling purposes. The carbide-promoting elements are: chromium, manganese, niobium, molybdenum, titanium, tungsten and vanadium.

Graphitisers

Not all the alloying elements tend to combine with carbon when in the presence of iron. Far from promoting the formation of carbides, such alloying elements as nickel, aluminium and silicon cause instability in any carbide present so that carbon may be precipitated out as free graphite. If any of these elements are required in appreciable amounts, then carbide-promoting alloying elements must also be present or the carbon content of the steel must be kept very low. For this reason it is not possible to have a high-carbon, high-nickel alloy steel.

Austenite stabilisers

You were introduced to austenite in *Engineering Materials*, Volume 1. It is normally only present as a solid solution of carbon in iron at high temperatures when the metal is molten. However, by adding suitable alloying elements, austenite can be present as a solid solution at room temperature. Reference to Fig. 1.1 shows that some alloying elements such as cobalt, copper, nickel and manganese raise the A_4 temperature whilst at the same time depressing the A_3 temperature. Therefore, when these elements are added to carbon steels they stabilise the γ-phase (austenite) and increase the temperature over which this phase remains stable, because most of these alloying elements do not tend to form carbides and the carbon remains in solid solution in the austenite. When the amounts of the alloying elements present stabilise the austenite to the extent that it is present at room temperature (e.g. austenitic stainless steel), the steel becomes ductile. It also *loses its ferromagnetic properties*.

Ferrite stabilisers

In *Engineering Materials*, Volume 1, you were also introduced to ferrite as being a very weak solution of carbon in iron existing at room temperature. Alloying elements such as aluminium, chromium, molybdenum, tungsten, silicon and vanadium have the opposite effect to those described previously and stabilise the ferrite. This is achieved by raising the A_3 temperature and depressing the A_4 temperature to form what is referred to as the

'γ-loop' (Section 1.3). Since these alloying elements have body-centred-cubic (BCC) crystals at room temperature they tend to stabilise the ferrite which also has BCC crystals. We will consider all these effects more fully in the next section of this chapter.

SELF-ASSESSMENT TASK 1.1

1. For revision purposes refer to *Engineering Materials*, Volume 1, and define the following terms:
   (a) ferrite
   (b) austenite
   (c) cementite
   (d) pearlite
2. Explain briefly what is meant by the following terms:
   (a) carbide former
   (b) graphitiser
   (c) austenite stabiliser
   (d) ferrite stabiliser
3. Explain why manganese is such an important alloying element in both plain carbon and alloy steels.

1.3 The effects of alloying elements

Alloying elements can influence the properties of steels *directly* or *indirectly*. For example, nickel is stronger than iron so its presence increases the strength of the steel *directly*. Its presence also refines the grain of the steel and this further increases the strength of the steel. Other alloying elements, such as chromium, influence the properties of steels *indirectly* by making them more susceptible to heat treatment. Therefore, alloying elements have two main effects:

- They can form tough but ductile solid solutions with the parent metal.
- They can form compounds with the parent metal (and with each other) promoting brittleness and hardness.

Fortunately most of the elements used in alloy steels form *substitutional solid solutions* to varying extents. The formation of solid solutions was discussed in *Engineering Materials*, Volume 1. The effect of such solid solutions is to increase the tensile strength, the impact strength and the ductility of the alloy formed. In the case of low-carbon steels, the α-phase (ferrite) is strengthened in this way without any corresponding loss in ductility. This increase in strength and toughness without loss of ductility is one of the more important effects associated with alloy steels.

Alloying elements can also change the transformation temperatures of the iron–carbon phase equilibrium diagram. Figure 1.1 shows how elements such as cobalt, copper, nickel and manganese can raise the A_4 temperature and lower the A_3 temperature compared with the corresponding temperatures for plain carbon steels. Most of the alloying elements that promote this effect have face-centred-cubic (FCC) crystal lattices at room temperature and, because the austenite in which these elements are substitutionally dissolved also has an FCC lattice, the alloying elements will oppose the transformation from the FCC γ-phase to the BCC α-phase. Thus these alloying elements stabilise the austenite and increase the temperature range over which it can exist.

Fig. 1.1 *Effect of alloying elements on ferrous metals*

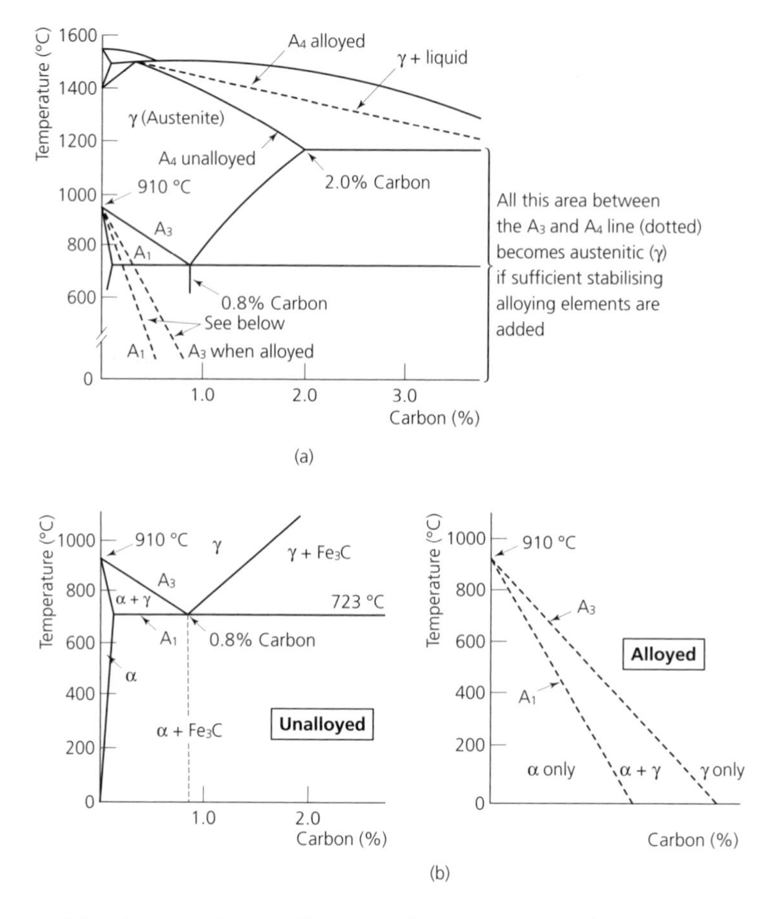

Since most of the elements that stabilise austenite do not react with the carbon present to form carbides, the carbon tends to remain in *solid solution* in the austenite. This further depresses the A_3 temperature. If the alloying elements are present in sufficient quantity, the γ-phase austenite can be stabilised right down to room temperature (e.g. austenitic stainless steel) and the alloy ceases to possess any ferromagnetic properties.

Figure 1.2 shows how other alloying elements such as aluminium, chromium, molybdenum, silicon, tungsten and vanadium can have the opposite effect. That is, they raise the A_3 temperature and lower the A_4 temperature to form what is described as the γ-loop. Since this group of alloying elements have BCC crystal lattices at room temperature, as does ferrite, this common lattice structure, coupled with the raising of the A_3 temperature, has the effect of stabilising and promoting the formation of the α-phase (ferrite). Chromium, molybdenum and tungsten, in particular, form stable carbides. The precipitation of carbon as metallic carbides still further promotes the transformation from austenite to ferrite.

Fig. 1.2 *Effect on ferrous metals of alloying elements that stabilise ferrite*

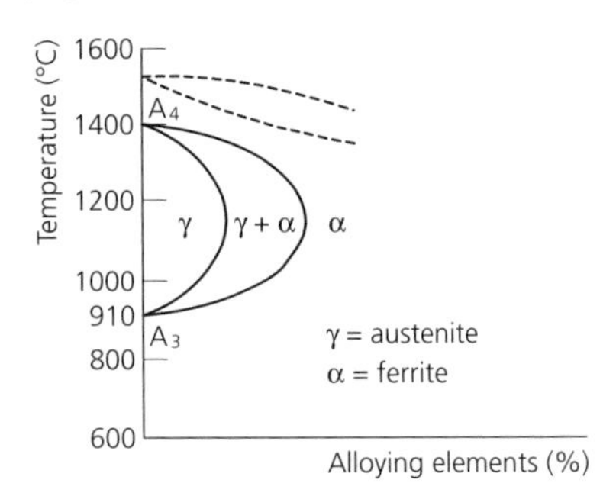

The hardness of alloy steels depends (as in plain carbon steels) on the formation of hard metallic carbides. It has been stated above that the alloying elements chromium, molybdenum and tungsten form very stable carbides that are harder than iron carbide; hence, these alloying elements are widely found in tool and die steels.

Another important effect of the carbide-stabilising alloying elements, and in particular tungsten, is the slowing down of the transformation rates; that is, the alloy does not have to be cooled as quickly as plain carbon steel to produce a given structure. This slowing down of the critical quenching speed enables alloy tool and die steels to be hardened by oil quenching, or even by air-blast quenching, with a corresponding reduction in the possibility of cracking and distortion. This induced sluggishness in the transformation rates also increases the temperature at which the steel may be used without loss of hardness. Steels that can be used at high operating temperatures without their 'temper being drawn' are referred to as having good 'hot-hardness' or good 'red-hardness'.

Although chromium improves the susceptibility of steel to heat treatment it has one major disadvantage, it promotes grain growth. Therefore, it is most important to heat-treat chrome steels at the lowest possible temperature and heat the steel for the minimum possible time if grain growth, and the brittleness associated with grain growth, is to be avoided. This is also a problem when attempting to weld such steels.

Fortunately, nickel and chromium tend to be complementary to each other when used as alloying elements. Nickel promotes fine grain but tends to unstabilise (graphitise) the carbides, whilst chromium promotes stable carbides but tends to cause grain growth. By careful control of the amount of nickel and chromium present in the alloy, it is possible to produce steels that have both a fine grain and stable carbides.

Generally, alloy steels contain less carbon than plain carbon steels. It has already been stated that alloy steels rarely contain more than 1 per cent carbon, and the reason for this is that the addition of any significant amount of an alloying element to a plain carbon steel reduces the *eutectoid composition* below the normal 0.83 per cent carbon. This effect is shown in Fig. 1.3. In this example, the addition of 2.5 per cent manganese reduces the eutectoid composition, at which the steel is wholly pearlitic, from 0.83 to 0.65 per cent carbon. This is also accompanied by a reduction in the A_1 and A_3 temperatures.

Fig. 1.3 *Displacement of the eutectoid composition*

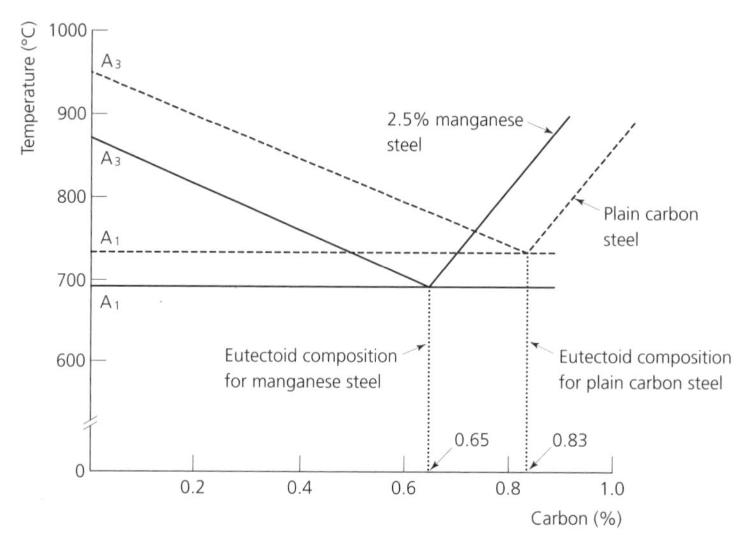

Finally, the addition of alloying elements to plain carbon steels can improve their corrosion resistance. Although not a particularly reactive metal, iron corrodes (rusts) readily in the presence of oxygen and moisture because the hydroxide film that forms on its surface is porous and offers no protection. Once corrosion commences, it tends to become spontaneous, the iron hydroxide (rust) coating reacting with the iron in the steel beneath it. Alloying elements such as aluminium, silicon, copper and chromium cause the formation of corrosion-resistant homogeneous oxide films to form on the surface of the steel if they are added in sufficient quantity.

SELF-ASSESSMENT TASK 1.2

1. Describe the difference between an alloying element that *directly* affects the properties of an alloy and an alloying element that *indirectly* affects an alloy. Give an example of each.
2. Describe the essential differences between:
   (a) interstial solid solutions
   (b) substitutional solid solutions
   (c) intermetallic compounds
3. Discuss the design decisions that have to be made when choosing between a plain carbon steel and an alloy steel in terms of:
   (a) material cost
   (b) processing costs
   (c) availability

1.4 The classification of steels

The most convenient way to classify the wide range of alloy steels available to the engineer is to group them according to application, and then form sub-groups according to their principal alloying elements. The main groups are:

- *Structural steels* These must not be confused with steels used for structural steelwork, but refers to those steels where strength is of paramount importance rather than, say, corrosion resistance or heat resistance.
- *Corrosion-resistant steels* These include the 'stainless' steels together with less expensive low-alloy steels.
- *Heat-resistant steels* These steels are used for such applications as the valves in automobile engines and for components for gas turbines and jet engines where low 'creep' characteristics are required.
- *Tool and die steels* These are hard and wear-resistant steels which retain their hardness at high operating temperatures and resist cracking and distortion during heat treatment.
- *Ferromagnetic steels* Most steels show some ferromagnetic properties, but this group of steels have been specially developed to exploit these properties and are discussed in detail in Sections 7.11–7.13 inclusive.

1.5 Structural steels

Having classified the wide range of alloy steels available to us, let's now look at each category in more detail.

1.5.1 *Manganese steels*

All steels contain small amounts of manganese to deoxidise the molten steel during manufacture. A small excess of up to 0.35 per cent beyond the amount necessary to ensure proper deoxidation is generally provided in order to combine with any residual sulphur present and prevent the formation of iron sulphide. Manganese also improves the rolling and forging qualities of steels. However, the true *manganese steels* contain larger amounts of the alloying element and fall into two groups:

1. Alloy steels containing from 11–14 per cent manganese.
2. Alloy steels containing from 1–2 per cent manganese.

The steels associated with group 1 also contain from 1.0–1.3 per cent carbon and, before heat treatment, are very hard, lacking in ductility. These steels have to be heat treated by quenching from 1050 °C, after which they become tough with a reasonable ductility and are sufficiently soft to be machinable. Steels in this group are rapidly hardened by cold working and form a very hard skin when subjected to abrasion. Therefore they have to be machined with very sharp cutting tools since, if any rubbing occurs during machining, the hard skin which forms can only be removed by grinding before machining can recommence. This resistance to abrasion makes the steel most suitable for such applications as railway points, dredger buckets and stone-crusher jaws.

The steels associated with group 2 contain only 0.25–0.55 per cent carbon and, although they exhibit similar properties to group 1, they do so to a much lesser extent. However, they are much less costly and low-carbon, low-manganese alloy steels are now widely used for automobile components in place of the more costly nickel–chromium steels. For example, a steel containing 0.2–0.45 per cent carbon, 1.2 per cent manganese, and 0.8 per cent silicon is sufficiently ductile and machinable for structural purposes, yet it will have a tensile strength of 560–910 MPa depending upon the heat-treatment to which it has been subjected.

1.5.2 *Nickel steels*

Plain nickel steels are widely used and fall into four main groups:

- Structural steels containing up to 6 per cent nickel.
- Corrosion-resistant steels containing up to 20–30 per cent nickel.
- Low-expansion steels containing 30–40 per cent nickel.
- High-permeability magnetic steels containing 50 per cent nickel or more.

In this section we will consider only the 'structural' alloys. The remaining alloys will be considered in later sections as and when appropriate.

Structural steels containing up to 6 per cent nickel are used for manufacturing components for machines and structures that are highly stressed. Typical alloys contain 0.1–0.55 per cent carbon, 0.3–0.8 per cent manganese and 0.4–6.0 per cent nickel. The steels in this group, which contain low percentages of carbon, are used for components that need to be case hardened. Since case hardening requires the steel to be heated into the α-phase for prolonged periods, grain growth is usually excessive. However, the presence of nickel reduces this grain growth and, after heat treatment, promotes a very tough, fine-grained core which greatly enhances the mechanical properties of the component.

As stated earlier in this chapter, nickel tends to promote graphitisation of the carbides in the steel and the manganese content has to be increased beyond that normally required for deoxidation in order to counteract this effect. Since nickel helps to prevent excessive grain growth at high temperatures, it enables fine-grain steels to be produced more easily.

Nickel also lowers the critical temperatures slightly and thus makes heat treatment less severe, reducing the chance of cracking and distortion. Figure 1.4 compares the properties of a 1 per cent nickel steel with a 3.5 per cent nickel steel after quenching in oil from 850 °C and tempering at various temperatures up to 700 °C.

1.5.3 *Nickel–chromium steels*

These are probably the most widely used of all alloy steels. The compositions most widely used are: 0.1–0.55 per cent carbon, 1.0–4.75 per cent nickel, 0.45–1.75 per cent chromium and 0.3–0.8 per cent manganese. The low-carbon steels (less than 0.25 per cent carbon) are used for case hardening, where the presence of chromium promotes a hard and wear-resistant case, whilst the nickel preserves a tough, fine-grained core.

Fig. 1.4 *Properties of nickel steels*

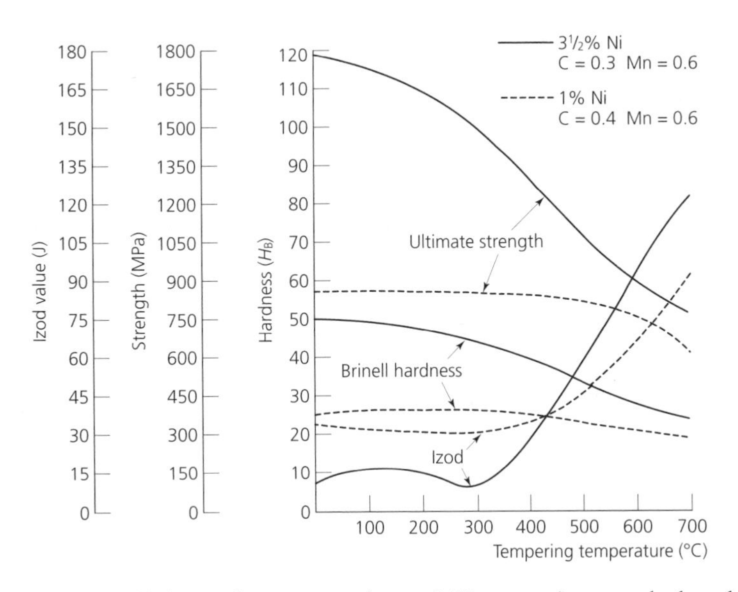

Alloys containing the higher carbon content (up to 0.55 per cent) are used where high-duty mechanical properties are required. Figure 1.5 shows the properties of a typical steel having a composition of 0.3 per cent carbon, 3.4 per cent nickel and 0.75 per cent chromium after quenching in oil from 850 °C and tempering at various temperatures up to 700 °C. The dip in the Izod curve that occurs when the tempering temperature lies between 250 and 450 °C should be noted. It is typical of these steels, and this range of temperatures must not be used for tempering and must be avoided in service. These alloys must be cooled rapidly from the tempering temperature or their impact strength will be low. This marked reduction in impact strength occurs when tempering is slow or when it lies between 250 and 450 °C. It is called *temper brittleness* and is indicated by the broken line in Fig. 1.5. It is also

Fig. 1.5 *Properties of a typical nickel–chromium steel*

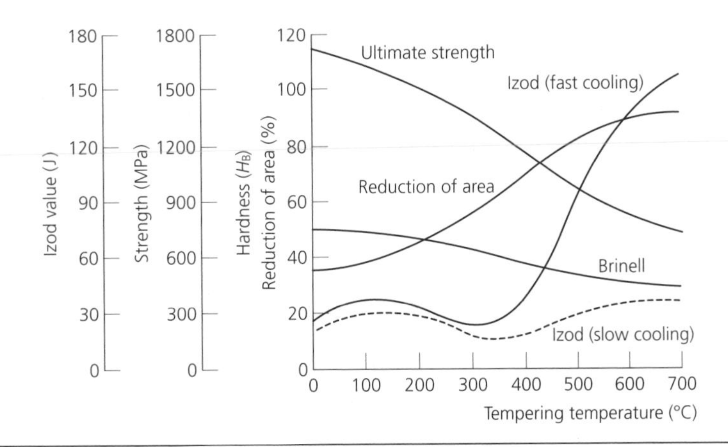

the cause of *weld decay* and nickel–chromium steels have to be 'proofed' by the addition of molybdenum, titanium or niobium if they are to be used in welded fabrications. When the nickel content exceeds 1 per cent and the chromium content exceeds 4 per cent, the alloy may be hardened by quenching in an air blast from a temperature just above the A_3 line. Such alloys are referred to as air-hardening steels and they possess outstanding properties for small, heavily loaded machine parts and gears.

1.5.4 *Nickel–chromium–molybdenum steels*

Although making the alloy more expensive, *temper brittleness* can be virtually eliminated by the addition of 0.3–0.6 per cent molybdenum to nickel–chromium steels. The addition of molybdenum also makes the mass effect less pronounced, allowing large components to be heat treated more easily. The property of molybdenum to prevent temper brittleness is very valuable since it enables large forgings to be cooled slowly without loss of strength and without the initiation of internal stresses that would be caused by more rapid cooling.

The addition of molybdenum to nickel–chromium steels also enables an increased amount of manganese to be used with no reduction in impact strength. The increase in manganese content allows the nickel content to be reduced and this, in turn, offsets the additional cost of the molybdenum. In fact, by increasing the manganese content substantially in place of the more expensive nickel, it is even possible to reduce the cost of the steel. Consider a steel alloy having a composition of, 0.35 per cent carbon, 1.6 per cent manganese, 2.0 per cent nickel and 0.6 per cent molybdenum. This will have properties almost identical to a more costly steel alloy whose composition is, 0.35 per cent carbon, 0.6 per cent manganese, 3.0 per cent nickel and 0.8 per cent chromium.

1.5.5 *Nickel–chromium–vanadium steels*

Vanadium is never used as the sole alloying element in steels but is used in conjunction with nickel or chromium or both. It is a very effective deoxidising element and thus, by eliminating or greatly reducing such impurities as the iron oxide content entrapped in the steel, it improves the mechanical properties generally and the resistance to *fatigue* in particular. Vanadium is an important addition to alloys that are subjected to repeated stress reversals, for example spring steels. Vanadium also intensifies the effect of the other alloying elements present and enables somewhat smaller quantities of these elements to be used without altering the mechanical properties of the steel. Its direct effect is to stabilise the carbides present and thus to harden the steel. Hence it is not normally used in excess of 0.2 per cent except in tool and die steels.

1.5.6 *Molybdenum steels*

Steels containing molybdenum as the sole alloying element (in addition to carbon and the small amount of manganese present in all steels) are also used for structural purposes. Like silicon steels, to be mentioned later, molybdenum steels give greater strength for a given ductility than is obtainable in plain carbon steels. Molybdenum steels containing 0.2–0.7 per cent molybdenum, 0.3–1.0 per cent manganese, 0.1–0.35 per cent silicon and 0.15–0.7 per cent carbon are used for rolled steel sections, forgings and castings. The tensile strengths for these steels range from about 378–462 MPa, whilst their elongation percentage can be kept as high as 20 per cent. The grain size in these steels can be controlled accurately and easily.

1.5.7 *Chromium–molybdenum steels*

In these alloys the chromium content may range from as low as 0.4 per cent to as high as 10 per cent, and the molybdenum content from 0.2–1.5 per cent. Steels containing low percentages of chromium and molybdenum have mechanical properties similar to other low-alloy structural steels. However, the steels with high percentages of chromium and molybdenum may be used as rolled steel sections, forgings and castings with tensile strengths up to 1120 MPa whilst still maintaining elongation values of 18–22 per cent. The high chromium content in these latter steels makes them highly corrosion resistant.

1.5.8 *Chromium steels*

Steels in which chromium is the principal alloying element can have tensile strengths up to 980 MPa, elongations of 10–15 per cent and relatively high impact strengths. Such steels may have 0.2–1.6 per cent chromium and up to 1.0 per cent manganese. Great care is required in their heat treatment as chromium promotes grain growth in steels. However, the high manganese content helps to offset this effect. Steels containing 5–10 per cent chromium and 1–4 per cent silicon are heat resistant and widely used for valves in automobile engines. These alloys are usually referred to as silicon–chromium steels.

1.5.9 *Silicon steels*

Small amounts of silicon are frequently used in many alloy steels, but the amount does not usually exceed 0.8 per cent. If the silicon content is increased to 1.25 per cent (together with 0.5–0.65 per cent carbon and 0.6–0.9 per cent manganese) a steel with a high resistance to fatigue is produced. Such steels are frequently used for springs when quenched in oil from 857 to 900 °C and tempered between 475 and 525 °C.

Steels containing 0.5–1.0 per cent silicon and 0.7–0.95 per cent manganese are now being used increasingly for structural purposes. These steels, like the molybdenum steels mentioned above, give greater strength for a given ductility than can be achieved in plain carbon steels. Also, the cost is lower than for molybdenum steels. A selection of structural steels is given in Table 1.1, together with their properties and some typical uses.

Table 1.1 *Structural (low-alloy) steels*

Compositions, properties and applications

BS 970 spec.	*Type of steel*	*Composition (%)* C	*Mn*	*Ni*	*Cr*	*Mo*	*Condition*	R_e *(MPa)*	R_m *(MPa)*	I *(J)*	A %	*Applications*
150M28	Low manganese	0.28	1.50	—	—	—	Normalised	355	590	—	20	A cheap medium-duty alloy for automobile components
530M40	Nickel–manganese	0.40	0.90	1.00	—	—	Quench hardened from 850 °C in oil, tempered at 600 °C	495	695	91	25	Crankshafts, connecting rods, axles and general components in the automobile and machine tool industries
608M38	Manganese–molybdenum	0.38	1.50	—	—	0.50	Quench hardened from 850 °C in oil, tempered at 600 °C	1000	1130	70	19	A lower-cost substitute for nickel–chrome–molybdenum steels for highly stressed components
653M31	Nickel–chromium	0.31	0.60	3.00	1.00	—	Quench hardened from 820 to 840 °C in oil, tempered at 600 °C	820	930	105	23	Highly stressed components such as: differential shafts, half-shafts, stub axles, connecting rods, pinion shafts, high-tensile studs and bolts
817M40	Nickel–chromium–molybdenum	0.40	0.55	1.50	1.20	0.30	Oil quenched from 840 °C and tempered at 600 °C		2010	27	14	Highly stressed gears for the automobile and machine tool industries
								990	1080	70	22	Highly stressed components where resistance to shock and fatigue is important
945M38	Manganese–nickel–chromium–molybdenum	0.38	1.40	0.75	0.50	0.20	Oil quenched from 840 °C, tempered at 600 °C	960	1040	85	21	A cheaper alloy than 817M40, but still suitable for highly stressed components where resistance to shock and fatigue is important

R_e = Yield stress, R_m = Tensile strength, *I* = Izod number, *A* = elongation %, Mn = manganese, C = carbon, Ni = nickel, Cr = chromium, Mo = molybdenum

1.6 Corrosion-resistant steels

These are alloy steels containing large amounts of nickel and chromium, and it is these alloying elements that promote the formation of homogeneous, corrosion-resistant oxide films on the surface of the metal.

1.6.1 *Nickel steel*

Alloys containing 0.4–0.5 per cent carbon and 20–30 per cent nickel are austenitic in structure even when cooled slowly. Besides being extremely tough, they are highly resistant to corrosion by sea-water, steam and hot gases. They also have low coefficients of thermal expansion and are used for steam turbine blades and internal combustion engine valves. Since they are austenitic, these alloys are also non-magnetic and use is often made of this property. To stabilise the carbides present the alloys always contain 1.4 per cent manganese and up to 0.5 per cent chromium. The only heat treatment required is cooling in air from 800 °C to render the alloy machinable. Alloys containing from 30–40 per cent nickel are also available, but are very expensive and are only used where a negligible coefficient of thermal expansion is required.

1.6.2 *Chromium–molybdenum steels*

These have already been introduced as high-strength structural steels. Alloys containing up to 10 per cent chromium and 1.5 per cent molybdenum combine very high strength and toughness with very high corrosion resistance, particularly to acids.

1.6.3 *Stainless steels*

There are several distinctive types of stainless steel and they all contain fairly large amounts of chromium and, sometimes, nickel. The chromium content ranges from 4–22 per cent and the nickel content from 0–26 per cent. Stainless steels can be categorised into four main groups:

- *Austenitic* stainless steels containing 15–20 per cent chromium and 7–10 per cent nickel.
- *High-austenitic* stainless steels containing 22–26 per cent chromium and 12–14 per cent nickel.
- *Martensitic* stainless steels containing 10–14 per cent chromium.
- *Ferritic* stainless steels containing 14–18 per cent or 23–30 per cent chromium.

Austenitic stainless steels are probably the most important group of stainless steels. Within this group lies the 18/8 stainless alloy which is the most widely used of all. Stainless steels in the 18/8 group are austenitic in structure, even when cooled slowly, and they are not responsive to heat treatment, except that they can be annealed from 1100 °C. After annealing they are very tough but ductile. They are hardened by cold working, resulting in increased tensile strength and decreased ductility. Austenitic stainless steels retain their properties at fairly high temperatures, whilst maintaining their toughness and impact resistance at very low temperatures. This range of operating temperatures is somewhat exceptional.

18/8 stainless steels have excellent corrosion-resisting properties, but are liable to suffer from intercrystalline corrosion after being heated to between 600 and 900 °C. This must be borne in mind when these steels are welded since the parent metal will be at this temperature on each side of the weld zone. In these circumstances, this form of intercrystalline corrosion is referred to as weld decay. This susceptibility can be reduced by quenching the steel from 1050 °C after welding or by the addition of such alloying elements as molybdenum or titanium (1.6 per cent) or both. Stainless steels that include these additional alloying elements are said to have been 'proofed' against weld decay.

Ferritic stainless steels are susceptible to sigma-phase formation if subjected to prolonged heating at 475 °C. The sigma phase, which is formed from ferrite in the presence of chromium, promotes hardness and brittleness in the steel. This leads to cracking and fatigue failure and is referred to as temper brittleness. Ferritic steels, containing a high chromium content, must be cooled rapidly from 650 °C after welding to prevent the onset of weld decay. The sigma phase can also occur in austenitic stainless steels following prolonged heating at 800 °C or slow cooling from this temperature.

Martensitic stainless steels can be quench hardened and tempered and are, therefore, used for stainless edge tools in the catering and food-processing industries and for domestic cutlery.

Properties

Stainless steels can be manipulated by all the usual engineering processes. They machine quite well provided the cutting tools used are kept sharp. For maximum ductility the carbon content should be kept low, whilst for cutlery and general engineering the carbon content is increased to 0.3–0.55 per cent and the nickel content is eliminated to ensure the promotion of hard, stable carbides. Table 1.2 lists some corrosion-resistant steels, together with their properties and some typical applications.

1.7 Heat-resistant steels

These alloy steels are designed to resist corrosion and oxidation by reactive gases at high temperatures, whilst retaining their strength at such temperatures. They must also be creep resistant and free from any tendency to temper brittleness, carbide precipitation and the formation of the sigma phase. Such steels contain up to 30 per cent chromium together with up to 3.5 per cent silicon. Nickel is also present to limit the grain growth at sustained high temperatures that would otherwise result from the high chromium content.

To maintain their strength at high temperatures, heat-resisting alloys are 'stiffened' by the addition of one or more of the following elements in small quantities: aluminium, carbon, molybdenum, titanium and tungsten.

1.8 Maraging steels

These are high-strength, high-alloy steels containing approximately 18 per cent nickel. Some typical compositions are shown in Table 1.3. The high nickel content has a

Table 1.2 *Corrosion- and heat-resistant steels*

Compositions, properties and applications

Type of steel	*Composition*						*Mechanical properties*				*Heat treatment*	*Applications*
	C	*Mn*	*Cr*	*Ni*	*Ti*	*Si*	R_m	R_e	A	H_B		
403S17 Ferritic	0.04	0.45	14.0	0.50	—	0.80	510	340	31	—	Condition soft; cannot be hardened except by cold work	Soft and ductile; can be used for fabrications, pressings, drawn components, spun components; domestic utensils
420S45 Martensitic	0.30	1.0	13.0	1.0	—	0.80	1470 1670	— —	— —	450 534	Quench from 950–1000 °C; temper 400–450 °C; temper 150–180 °C	Corrosion-resistant springs for food-processing and chemical plant; corrosion-resistant cutlery and edge tools
302S25 Austenitic	0.1	1.0	18.0	8.50	—	0.80	61 896	278 803	50 30	170 —	Condition soft solution treatment from 1050 °C	18/8 stainless steel widely used for fabrications and domestic and decorative purposes
321S20 Austenitic weld-decay resistant	0.1	0.80	18.0	8.50	1.60	0.80	649 803	278 402	45 30	180 225	Condition soft solution treatment from 1050 °C; can only be work hardened	18/8 which can be safely fabricated by welding, used for brewing, food-processing, and chemical plant
401S45 Valve steel	0.4	0.50	8.0	0.5	—	3.0	—	—	—	225 (min)	Quench harden in oil from 1030 to 1060 °C; temper at 750–850 °C	Heat-resistant steel of relatively low cost; general-purpose steel
349S54 Valve steel	0.5	10.0	22.0	4.50	—	0.25	—	—	—	321 (min)	Soften by solution treatment at 1160–1190 °C; harden by precipitation treatment at 750–850 °C for 6 to 15 hours	A high-quality, high cost valve steel suitable for hostile environments, e.g. furnace and chemical-plant components

C = carbon, Mn = manganese, Cr = chromium, Ni = nickel, Ti = titanium, Si = silicon, R_m = Tensile stress (MPa), R_e = Yield stress (MPa), A = elongation %, H_B = Brinell hardness considerable graphitising effect so that the carbon content must not exceed 0.3 per cent. It has been found that balanced additions of cobalt and molybdenum to iron–nickel martensite gave a combined age-hardening effect far greater than when these alloying elements are used separately. The ability to *age harden* the martensite gave this group of steels its name – *maraging steels*. Furthermore, it was found that the iron–nickel–cobalt–molybdenum matrix was amenable to supplemental age hardening by small additions of aluminium and titanium. Because of the high cost of the alloying elements in maraging steels, they are only used where their special properties can be fully exploited. Table 1.4 lists the advantages of nickel maraging steels over conventional alloy steels, whilst Table 1.5 lists some typical applications. The heat treatment of maraging steels by solution and precipitation processes are described in Section 2.10. The welding of maraging steels is discussed in Section 5.15.

Table 1.3 *Maraging steels: 18 per cent Ni-Co-Mo*

Composition ranges by weight per cent*

	Wrought				*Cast*
Grade	*18Ni1400*	*18Ni1700*	*18Ni1900*	*18Ni2400*	*17Ni1600*
Nominal 0.2% proof stress					
N/mm^2 (MPa)	1400	1700	1900	2400	1600
$tonf/in^2$	90	110	125	155	105
10^3 lbf/in^2	200	250	280	350	230
kgf/mm^2	140	175	195	245	165
hbar	140	170	190	240	160
Ni	17–19	17–19	18–19	17–18	16–17.5
Co	8.0–9.0	7.0–8.5	8.0–9.5	12–13	9.5–11.0
Mo	3.0–3.5	4.6–5.1	4.6–5.2	3.5–4.0	4.4–4.8
Ti	0.15–0.25	0.3–0.5	0.5–0.8	1.6–2.0	0.15–0.45
Al	0.05–0.15	0.05–0.15	0.05–0.15	0.1–0.2	0.02–0.10
C max.	0.03	0.03	0.03	0.01	0.03
Si max.	0.12	0.12	0.12	0.10	0.10
Mn max.	0.12	0.12	0.12	0.10	0.10
Si + Mn max.	0.20	0.20	0.20	0.20	0.20
S max.	0.010	0.010	0.010	0.005	0.010
P max.	0.010	0.010	0.010	0.005	0.010
Ca added	0.05	0.05	0.05	none	none
B added	0.003	0.003	0.003	none	none
Zr added	0.02	0.02	0.02	none	none
Fe	Balance	Balance	Balance	Balance	Balance

*The composition ranges given are those originally developed by Inco which broadly cover current commercial practice. Slight changes in these ranges have been made in some national and international specifications.
Source: Inco Europe Ltd.

Table 1.4 *Advantages of nickel maraging steels*

Excellent mechanical properties	*Good processing and fabrication characteristics*	*Simple heat treatment*
1. High strength and high strength-to-weight ratio 2. High notched strength 3. Maintains high strength up to at least 350 °C 4. High impact toughness and plane strain fracture toughness	1. Wrought grades are amenable to hot and cold deformation by most techniques. Work-hardening rates are low 2. Excellent weldability, either in the annealed or aged conditions. Preheat not required 3. Good machinability 4. Good castability	1. No quenching required. Softened and solution treated by air cooling from 820 to 900 °C 2. Hardened and strengthened by ageing at 450–500 °C 3. No decarburisation effects 4. Dimensional changes during age hardening are very small – possible to finish machine before hardening 5. Can be surface hardened by nitriding

Source: Inco Europe Ltd.

Table 1.5 *Applications of nickel maraging steels*

Aerospace	*Tooling and machinery*	*Structural engineering and ordnance*
Aircraft forgings (e.g. undercarriage parts, wing fittings) Solid-propellant missile cases Jet-engine starter impellers Aircraft arrestor hooks Torque-transmission shafts Aircraft ejector release units	Punches and die bolsters for cold forging Extrusion press rams and mandrels Aluminium die-casting and extrusion dies Cold-reducing mandrels in tube production Zinc-base alloy die-casting dies Machine components: gears index plates lead screws	Lightweight portable military bridges Ordnance components Fasteners

Source: Inco Europe Ltd.

1.9 Tool and die steels

Plain carbon steels with a carbon content between 0.7 and 1.5 per cent make excellent cutting tools for low-strength materials, such as wood, where the keen edge attainable with such steels is a distinct advantage for hand tools. However, for machining metal, plastic or wood under modern production conditions, plain carbon steels are no longer adequate as cutting tool materials. Quench-hardened high-carbon steels are very brittle and have to be tempered to improve their toughness. Unfortunately, tempering also reduces their hardness and wear resistance. Further, it is the ease with which the temper of plain carbon steel can be 'drawn' (i.e. the steel becomes soft) that renders such steels unsuitable for the high-speed machining of modern high-duty alloys because of the high temperatures generated in the cutting zone.

The addition of such alloying elements as chromium, cobalt, manganese, molybdenum, tungsten and vanadium makes tool and die steels harder, more wear resistant, more shock resistant, less liable to shrink and warp and better able to operate at high temperatures. For example, correctly hardened high-speed steel can retain its hardness and continue cutting at temperatures approaching dull-red heat. Most of the alloying elements used in tool and die steels are refractory metals with very high melting points. They also form very hard stable carbides and, having body-centred-cubic crystal lattices, they limit the range of temperatures over which austenite can exist, thus stabilising the ferrite and the hard tetragonal martensite.

The heat treatment of alloy steels will be dealt with in detail in Chapter 2. However, at this point, it is sufficient to say that the low-alloy tool and die steels are quench hardened in oil from temperatures only slightly above those used for plain carbon steels of equivalent carbon content. They are tempered at similar temperatures to plain carbon steels. The more heavily alloyed steels have much slower transformation rates and can be air hardened (air-blast quenched) for thinner sections or oil quenched for heavier sections. In order that the full advantage can be taken of the ability of the high-alloy tool and die steels to retain their hardness at high operating temperatures, the maximum amounts of tungsten and molybdenum must be present in solid solution in the austenite before quenching. For this reason very high hardening temperatures are required (as much as 1300 °C in some instances) and great care has to be taken to avoid grain growth. Despite the large amount of carbide-stabilising alloying elements present, a substantial amount of austenite still remains after the initial quench hardening. To transform the retained austenite into martensite a secondary hardening treatment is required. This involves quenching the already hard steel from 550 °C. With some steels it is necessary to repeat the secondary hardening treatment two or even three times before the transformation of austenite into martensite is complete. This secondary hardening process must not be confused with tempering. Tempering increases toughness at the expense of hardness, whilst secondary hardening increases both hardness and toughness. Once the martensite has been formed the steel can be used at temperatures up to 700 °C before tempering and softening sets in. This is because of the sluggishness of the transformations in such heavily alloyed steels. Cutting tool materials will be considered further in Chapter 10.

EXERCISES

1.1 Discuss the relative advantages and limitations of alloy steels compared with plain carbon steels.

1.2 Select **five** of the following alloying elements and, in each instance, explain the effect of adding the alloying element to a carbon steel. Chromium, cobalt, lead, manganese, molybdenum, nickel, phosphorus, silicon and tungsten.

1.3 Explain why nickel and chromium are frequently used together in alloy steels.

1.4 Explain what is meant by the term *temper brittleness* when related to alloy steels. What causes this phenomenon and how it can be avoided?

1.5 Explain, with examples, how alloying elements can change the transformation temperatures for steels.

1.6 (a) Compare the properties, composition and applications of the following groups of stainless steel:

(i) autensitic
(ii) ferritic
(iii) martensitic

(b) Describe the precautions that must be taken when heat treating and welding stainless steels to avoid *temper brittleness* and *weld decay*.

1.7 Compare the essential differences between a stainless steel and a heat-resisting steel in terms of composition and properties.

1.8 Compare the essential differences between a *maraging* steel and a conventional alloy steel in terms of composition, properties and heat treatment.

1.9 Referring to manufacturers' literature, compare the properties, composition and applications of typical alloy steels from the following groups:

(a) nickel–chromium steels
(b) nickel–chromium–molybdenum steels
(c) nickel–chromium–vanadium steels

1.10 Explain how elements such as cobalt, molybdenum and tungsten improve the hardness of alloy steels when operating at elevated temperatures.

2 The heat treatment of steels

The topic areas covered in this chapter are:

- Non-equilibrium transformations.
- Construction of time–temperature transformation diagrams.
- Interpretation of time–temperature transformation diagrams.
- Quenching.
- Hardenability.
- Tempering.
- Martempering.
- Austempering.
- Solution and precipitation treatment of maraging steels.

2.1 Non-equilibrium transformations

The heat treatment of plain carbon steels was introduced in *Engineering Materials*, Volume 1, by relating the processes of full annealing, sub-critical annealing, normalising and quench hardening to the iron–carbon phase equilibrium diagram, as shown in Fig. 2.1. These transformations can only occur if heating and cooling proceeds sufficiently slowly so that all the diffusion processes associated with the transformations are completed – that is, for all the transformations to achieve equilibrium. However, most heat-treatment processes involve heating and cooling the steel more rapidly so that *equilibrium is not achieved* and this leads to the formation of microstructures that cannot be forecast from the iron–carbon phase equilibrium diagram.

Reference to Fig. 4.8 in *Engineering Materials*, Volume 1, shows that, depending upon whether the steel is being heated or cooled, stable austenite exists at temperatures above the Ac_1/Ar_1 line, and that the steel is fully austenitic above the Ac_3–Ac_{cm}/Ar_3–Ar_{cm} lines. Further, carbon dissolves interstitially in γ-iron to form the solid solution known as austenite and this enables the carbon to diffuse quickly throughout the iron so that coring is negligible. However, despite the ease and rapidity with which diffusion of the carbon in γ-iron occurs under equilibrium conditions, if the austenite is cooled very quickly from above its upper critical temperature to a much lower temperature (say 200 °C) there is insufficient time for total diffusion to occur. Rapid cooling results in a correspondingly

Fig. 2.1 *Heat-treatment temperatures for plain carbon steels related to the iron carbon phase equilibrium diagram*

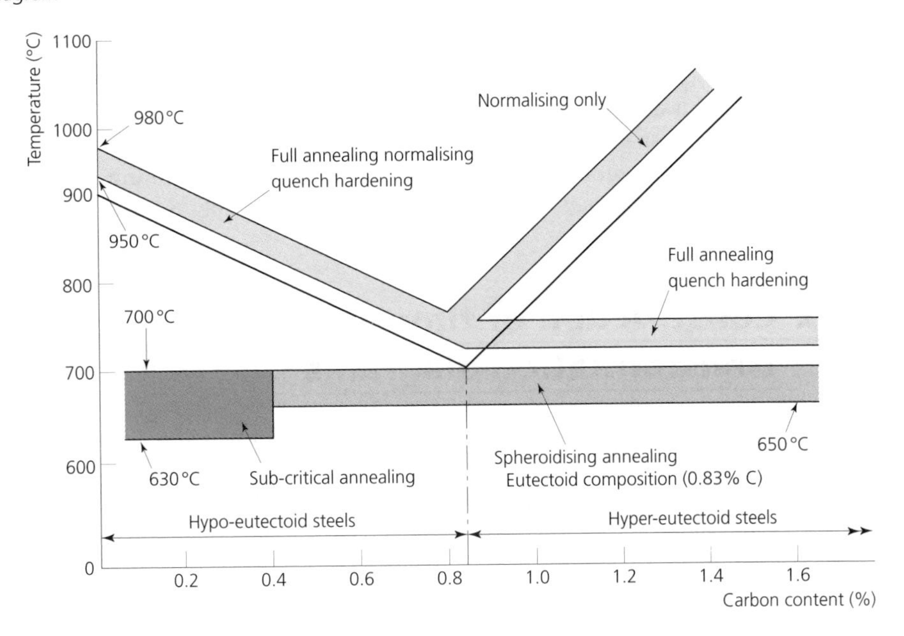

rapid transformation from the face-centred-cubic structure of γ-iron to the body-centred-cubic structure of α-iron *before* the carbon can diffuse out of solution and form iron carbide, thus causing supersaturation of the body-centred structure. This, in turn, causes such distortion of the crystal lattice structure that slip (see Section 8.2) becomes virtually impossible and the steel exhibits the property of hardness. In fact, such sudden quenching produces a solid solution of carbon in body-centred-cubic crystals that is over one thousand times supersaturated and is thus very unstable. Under a microscope a polished and etched specimen of plain, high-carbon steel, quench hardened from above the Ac_{cm} line, shows a structure of acicular (needle-like) crystals called *martensite* after the metallurgist, Martens, who first identified it.

- If the quenching is less severe so that the transformations can proceed isothermally (constant temperature), then some iron carbide can form and precipitate out.
- If the temperature of the quenching bath permits the transformations to occur at 250 °C the particles formed will be those of *lower bainite*. These are very fine and can only be seen under a microscope of very high magnification. Lower bainite is similar in appearance to martensite but rather less hard and appreciably tougher.
- If the temperature of the quenching bath permits the transformations to occur nearer 550 °C, the particles formed are called *upper bainite*. These particles are less fine and more 'feathery' in appearance than lower bainite. They are also softer and very much tougher.
- If the temperature of the quenching bath permits the transformations to occur above 550 °C but below the A_1 line (723 °C), then pearlite will be formed with a further increase in toughness and loss of hardness.

Thus it becomes clear from the above comments that the microstructure produced is directly related to two factors, namely:

- The temperature at which the transformation occurs.
- The time taken during which the transformation occurs (i.e. the transformation rate).

The effect of time and temperature on the non-equilibrium transformations can be better explained using *time–temperature transformation (TTT) diagrams* than by using the iron–carbon phase equilibrium diagram. Time-temperature transformation diagrams are also known as *isothermal transformation diagrams*, or simply *S-curves* from the shape of the diagram.

2.2 Construction of time–temperature transformation diagrams

Let's now consider how time–temperature transformation diagrams can be constructed. Unlike the iron–carbon phase equilibrium diagram, which is applicable to all plain carbon steels, a time–temperature transformation curve can refer to only one steel of a particular composition at a time. Thus, if a comparison is to be made between several different steels, each steel will have to be represented by its own particular time–temperature transformation diagram. These diagrams are equally applicable to both plain carbon steels and alloy steels. However, they are mainly used for alloy steels where they are essential for predicting the outcome of the heat treatment of such steels. This is because alloy steels cannot be referred to the iron–carbon phase equilibrium diagram.

However, for simplicity, let's first consider a time–temperature transformation diagram for a plain carbon steel, as shown in Fig. 2.2. To produce this diagram, the following procedure may be adopted.

Fig. 2.2 *Typical time–temperature transformation (TTT) curves for a plain carbon steel*

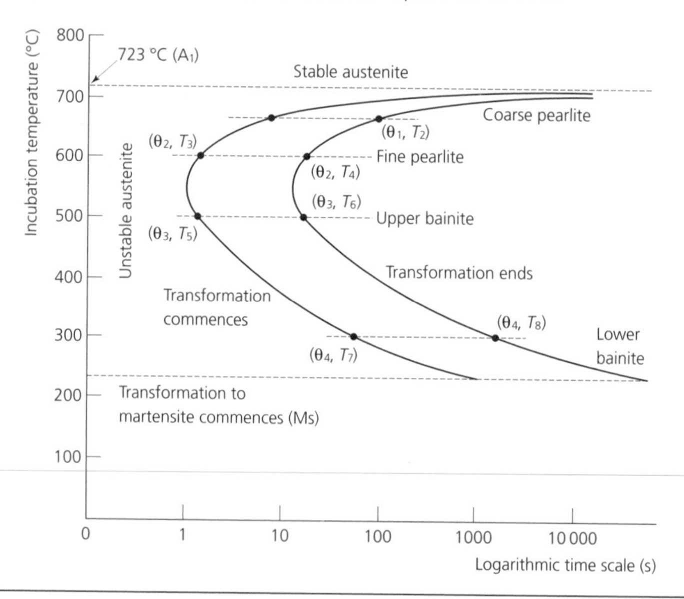

- A large number of specimens, approximately 12 mm diameter by 1.5 mm thick, are produced from a sample of the steel being examined.
- These specimens are suspended in a salt bath at a temperature just above that required to ensure that the steel is fully austenitic. This is the 'austenising bath' in Fig. 2.3.
- A suitable number of specimens are transferred to a second salt bath. This is referred to as the 'incubation' bath in Fig. 2.3. For a plain carbon steel, this furnace is held at a predetermined temperature below 723 °C. For example, a temperature of 250 °C would be suitable if a structure of lower bainite is required. In this bath the transformation of the austenite to carbide takes place, dependent upon the time the specimens remain in the bath.
- The specimens are then removed from the incubation bath one by one at predetermined times and quenched in water, as shown in Fig. 2.3. This final quench halts any transformations that were taking place in the incubation bath, and converts any residual austenite into martensite.
- The quenched specimens are polished, etched and examined microscopically to assess the extent to which the transformations have occurred during incubation. Figure 2.4

Fig. 2.3 *Heat treatment sequences for producing time–temperature transformation diagrams*

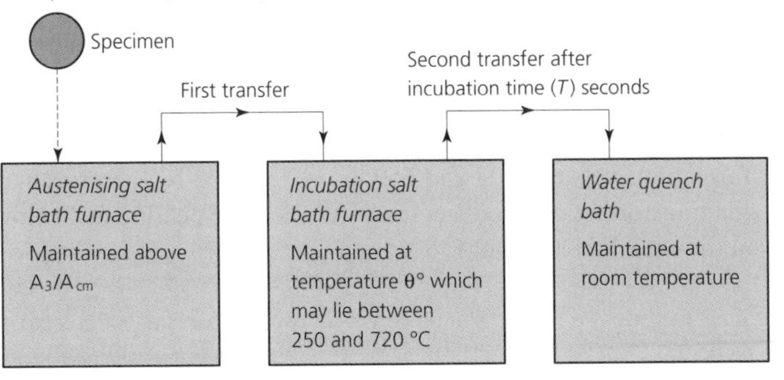

Fig. 2.4 *Typical incubation transformations*

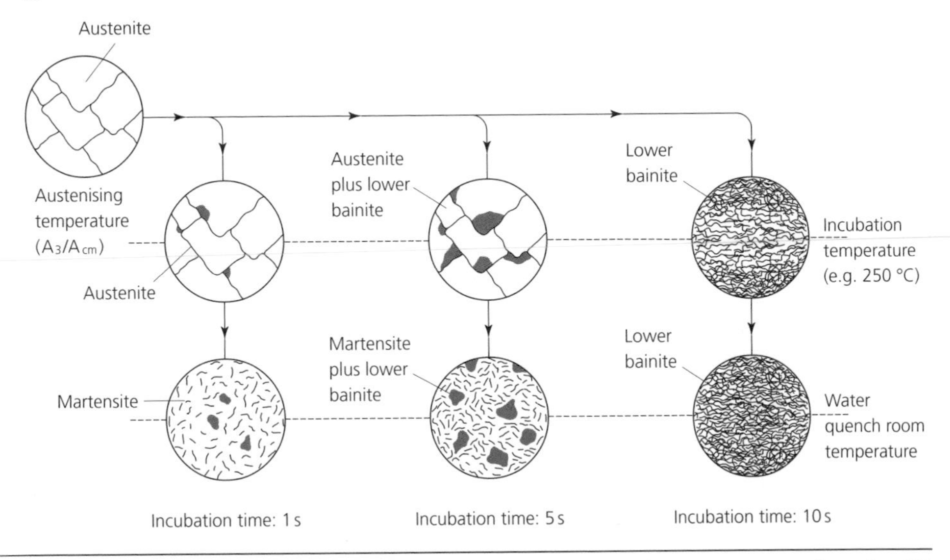

shows the results of this sequence of treatments. The greater the time interval, the greater will be the degree of transformation for any given incubation temperature.

- This sequence of events is repeated for a range of temperatures from 250 °C to just below 723 °C (θ_1, θ_2, θ_3,...) and sets of results are obtained similar to those shown in Fig. 2.2 except that the time intervals (T_1, T_2, T_3,...) will be different. All these results are then plotted on a common set of axes and a time–temperature transformation diagram is produced, similar to that shown in Fig. 2.2.

2.3 Interpretation of time–temperature transformation diagrams

Figure 2.5 shows the time–temperature transformation diagram for a plain carbon steel of eutectoid composition. Since the steel is of eutectoid composition its structure will be entirely stable austenite above 723 °C. However, below this temperature the austenite will become increasingly unstable and the two curved lines indicate the times taken for the transformations to begin and end as previously explained.

Just below 723 °C there is considerable inertia in the transformation process as there is little instability in the austenite. However, as the temperature falls, the instability in the austenite increases so that, at 550 °C, the time taken before the transformation commences is at a minimum, as is the time for the completion of the transformation. The transformations between 723 and 550 °C are initiated by the precipitation of iron carbide, and the transformation products range from coarse pearlite just below 723 °C to upper bainite at (fine pearlite) at about 555 °C.

Fig. 2.5 *Time–temperature transformation curves for plain carbon steel of eutectoid composition (0.83 per cent carbon)*

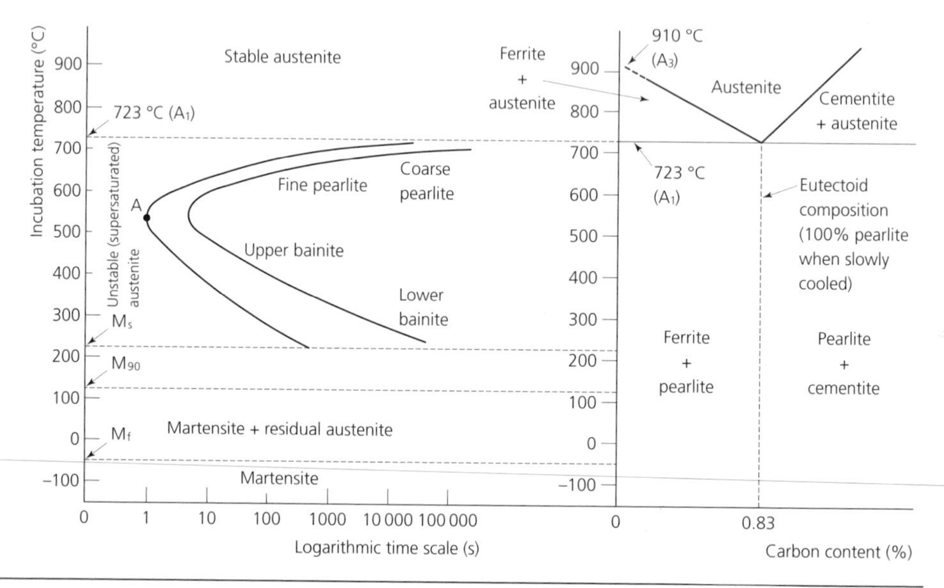

Although the austenite becomes increasingly unstable as the temperature falls below 550 °C, the lower temperature causes the diffusion of carbon in the iron to become increasingly more sluggish. This latter factor has more influence than the increasing instability of the austenite below 550 °C and the time taken for the transformations to commence and finish increases again. This is shown in Fig. 2.5. Whereas the transformations above 550 °C are initiated by the precipitation of iron carbide and result in a structure of dark, feathery upper bainite, below 550 °C the transformations are initiated by the precipitation of ferrite and result in a structure of acicular lower bainite at 220 °C.

Let's consider what happens if the temperature of the austenised steel is lowered sufficiently rapidly by severe quenching so that the transformation into bainite is avoided. Such rapid quenching would result in the austenite being transformed directly into martensite. Note that the 'M' lines indicate the amount of martensite present. In Fig. 2.5 this would mean cooling the steel in less time than that indicated by the point 'A' on the diagram (less than one second). At M_s the martensite only just commences to form and very little will be present, whilst at M_{90} the transformation into martensite is 90 per cent complete. Total transformation is not achieved unless the steel is quenched to M_f (–50 °C) in less than one second (time 'A'). Thus some retained austenite is always present when the steel is quenched to room temperature.

Let's now consider Fig. 2.6, which shows the time–temperature transformation diagram for a 0.4 per cent (hypo-eutectoid) plain carbon steel. Reference to the iron–carbon phase equilibrium diagram shows that for all hypo-eutectoid steels there is a zone between the upper (A_3) and lower (A_1) critical temperatures where both ferrite and stable austenite are present. Thus there is an additional zone on the time–temperature transformation diagram for the ferrite transformations.

Fig. 2.6 *Time–temperature transformation curves for a plain carbon steel of 0.4 per cent carbon content*

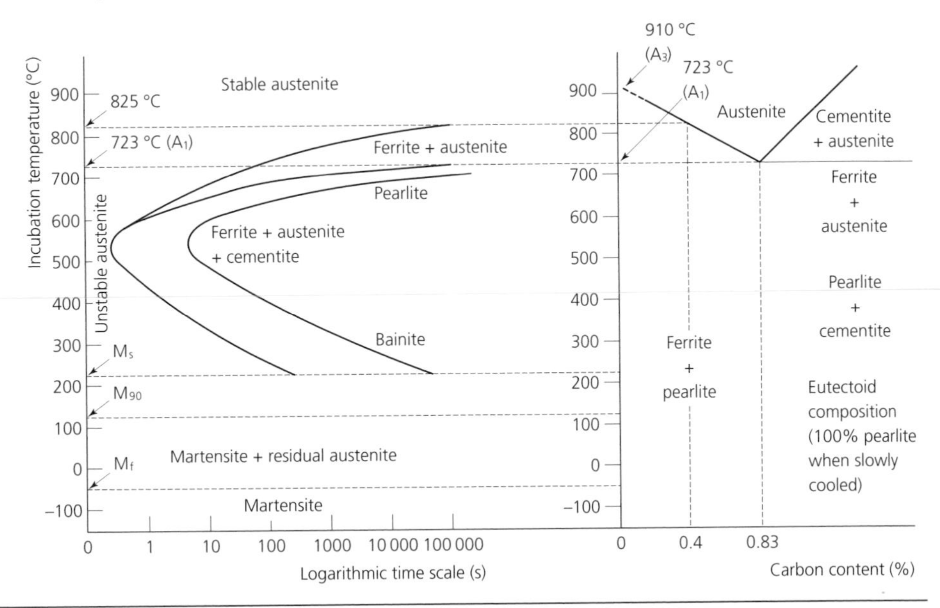

Also, the transformations are shifted to the left of the diagram so that they commence more quickly and finish more quickly. This leftward shift applies to all hypo-eutectoid steels and the transformation times for such steels are so short that the need for drastic cold-water quenching becomes apparent. For steels below 0.3 per cent carbon the leftward shift is such that the transformation curve touches zero time. Thus it is impossible to cool the steel sufficiently quickly to achieve a totally martensitic structure and some ferrite will always be present (see Fig. 2.9).

The rapid cooling required to control the transformations in plain carbon steels can result in cracking and distortion. However, the addition of such alloying elements as nickel, chromium and molybdenum shifts the curves to the right and reduces the cooling rate necessary to produce a martensitic structure. Therefore, these alloying elements improve the hardenability of a steel, as discussed earlier in this chapter. Some typical time–temperature transformation diagrams for simple alloy steels are shown in Fig. 2.7. Except for the 1 per cent nickel steel, quenching in oil is adequate and there is less chance of cracking and distortion (see also Fig. 2.10).

Fig. 2.7 *Some typical time–temperature transformation curves for alloy steels*

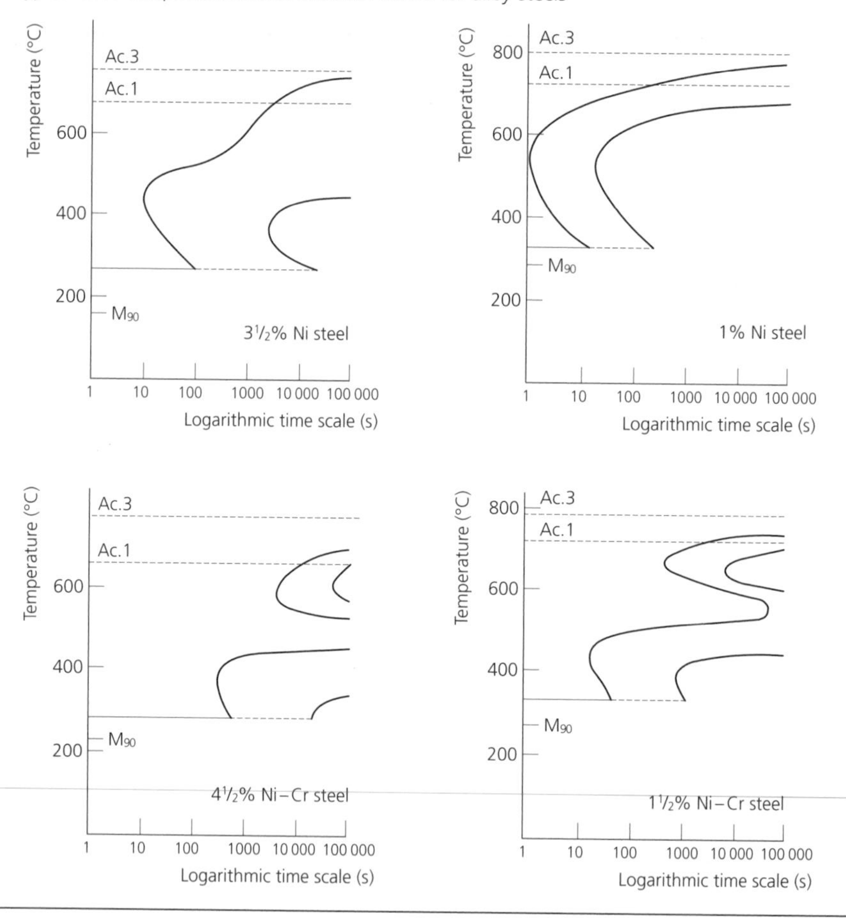

2.4 Quenching

In *Engineering Materials*, Volume 1, it was stated that the faster a component made from a plain carbon steel was quenched to room temperature, the harder it became for a given carbon content. The truth of this statement was proved in Section 2.3, where it was shown that:

- For a martensitic structure to be formed, cooling from the austenising temperature has to be too rapid for the bainite transformations to commence.
- The final temperature at the end of the quenching process has to be sufficiently low to ensure the transformation from unstable austenite into martensite.

Under production conditions, 100 per cent transformation into martensite is never achieved in plain carbon steels. Firstly, the change from austenite (γ-iron) to martensite (α-iron) is accompanied by structural changes in the steel (FCC $\rightarrow$ BCC) that results in volumetric changes. Except for very thin components, these changes will occur at the outer layers of the component some time before they occur at the core of the component where cooling is slowest. This results in cracking and distortion of the component. Secondly, it is impractical to maintain the quenching bath at the M_f temperature ($-50\,°C$) since all liquid-quenching media are frozen at this temperature.

Thus quenching becomes a compromise between minimising the chance of cracking and/or distorting the work and achieving the maximum practical transformation of austenite into martensite. To this end, we will now consider the process of quench hardening in greater detail.

The time–temperature transformation diagrams discussed in Section 2.2 were derived by quenching the austenised steel in an incubation bath, whose temperature lay between 723 and 250 °C, and holding the specimen at the incubation bath temperature whilst transformation of the metal structure took place. The transformations therefore took place at a constant temperature – that is, *isothermally*. This does not happen when steels are quenched in a liquid such as water or oil at room temperature. Under these conditions, cooling occurs continuously from the austenising temperature down to or slightly above room temperature. (The heat energy transfer will raise the temperature of the quenching bath.) Therefore, any transformations that take place will occur whilst cooling proceeds, and not isothermally.

The effects of quench cooling on the TTT diagram are shown in Fig. 2.8. Again, a steel of eutectoid composition has been chosen for simplicity since, under equilibrium conditions, the austenite changes at the A_1 temperature (723 °C) instantaneously, whereas hypo- and hyper-eutectoid steels change over a range of temperatures, accompanied by various transformation products. Cooling curves should not be applied to isothermally derived TTT diagrams, but they can be applied to modified TTT diagrams as shown in Fig. 2.8.

2.4.1 *Curve 1*

Curve 1 is the result of severe water quenching. The cooling curve misses the 'nose' of the transformation curve and therefore no bainite will be formed. The unstable austenite will

Fig. 2.8 *Continuous cooling curves for a plain carbon steel of eutectoid composition (0.8 per cent carbon)*

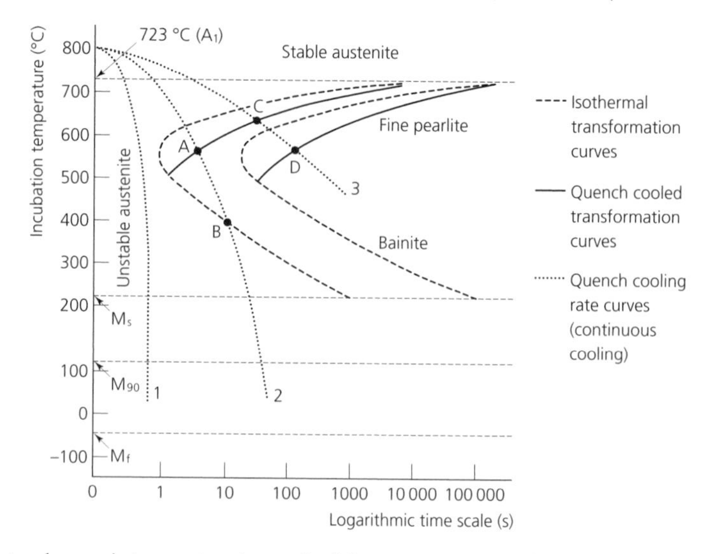

commence to change into martensite at the M_s temperature and the percentage change will increase until the metal reaches the temperature of the quenching bath. Since this will normally lie between 0 and 100 °C, some 90 per cent martensite will be present together with some residual austenite. To achieve this condition quenching has to be more rapid than the critical cooling rate. The critical cooling rate is defined by the cooling curve that just grazes the nose of the appropriate transformation curve for a given steel.

2.4.2 *Curve 2*

Curve 2 is the result of a less rapid quenching and does not achieve the critical cooling rate. Thus some transformations into bainite will occur between A and B in Fig. 2.8 and the remaining austenite will commence to transform at the M_s temperature. This latter transformation will cease when the temperature of the quenching bath is reached. Thus the final composition of the steel will contain a mixture of martensite together with some softer bainite and even softer residual austenite. Hence the hardness of *curve 1* will not be achieved. A thick component could well achieve the conditions of *curve 1 at its surface*, whilst only achieving those of *curve 2 at its core*, where cooling proceeds more slowly, and this effect will be discussed more fully in Section 2.5.

2.4.3 *Curve 3*

Curve 3 is the result of slow cooling as, for instance, when normalising a component. In Fig. 2.8, there will be time for complete transformation of the unstable austenite into fine pearlite together with some upper bainite between C and D depending upon the section thickness and rate of cooling. Although the temperature must pass through the M_s point, no martensite will be formed since there has been sufficient time for all the unstable austenite to transform completely into pearlite and bainite and there is no residual austenite to transform into martensite.

Figure 2.9 shows the curve for a low-carbon steel. It can be seen that the nose of the cooling curve cuts the zero time point of the diagram so that no critical cooling curve can exist. Thus there can be little direct transformation from unstable austenite into martensite, as most of the austenite transforms directly into pearlite and bainite between A and B and only very limited hardening can occur. No attempt is made to quench harden such steels in practice.

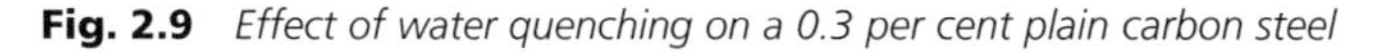

Fig. 2.9 *Effect of water quenching on a 0.3 per cent plain carbon steel*

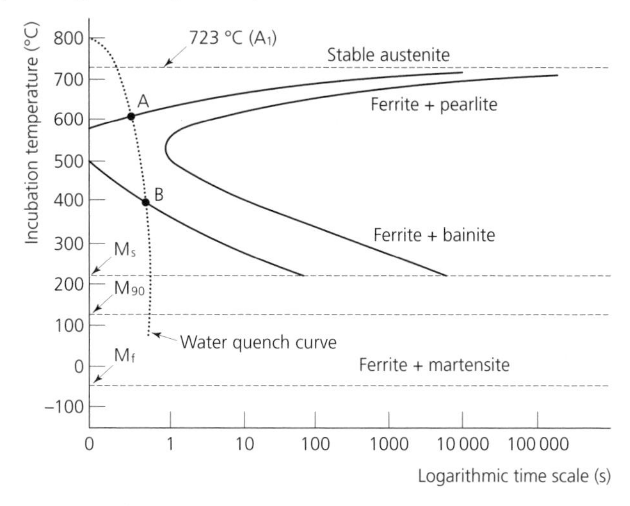

It has already been stated that the effect of adding alloying elements such as nickel, chromium and molybdenum can shift the curve of the TTT diagram to the right (Section 2.3). Figure 2.10 shows how a 4.5 per cent nickel–chromium steel can be quenched in oil and still achieve a direct transformation from unstable austenite directly into

Fig. 2.10 *Effect of oil quenching on a 4.5 per cent nickel–chromium steel*

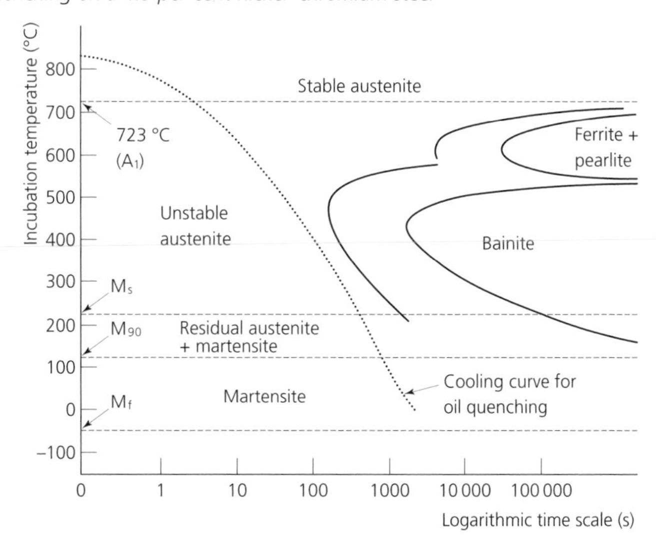

martensite. Although the cooling curve is less steep than for water quenching, it is still to the left of the 'nose' of the curve and, therefore, exceeds the critical cooling rate. Obviously, there is much less chance of cracking and distortion occurring when steels can be satisfactorily hardened at a slower cooling rate.

SELF-ASSESSMENT TASK 2.1

1. Explain what is meant by a time–temperature transformation (TTT) diagram and how such a diagram can be constructed.
2. Explain how such a diagram can be used to predict the outcome of a heat-treatment process for a given alloy steel.

2.5 Hardenability

It has already been explained that the hardness of a plain carbon steel depends upon its carbon content and the rate of cooling from the hardening temperature for a given steel. When a thick component is quenched from its hardening temperature it will take longer for the inner core of the workpiece to cool than for the surface layers that are in contact with the quenching medium (water or oil). This leads to a variation in hardness across the section of the material component, as shown in Fig. 2.11(a). This variation in hardness is referred to as *mass effect*.

Since plain carbon steels have a high critical cooling rate it follows that large sections cannot be fully hardened throughout, and this is shown in Fig. 2.11(b). Thus plain carbon steels have *poor hardenability*. However, a 3 per cent nickel steel containing only 0.3 per cent carbon has a lower critical cooling rate and will harden uniformly across comparatively thick sections. An alloy steel of this composition is said to have *good hardenability*.

Fig. 2.11 *Mass effect (hardenability): (a) cross-section hardenability of a plain carbon steel; (b) effect of hardenability on structure*

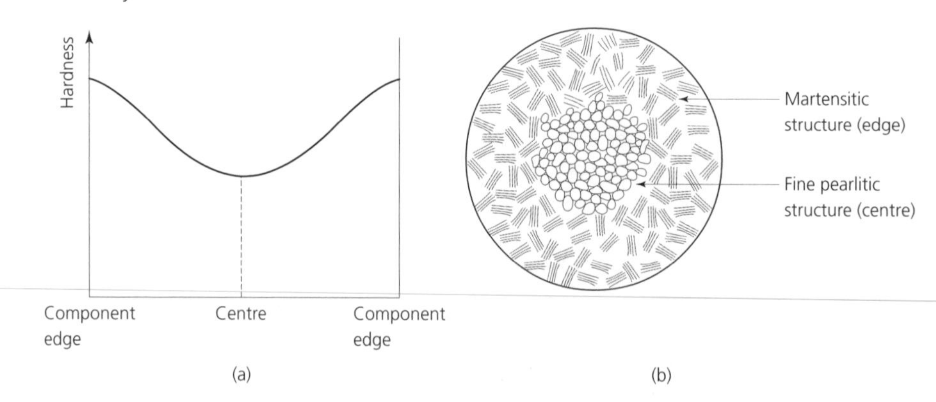

Hardness and hardenability should not be confused. It has already been stated that a steel containing 1.0 per cent carbon has poor hardenability compared with a 3 per cent nickel steel containing only 0.3 per cent carbon. Nevertheless, because of its higher carbon content, the 1.0 per cent plain carbon steel will have a very much higher surface hardness.

Lack of uniformity of structure and hardness in steels with a poor hardenability can seriously affect their mechanical properties. For this reason it is necessary to specify the maximum diameter of bar (*ruling section*) for which the stated mechanical properties can be achieved under normal heat-treatment conditions. Examples of how the ruling section can affect the mechanical properties of a carbon steel and an alloy steel are shown in Table 2.1.

Table 2.1 *Effect of ruling section on mechanical properties*

BS 970 Spec.	*Condition*	*Limiting ruling section (mm)*	*Tensile strength (MPa)*	*Minimum elongation (%)*
070M55 (carbon steel)	Hardened and tempered	19	850 → 1000	12
		63	770 → 930	14
		100	700 → 850	14
835M30 (alloy steel)	Hardened and tempered	63	1080 → 1240	11
		100	1000 → 1160	12
		150	930 → 1080	12
		250	850 → 1000	13

One of the main reasons for adding alloying elements such as nickel and chromium to steel is to reduce the mass effect and to increase the ruling section for which the required properties can be achieved. Figure 2.12 shows the effect of the ruling section on the transformations in a typical plain carbon steel. The diameter of steel (1) is less than the ruling section and both surface layers and core transform to martensite. The diameter of steel (2) is greater than the ruling section thus, although its surface layers will still be martensitic, its core will be pearlitic

Fig. 2.12 *Effect of ruling sections on transformations*

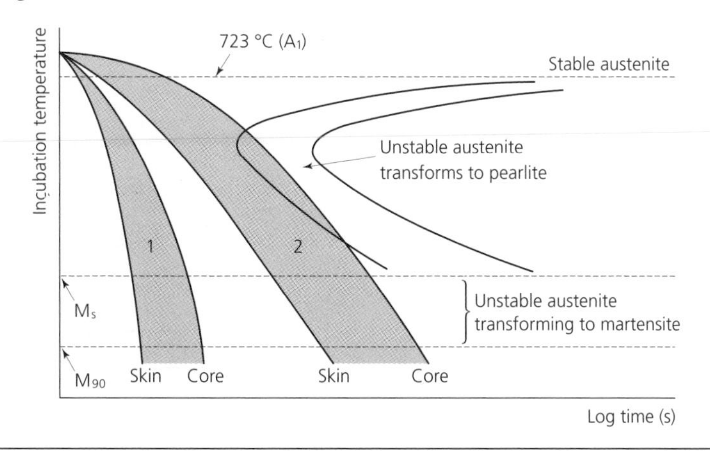

2.5.1 *The Jominy end-quench test*

This test is used to determine the hardenability of steels. It involves heating a specimen to just above its upper critical temperature so that it is fully austenitic, and then quenching it by spraying a jet of water against its lower end, as shown in Fig. 2.13. This figure also shows details of the specimen and the test. The specimen cools very rapidly at the quenched end and progressively less rapidly towards the opposite (shouldered) end. When cold, a flat is ground along the side of the specimen and its hardness is tested every 3 mm from the quenched end. The hardness readings are plotted against distance from the quenched end to give hardenability curves, as shown in Fig. 2.14. You can see that the hardness is more uniform for the alloy steel than it is for the plain carbon steel, thus the alloy steel has better hardenability. This is purely an empirical test and there is no mathematical relationship between the results of the Jominy end-quench test and the ruling section for any particular steel.

Fig. 2.13 *Jominy end-quench test*

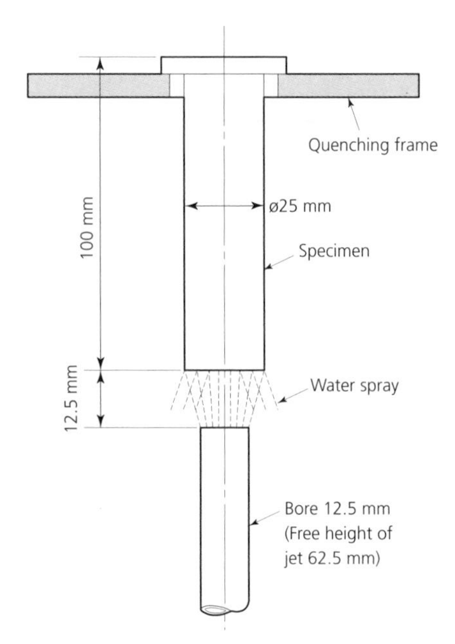

Fig. 2.14 *Typical hardenability curve*

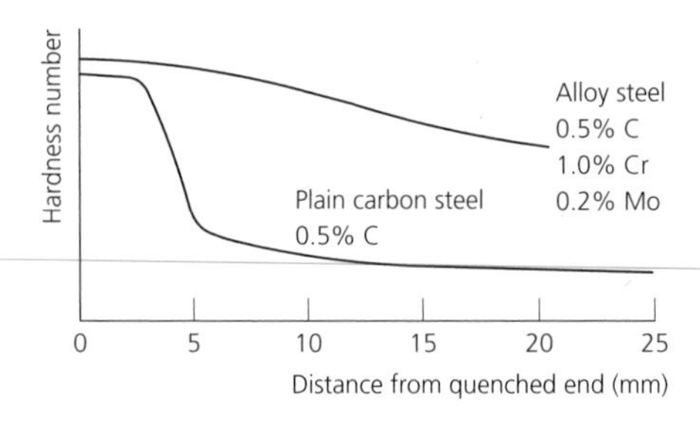

2.6 Tempering

Quench-hardened plain carbon steels and low-alloy steels are brittle and hardening stresses are present. In such a condition they are of little practical use and must be reheated (tempered) to relieve the stresses and to reduce the brittleness. Tempering causes the transformation of martensite into the less brittle structure now to be described. Unfortunately any increase in toughness as a result of tempering is accompanied by some decrease in hardness.

Tempering always tends to transform unstable martensite back to the stable pearlite of the equilibrium transformations. This is because tempering causes the dissolved carbon atoms to precipitate out as iron carbide particles. These particles increase in size as the tempering temperature increases. At temperatures between 100 and 200 °C the iron carbide which forms is not the normal Fe_3C (cementite) composition, but ε-carbide (epsilon-carbide) which is different in composition. This leaves the remaining martensite with a reduced carbon content of 0.3 per cent. From Fig. 2.15 it can be seen that, at first, there is a slight increase in hardness due to the presence of the hard ε-carbide. However, as the temperature rises, the hardness falls off as the unstable martensite starts to transform to pearlite. At 400 °C the ε-carbide starts to transform to the more usual Fe_3C composition, and the residual low-carbon (0.3 per cent) martensite starts to transform into ferrite accompanied by a reduction in hardness with a corresponding increase in toughness and ductility.

Fig. 2.15 *Effect of tempering on hardness for a 0.83 per cent carbon steel*

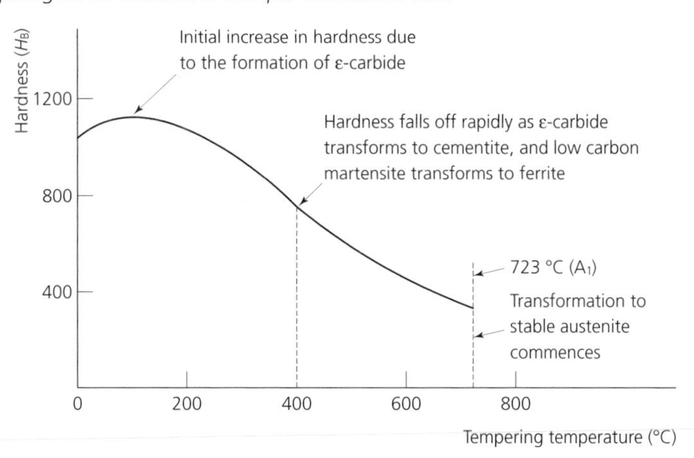

As the tempering temperature continues to rise towards the A_1 temperature (723 °C) the precipitation of iron carbide (Fe_3C) results in a structure similar to that produced by *spheroidising annealing*. In fact, quench hardening followed by high-temperature tempering is often used in place of spheroidising annealing since it is a faster process and gives a more uniform dispersion of the iron carbide.

Plain carbon and low-alloy steels are usually tempered below 300 °C where hardness and wear resistance are of primary importance. Examples of tempering temperatures for such steels are listed in Table 2.2. At these tempering temperatures the structure is of a fine pearlite called *troostite*. To differentiate the troostite of tempering from the troostite of quenching, the former is called *secondary troostite* (or just 'troostite'), whilst the latter is called *primary troostite* (or 'bainite'). However, the term troostite is falling into disuse and nowadays it is more usual to use the terms 'tempered martensite' and 'bainite'.

Table 2.2 *Steel tempering temperatures*

Surface appearances and applications

*Colour**	*Equivalent temperature (°C)*	*Application*
Very light straw	220	Scrapers; lathe tools for brass
Light straw	225	Turning tools; steel-engraving tools
Pale straw	230	Hammer faces; light lathe tools
Straw	235	Razors; paper cutters; steel plane blades
Dark straw	240	Milling cutters; drills; wood-engraving tools
Dark yellow	245	Boring cutters; reamers; steel-cutting chisels
Very dark yellow	250	Taps; screw-cutting dies; rock drills
Yellow–brown	255	Chasers; penknives; hardwood-cutting tools
Yellowish brown	260	Punches and dies; shear blades; snaps
Reddish brown	265	Wood-boring tools; stone-cutting tools
Brown–purple	270	Twist drills
Light purple	275	Axes; hot setts; surgical instruments
Full purple	280	Cold chisels and setts
Dark purple	285	Cold chisels for cast iron
Very dark purple	290	Cold chisels for iron; needles
Full blue	295	Circular and band saws for metals; screwdrivers
Dark blue	300	Spiral springs; wood saws

*Appearance of the oxide film that forms on a polished surface of the material as it is heated.

2.7 Martempering

This is a process for hardening alloy steels without the risk of distortion and cracking that is present when quench hardening in water or oil. The process consists of heating the steel to its austenising temperature and then quenching it in a salt bath furnace maintained at just above the M_s temperature for the steel. The steel is maintained at this lower temperature until the structure is uniformly heated throughout and the transformation to martensite is complete. At this point the work is removed from the low-temperature salt-bath furnace and it is allowed to cool naturally to room temperature. The transformations are shown in Fig. 2.16 where it can be seen that both the surface of the metal and its core pass through the M_s to M_{90} range at the same time. This uniform transformation of both the surface and the core of the material, coupled with a relatively mild quench into a salt-bath furnace rather than into oil or water at room temperature, results in the risk of distortion and cracking (due to internal stresses) being reduced to a minimum. A uniform martensitic structure will have been achieved throughout the work and this has to be tempered as for any other hardened steel.

Fig. 2.16 *The martempering process*

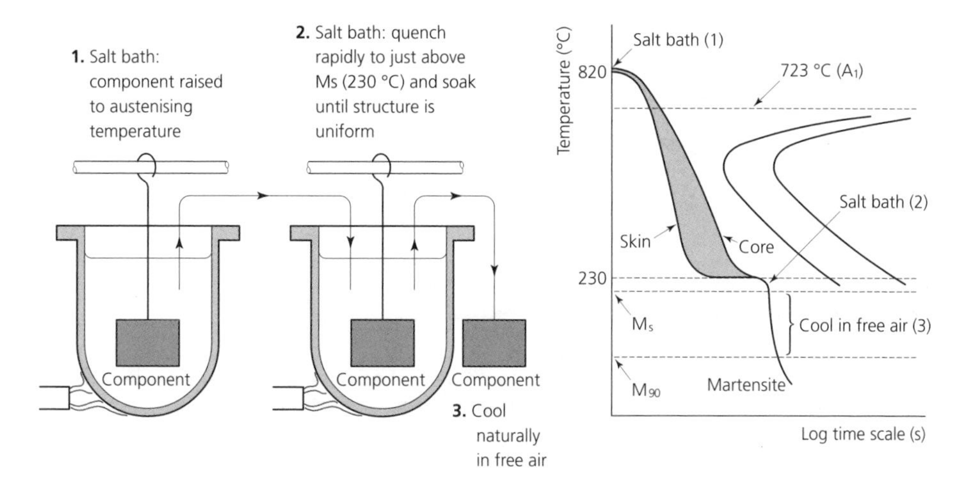

2.8 Austempering

This process is identical to that used in producing specimens when constructing time–temperature transformation diagrams. Reference to Fig. 2.17 shows that the work is first raised to above its austenising (A_3) temperature and then quenched in a salt bath at the required incubating temperature above the M_s temperature. The work remains in the incubating salt bath until both the surface and the core of the work have transformed into lower bainite. Once the transformations are complete, the work can cool down to room temperature naturally. The relatively high temperature of the quenching bath and the slow subsequent cooling reduces the possibility of cracking and distortion due to internal stresses. Austempering is a true isothermal process.

The mechanical properties of the lower bainite structure produced by austempering are similar to those for tempered martensite. Whilst austempering is widely used for the heat treatment of alloy steel components, it can only be used on plain carbon steels of high carbon content and small cross-section (less than 10 mm thick). This is due to the fact that it is difficult to cool a plain carbon steel quickly enough in the incubation salt bath for the cooling curve to avoid the 'nose' of the transformation curve. The lower critical cooling rates (better hardenability) of alloy steels overcome this problem.

Fig. 2.17 *The austempering process*

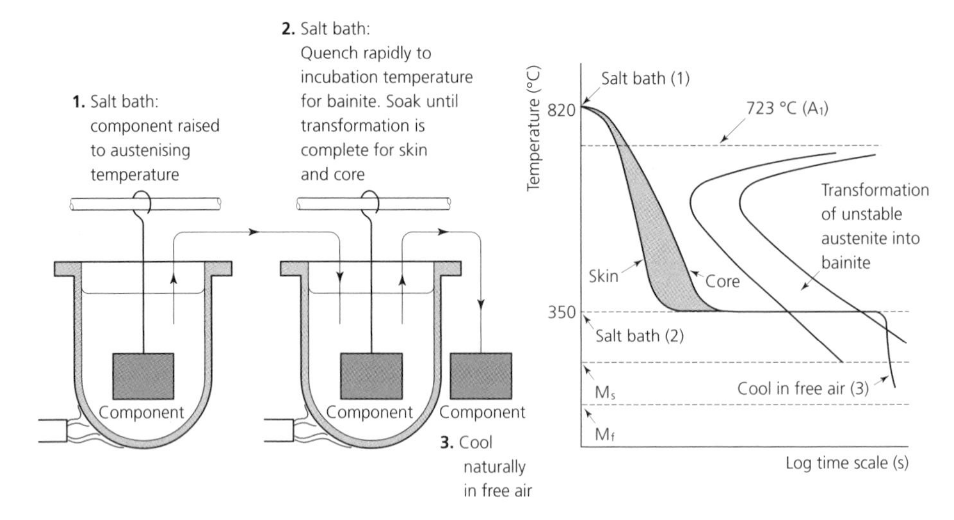

SELF-ASSESSMENT TASK 2.2

1. Describe how you would harden and temper cold chisels made from 0.8 per cent carbon steel on a batch production basis. Pay particular attention to the hardening and tempering temperatures and the quenching bath used.
2. Research the manufacturers' literature for the composition of a typical die-steel alloy for sheet metal press tools and discuss how such a steel should be hardened and tempered.
3. Research the manufacturers' literature for the composition of a typical high-speed steel and discuss how a lathe tool made from such a steel should be hardened.

2.9 Solution and precipitation treatment of maraging steels

Such steels are annealed by solution treatment to absorb the precipitated compounds, in a similar manner to aluminium–copper alloys, at a temperature of 820 °C for a time dependent upon the section thickness. This is usually about 15–30 minutes for thin sections and up to 1 hour per 25 mm for thick sections. Air cooling provides an adequate quenching rate because of the high alloy content, and the structure will become martensitic. This is not the hard, brittle tetragonal martensite associated with high-carbon steels and alloy tool steels, but a BCC martensite that is softer and tougher and capable of being machined and flow formed.

After processing the alloy can be precipitation age hardened, again in a similar manner to aluminium–copper alloys, by heating the alloy to 480 °C for 3 hours. The effects of over-ageing are slight, even after 200 hours. The exception is the 18Ni2400 grade that needs to be aged for 12 hours, although this time can be reduced by ageing at 540 °C. Age-hardening precipitates intermetallic compounds such as $TiNi_3$ and it is these compounds that give maraging steels their characteristic high strength and toughness.

EXERCISES

2.1 Explain why the iron–carbon phase equilibrium diagram is unsuitable for predicting the changes that take place during the quench hardening of a plain carbon steel.

2.2 (a) Describe the essential difference between hardness and hardenability.
(b) Explain what is meant by the term 'ruling section'.

2.3 In terms of hardness and hardenability, explain the essential differences between:

(a) a 1 per cent plain carbon steel, and
(b) a 0.3 per cent carbon 3.0 per cent nickel alloy

2.4 Sketch a time–temperature transformation curve for a high-carbon steel and explain how it can be used to determine the critical cooling rate for that steel and how it can be used to predict the changes that will take place during cooling.

2.5 With reference to a time–temperature transformation curve for a plain carbon steel of eutectoid composition, explain the significance of the M_s and M_f temperatures.

2.6 Explain why quench-hardened plain carbon and alloy steels need to be tempered before use.

2.7 Compare and contrast the following processes for hardening plain carbon and alloy steels in terms of their relative advantages and limitations.

(a) austempering
(b) martempering

2.8 Compare the advantages and limitations of either austempering or martempering with conventional quench hardening followed by tempering.

2.9 Compare and contrast the heat treatment of maraging steels with the heat treatment of conventional alloy steels.

2.10 Compare and contrast the hardening of a high-speed steel with the hardening and tempering of a plain carbon steel. Pay particular attention to the difference between secondary hardening and tempering, and the precautions that must be taken to avoid temper brittleness, grain growth and cracking.

3 Welded and brazed materials

The topic areas covered in this chapter are:

- Principles of fusion welding.
- Oxy-fuel gas welding.
- Manual metallic arc welding.
- TIG and MIG and submerged-arc welding.
- Welding defects.
- Electric resistance welding.
- Braze (bronze) welding.
- Brazing and silver soldering.
- Effects of welding and brazing on material properties.
- Comparison of joining techniques.

3.1 Fusion welding

Fusion welding produces a bond between two components by the melting and solidification of the parent metal and filler metal at the joint interface. The principle of fusion welding is illustrated in Fig. 3.1, where it can be seen that the edges of the 'V' are melted and fused together with the molten filler metal. There are two fundamental methods of fusion welding and they are named after the heat source used, as shown in Fig. 3.2.

Fig. 3.1 *Fusion welding: (a) before – a single 'V'-butt weld requires extra metal; (b) after – the edges of the 'V' are melted and fused together with the molten filler metal*

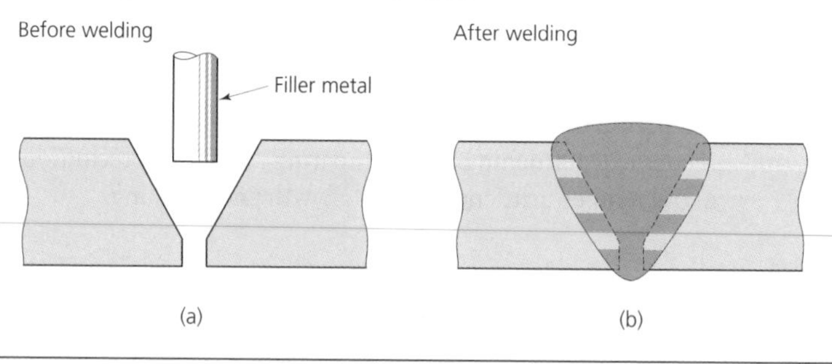

Fig. 3.2 *Comparison of (b) oxyacetylene welding and metallic arc welding*

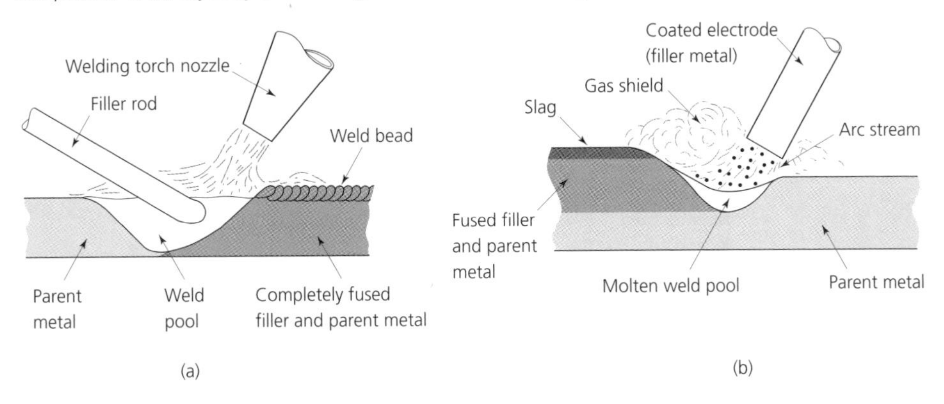

- Oxy-fuel gas (usually oxyacetylene) in which a very high temperature flame is used to melt the metal being joined as well as melting the filler material.
- Manual metallic arc in which an electric current forms a very high-temperature 'arc' (prolonged spark) between the metals being joined and an electrode made from suitable filler material.

Let's now examine some typical welding processes and find out how such processes can affect the structure and properties of the metals being joined.

3.2 Oxy-fuel gas welding

The principle of oxy-fuel gas welding is shown in Fig. 3.2(a). The flame from the welding torch provides the heat energy to melt (fuse):

- The filler rod that is made from a similar metal to that of the components being joined.
- The edges of the parent metal.

Upon cooling, the joint solidifies and the welded joint is complete. During welding the parent metal and the filler rod have been heated from room temperature to their melting temperatures and then cooled down again to room temperature relatively slowly. Obviously this will affect the grain structure of the metal not only at the edges of the metal where fusion occurs (the weld zone), but also for some distance back along the metal as it becomes heated by conduction.

For the production of satisfactory welded joints, it is essential to use the correct flame conditions. The appearance of the three basic flames is shown in Fig. 3.3.

3.2.1 *Neutral flame*

For most applications neutral flame conditions are used and this type of flame is shown in Fig. 3.3(a). The flame is easily recognised by its clearly defined U-shaped white inner cone at the tip of the nozzle. It is produced when approximately equal volumes of oxygen and acetylene are mixed in the torch, and the combustion reactions take place in two stages.

Fig. 3.3 *Oxyacetylene welding flame conditions: (a) the neutral flame; (b) the oxidising flame; (c) the carburising flame*

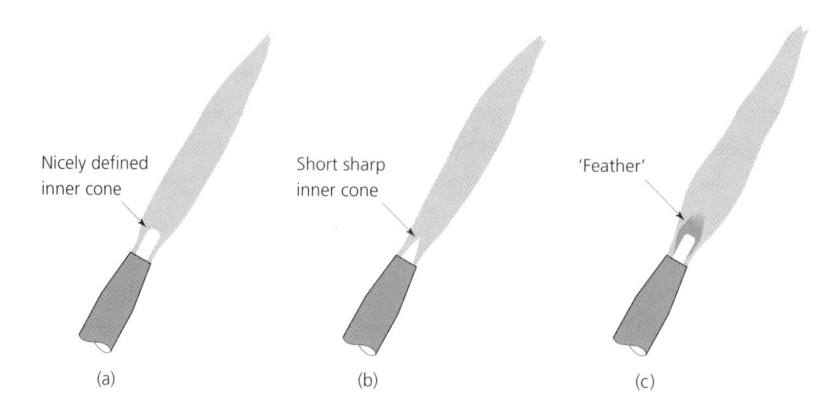

The primary reaction produces carbon monoxide and hydrogen, which are both reducing gases. The secondary reaction for complete combustion takes some additional oxygen from the surrounding atmosphere to produce carbon dioxide and water vapour. This secondary reaction reduces the possibility of the molten weld pool becoming oxidised. Further, with complete combustion, there is no free carbon to be absorbed by the molten metal. Thus with a neutral flame virtually no change will take place in the composition of the weld metal.

3.2.2 *Oxidising flame*

This type of flame is shown in Fig. 3.3(b). It is obtained by first setting the torch for a neutral flame and then reducing the acetylene supply slightly at the torch. The oxidising flame is easily recognisable by the shorter and sharply pointed conical inner cone. Combustion is noisier as the flame tends to 'roar'.

For complete combustion, one volume of acetylene requires two and a half volumes of oxygen. Thus, with the neutral flame, one and a half volumes of oxygen are taken from the atmosphere during the secondary reaction but, with the oxidising flame, all the oxygen necessary for combustion is taken from the cylinder and none is taken from the atmosphere. Therefore, the hot weld metal will be rapidly oxidised by the atmospheric oxygen to which it is exposed.

When welding carbon steels with an oxidising flame, the excess oxygen tends to combine with the carbon in the metal to form bubbles of carbon monoxide in the molten weld pool. This leads to degradation of the joint resulting from decarburisation of the weld metal and gas porosity. Iron oxides will also be formed and will become trapped in the joint as undesirable and weakening inclusions. Since the flame temperature is at a maximum with an oxidising flame, the metal may be overheated, resulting in oxidation of the grain boundaries (burning). This again leads to weakening of the joint. *For general welding, the oxidising flame must be avoided.* The only time that an oxidising flame is advantageous is when welding some brass, bronze and aluminium alloys.

3.2.3 *Reducing or carburising flame*

This flame is shown in Fig. 3.3(c) and is characterised by the 'feather' of incandescent carbon particles between the inner cone and the outer envelope. It is obtained by first setting the torch to neutral flame conditions and then slightly increasing the volume acetylene at the torch. When the ratio of oxygen to acetylene becomes less than one to one, a 'reducing' (carburising) flame condition is produced. This results in incomplete combustion that produces some free carbon. Thus ferrous weld metals become heavily carburised and this results in major changes in the composition, hardness and ductility of the weld metal.

- The carbon content of the molten ferrous weld metal is increased by diffusion in a manner similar to that occurring during case hardening.
- If the critical cooling rate of the carburised metal is exceeded and the carbon content of the carburised metal exceeds 0.4 per cent, hardening will occur accompanied by a sharp reduction in ductility and toughness.

These effects can be used for the flame hardening of machine slide ways by the Shorter process (see *Engineering Materials*, Volume 1). When carbon is dissolved in iron the melting temperature of the iron is lowered, and since carbon pick-up from the carburising flame is greatest at the surface of the metal, it follows that under these conditions the surface of the metal will melt before the body of the metal. This effect is exploited when cladding low- and medium-carbon steels with a hard surfacing material such as stellite.

3.3 Chemical reactions (gas welding)

The high temperatures encountered during welding not only cause structural changes in the metal, but they also promote chemical reactions between the metal being welded, the products of combustion, and atmospheric gases. The higher the temperature is raised, the more rapid becomes the rate of absorption of the contaminating gases and the rate of reaction with these gases, particularly when the metal is molten.

Oxygen tends to react with the parent metal and filler metal to form metallic oxides. Metals which oxidise readily are likely to be difficult to weld and the use of protective fluxes may be necessary. Oxidised welds are undesirable and can be identified by the appearance of the surface, which is irregular and pitted. Oxidation can result in:

- Fusion becoming difficult.
- Inclusions which may weaken the joint if the oxides formed are absorbed into the weld pool and remain entrapped in the solidified metal.
- Lack of ductility in the joint (brittleness).

Nitrogen dissolves in many molten metals and alloys, and may react with some of the constituents. If, when welding some steels, nitrogen is allowed to enter into the weld pool, the resulting joint will be porous and lacking in ductility.

Hydrogen can cause problems with many molten metals such as steel, aluminium and tough pitch copper as it is a powerful reducing agent. For example, tough pitch copper relies on the presence of particles of copper oxide to strengthen and harden the metal. Such particles would be reduced to metallic copper and water vapour by the hydrogen. The presence of hydrogen causes *gas porosity* that renders the joint weak and unsuitable for pipe joints and pressure vessels. Also, the removal of oxides by the hydrogen can directly reduce the strength and hardness of the metal. The water vapour formed during the reduction of the oxides can also cause gas porosity and this further weakens the joint. The main sources of hydrogen are the products of combustion of the fuel gas which is a hydrocarbon or, in the case of electric arc welding, from the electrode coating.

Water vapour tends to dissociate into hydrogen and oxygen when in contact with the molten metal. These nascent gases are highly reactive and exacerbate the effects described above.

3.4 The oxidation of welds

Some metals have such a high affinity for oxygen that the oxides on the surface of the metal reform as fast as they are removed. Although, as previously described, oxidised welds are generally unsatisfactory, this affinity for oxygen can be used to advantage in certain welding operations. For example, manganese and silicon, elements common to plain carbon steels, readily react with oxygen when the steel is in a molten condition. The reactions form a thin layer of *liquid slag* that protects the weld pool from further oxidation. It also prevents the formation of gas pockets (cavities) in the joint. Steel welding wires (filler rods) have a high silicon and manganese content to ensure that the protective slag is formed.

When welding other metals and alloys, removal and dispersion of the oxide film is not so easily achieved. Practical difficulties occur in welding when:

- The surface oxide forms a tenacious film.
- The oxide film has a very much higher melting point than the parent metal.
- The oxide film reforms very rapidly.

Fluxes are substances used to prevent oxidation and other undesirable chemical reactions from adversely affecting the quality of the joint. Table 3.1 lists some common fluxes and their applications when gas welding. The general requirements of fluxes are:

- To assist in removing the oxide film present on the surface of the parent metal in the joint zone.
- To prevent the oxide film reforming until the welded joint has been completed.
- To assist in removing any oxides which occur during welding by forming a 'fusible slag' that floats to the surface of the weld pool and does not interfere with the deposition and fusion of the filler material.
- To protect the weld pool from atmospheric oxygen and prevent the absorbtion and reaction of other gases (products of combustion), without obscuring the welder's vision, or hampering the manipulation of the molten pool.
- To lower the melting temperature of the oxide film on the parent metal below that of the parent metal itself. The flux should also have a lower melting temperature than the parent metal.

Table 3.1 ***Applications of fluxes for gas welding***

Metal or alloy	*Flux*	*Remarks*
Mild steel	—	No flux is required because the oxide produced has a lower melting point than the parent metal. Being less dense it floats to the surface of the molten weld metal as scale which is easily removed after welding. Use a neutral flame
Copper	Borax base with other compounds	If Borax alone is used, a hard scale of copper borate is formed on the surface of the weld which is difficult to remove. Use a neutral flame
Aluminium and aluminium alloys	Contains chlorides of lithium and potassium	Aluminium fluxes absorb moisture from the atmosphere, i.e. they are hygroscopic. Always replace the lid firmly on the container when the flux is not in use. The flux residue is very corrosive. On completion of the weld it is essential to remove all traces of this residue. This can be accomplished by scrubbing the joint area with a 5% nitric acid solution or hot soapy water. Use carburising flame
Brasses and bronzes	Borax type containing sodium borate with other chemicals	The flux residue is a hard glass-like compound which can be removed by chipping and wire brushing. Use an oxidising flame
Cast iron	Contains borates, carbonates, and bicarbonates plus other slag-forming compounds	Oxidation is rapid at red heat, and melting point of the oxide is higher than that of the parent metal. For this reason it is important that the flux combines with the oxides to form a slag which floats to the surface of the weld pool and prevents further oxidation

SELF-ASSESSMENT TASK 3.1

1. Explain what is meant by the term 'fusion welding', and explain how fusion welding differs from any of the soldering processes.
2. Describe the essential differences between reducing, oxidising and neutral welding flames, stating the circumstances in which they would be used.
3. Explain the need for, and essential requirements of, a flux, and describe the circumstances in which such a flux would be needed when oxyacetylene welding.

3.5 Manual metallic arc welding

The basic principles of manual metallic arc welding are shown in Fig. 3.4. The electric arc is 'struck' between the electrode and the metal being welded. The heat of the initial spark ionises the air gap and produces a conductive path for the electric current to flow between the electrode and the work. This enables the electric arc to be sustained. For safety and for maximum heat energy, a low-voltage, high-current density arc is used when welding.

Fig. 3.4 *Basic principles of electric arc welding*

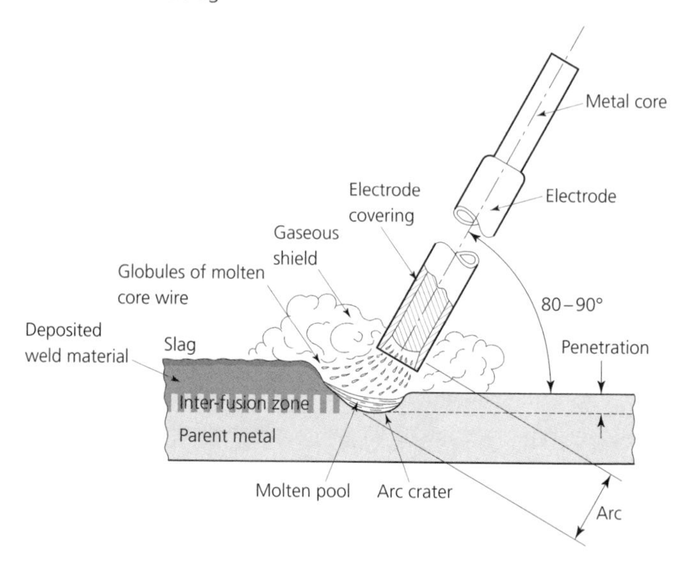

The heat of the arc is concentrated on the edges of the two pieces of metal being joined. This causes the metal edges to melt and additional molten metal, transferred across the arc from the electrode, also melts and acts as a filler rod. The molten metal from the electrode (the *arc stream*) is projected forcibly across the arc gap rather than flowing by gravity alone. This allows welding to take place vertically and overhead as well as along horizontal joints. As soon as the arc is struck, the tip of the electrode begins to melt thus increasing the gap between the electrode and the work. It is necessary, therefore, to feed the electrode towards the work to maintain an arc gap approximately 3 mm in length. The electrode is moved at a uniform rate along the joint, melting the metal to be joined and adding additional metal into the joint as it passes.

The majority of electrodes used in manual metallic arc-welding processes are *coated* electrodes. These consist of a core wire of closely controlled composition surrounded by a concentric coating of a solid flux that will melt uniformly with the core wire. The flux forms a partly vaporised and partly molten screen around the arc stream and protects it from contamination by atmospheric gases. The flux coating that surrounds the electrode has several important functions apart from protecting the arc stream. The more important of these are as follows:

- The flux coating forms a liquid slag over the weld pool which:
  (a) protects the solidifying weld metal from further contamination by atmospheric gases
  (b) prevents over-rapid cooling of the weld metal
  (c) controls the contour of the completed weld
  (d) picks up and removes impurities and inclusions from the weld pool
- It facilitates striking the arc and enables it to burn in a stable manner. This enables an alternating welding current to be used without the difficulties associated with uncoated wire electrodes.
- It serves as an insulator for the core wire when welding in deep grooves.
- It directs the arc and the globules of molten electrode into the weld pool, as shown in Fig. 3.4.
- It increases the rate of melting (metal deposition) and so speeds up the welding process.
- The coating can be manufactured with additions that will replace any alloying constituents in the electrode wire or the parent metal that may be lost during the welding process.
- It improves the penetration and enhances the strength of the weld.
- It can be formulated to increase or decrease the fluidity of the slag for special purposes. For example, a less fluid slag is desirable when welding overhead.

The coating of the core wire of an electrode consists of a mixture of a number of constituents. The more important of these constituents are listed in Table 3.2 together with

Table 3.2 *Properties of electrode-coating materials*

Constituent	*Remarks*
Titanium dioxide	Available in the form of natural sands as **Rutile** containing 96% titanium oxide. Forms a highly fluid and quick freezing slag. Is a good ionising agent
Cellulose	Provides a reducing gas shield for the arc. Increases the arc voltage
Iron oxide *and* manganese oxide	Used to adjust the fluidity and the properties of the slag
Potassium aluminium silicate	Is a good ionising agent, also gives strength to the coating
Mineral silicates *and* asbestos	Provides slag and adds strength to the coating
Clays *and* gums	Used to produce the necessary plasticity for extrusion of the coating paste
Iron powder	Increases the amount of metal deposited for a given size of core wire
Calcium fluoride	Used to adjust the basicity of the slag
Metal carbonates	Provides a reducing atmosphere at the arc. Adjusts the basicity of the slag
Ferro-manganese *and* ferro-silicate	Used to deoxidise and supplement the manganese content of the weld metal

the reasons for their inclusion. By varying the composition of the coating it is possible to produce a variety of electrode types. In practice, most electrodes conform to the six main types specified in BS 1719.

3.6 Shielding gases (arc welding)

As an alternative to using coated electrodes, shielding gases may be used to prevent oxidation of the weld pool. Although processes using shielding gases have higher material costs, labour costs are reduced because descaling is no longer required. Let's now look at the two most commonly used processes.

3.6.1 *Tungsten inert gas welding (TIG)*

This process employs a tungsten electrode that is not consumed. The atmospheric gases – oxygen and nitrogen – are excluded from the weld pool by a blanket of an inert gas such as *argon* or *helium* that will not react with the molten metal. The principle of TIG welding is shown in Fig. 3.5(a). You can see that the arc is struck between the tungsten electrode and the parent metal and, unlike other arc-welding processes, a separate filler rod is used in a similar way to gas welding.

Fig. 3.5 *Shielding gases in arc welding: (a) TIG welding; (b) MIG welding – the atmosphere is excluded from the weld by shielding with an inert gas; no chemical reactions take place*

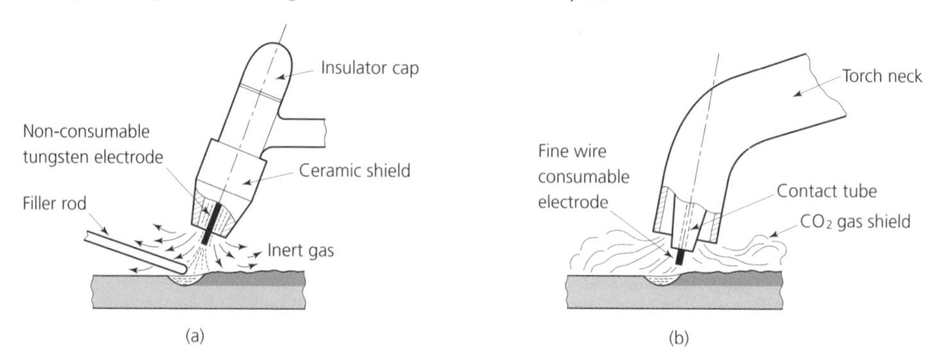

3.6.2 *Metal inert gas welding (MIG)*

This is a semi-automatic process. The electrode is a bare wire that is continuously fed from a drum into the welding gun by means of an automatic electrode wire drive unit that senses the potential difference across the arc and maintains a constant length of arc gap. The shielding gases used may be *argon*, *helium* or *carbon dioxide*. The use of carbon dioxide considerably reduces the operating costs of the process and, when used, the process is often called 'CO_2 welding'. Unfortunately, carbon dioxide is not an inert gas and cannot be used with some non-ferrous metals and alloys. When it passes through the arc it tends to break down into carbon monoxide and oxygen. (The effect of these gases on the strength of welded joints was discussed in Section 3.4.) To ensure that the liberated oxygen does not contaminate the weld metal, deoxidising alloying elements are incorporated into the welding wire. These deoxidising elements combine with the oxygen to form a very thin and neat protective layer of slag on the surface of the completed weld. This layer of slag does not have to be removed. Figure 3.5(b) shows the principle of MIG welding.

3.7 Submerged arc welding

This is an automatic process used for arc welding thick plate where very heavy welding currents are required. The end of the bare wire electrode is submerged under powdered flux, thus preventing splatter and ultraviolet radiation that could be a hazard to persons working nearby. The principle of this process is shown in Fig. 3.6. The tube delivers the powdered flux to the joint from a hopper and surplus flux is removed by a vacuum cleaning unit and recycled to the hopper. The electrode wire is automatically fed into the joint to keep the arc constant. Much of the flux melts and rises to the top of the molten weld metal as a protective slag. The automatic flux supply and electrode feed mechanism is traversed along the joint on a small power driven 'tractor'. The process is most suitable for straight line welding and is widely used in ship construction and other heavy-duty, large-scale applications.

Fig. 3.6 *Submerged arc welding*

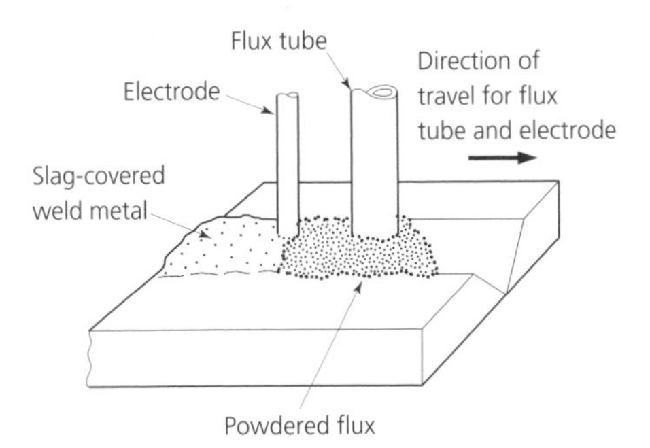

3.8 Welding defects

Some of the main factors to be considered when assessing the quality of a weld are:

- Shape of profile.
- Uniformity of surface and freedom from surface defects.
- Degree of undercut.
- Smoothness of join where weld is recommenced.
- Penetration of bead and degree of root penetration.
- Degree of fusion.
- Non-metallic inclusions and gas cavities.

Figure 3.7 shows a correctly formed 'V'-butt weld and a 'T'-joint fillet weld. The effect of the *heat-affected zone* on the joint strength will be considered in Sections 3.13 to 3.18 inclusive.

Fig. 3.7 *Correctly formed welds: (a) 'V'-butt weld; (b) 'T'-joint fillet weld*

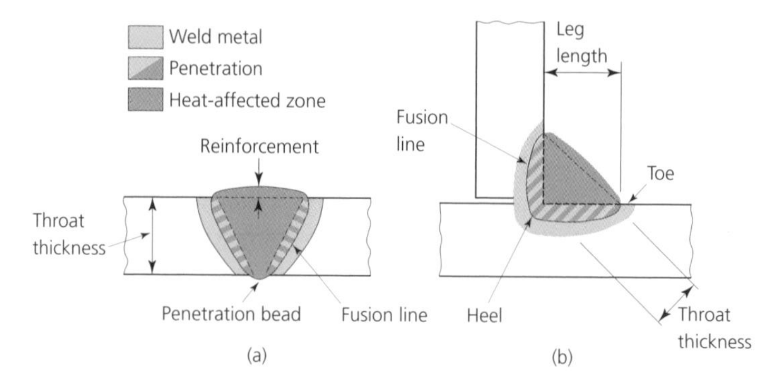

3.8.1 *Undercutting*

This is the term used to denote either the burning away of the side walls of the joint recess, or the reduction in parent metal thickness at the line where the weld bead is joined to the surface, as shown in Fig. 3.8. The formation of undercuts is particularly undesirable because they tend to weaken the joint by reducing the cross-sectional area and producing points of stress concentration. In multi-run welds there is also a danger of slag entrapment occurring.

Fig. 3.8 *Undercutting: (a) 'T'-joint fillet weld; (b) square butt weld*

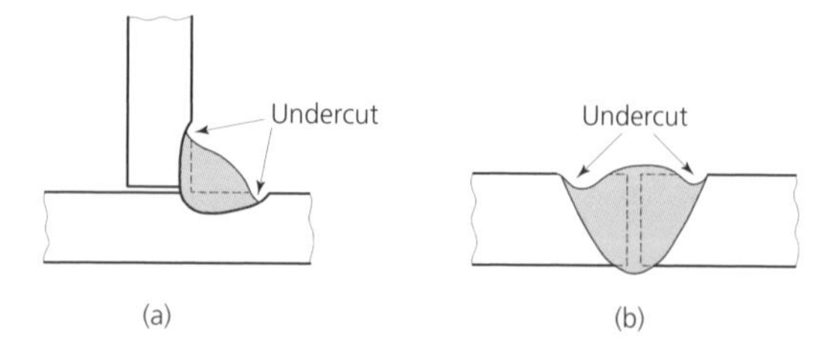

3.8.2 *Smoothness of joint when weld is recommenced*

Whenever a welding run has to be interrupted – for example, when changing an electrode or a filler rod – it is very important that when welding is recommenced the join should be as smooth as possible. The join should show no pronounced hump or crater in the weld surface, otherwise the join ends of the weld runs are liable to have poor strength. This is caused by crater cracks producing stress concentrations, and overlaps causing lack of fusion.

3.8.3 *Surface defects*

Surface defects in welded joints are generally due to the use of unsuitable materials and/or incorrect techniques. The weld surface should be free from porosity, cavities and either burnt-on scale (when gas welding) or trapped slag (when arc welding).

3.8.4 Penetration

One of the more common faults in welded joints is the lack of penetration. Examples of welds with unsatisfactory penetration are shown in Fig. 3.9. Lack of penetration results in a reduction of the cross-sectional area of the joint and corresponding lack of strength.

Fig. 3.9 *Incorrect penetration*

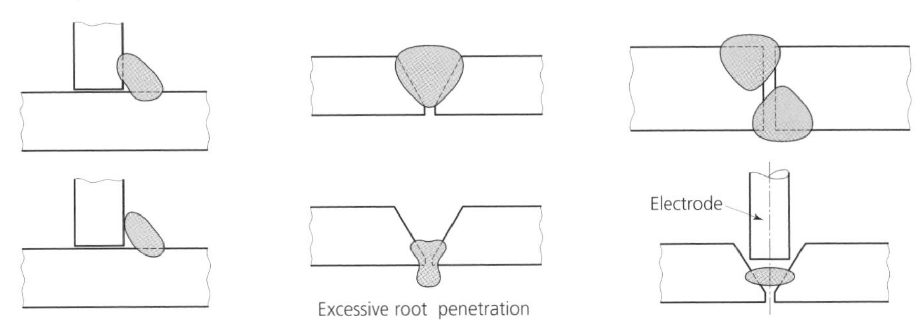

3.8.5 *Lack of fusion*

Lack of fusion is the failure to fuse together adjacent layers of weld or adjacent weld metal and parent metal, as shown in Fig. 3.10. This condition is caused by failure to raise the metal to its melting point either because of lack of heat energy, or because of an insulating layer of oxide or other impurities on the joint surfaces of the parent metal.

Fig. 3.10 *Lack of fusion: (a) 'T'-joint fillet weld; (b) 'V'-butt weld*

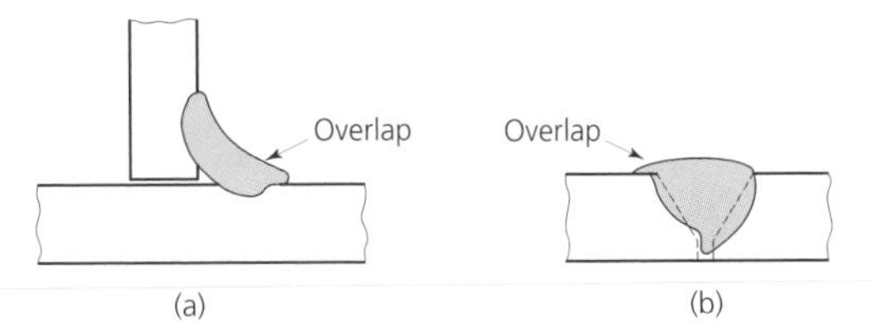

3.8.6 Inclusions

Inclusions may be slag or other foreign matter entrapped in the weld metal. These inclusions originate from the slag formed by the electrode coating, from badly prepared and dirty joint surfaces, or from mill-scale on the surface of the parent metal. They may also originate from atmospheric contamination. Slag inclusion in multi-run welds may result from inadequate removal of slag after the initial root run so that it becomes entrapped by the subsequent runs. Figure 3.11 shows some typical problems involving slag inclusions. In all instances inclusions are the result of faulty technique and lack of attention to detail.

Fig. 3.11 *Inclusions*

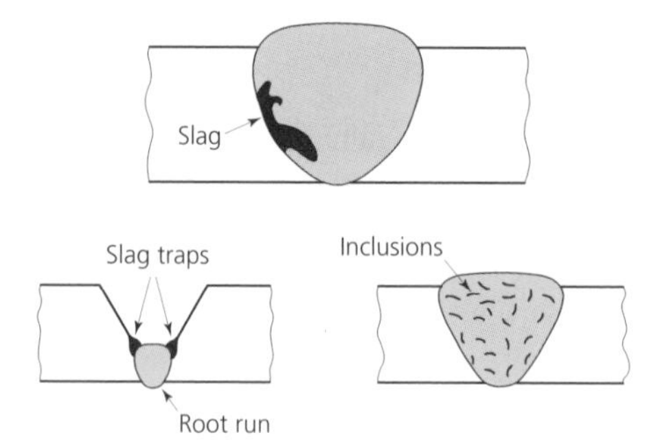

3.8.7 Porosity

This consists of groups of small cavities caused by gas entrapped in the weld metal, as shown in Fig. 3.12. Large cavities just below the surface or in the surface of the joint are called 'blow holes'. Table 3.3 lists some typical causes of porosity.

Fig. 3.12 *Porosity*

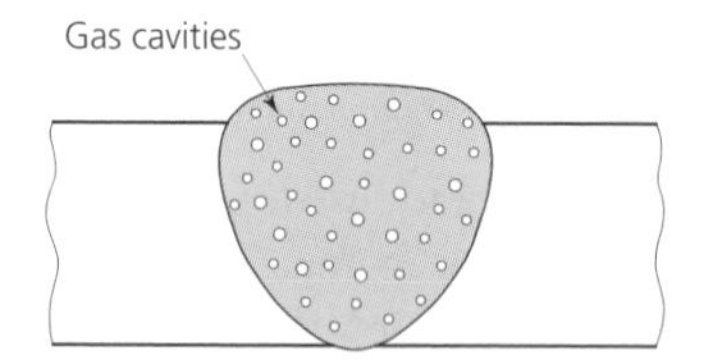

Table 3.3 *Common causes and remedies of porosity*

Cause	*Remedy*
1. High rate of weld freezing	Increase the heat input
2. Oil, paint, or rust on the surface of the parent metal	Clean the joint surfaces
3. Improper arc length, current, or manipulation	Use proper arc length (within recommended voltage range), control welding technique
4. Heavy galvanised coatings	Remove sufficient zinc on both sides of the joint
5. Use of oxidising flame	Use neutral flame for most welding operations

SELF-ASSESSMENT TASK 3.2

1. Describe the essential differences between oxy-fuel gas welding and manual metallic arc welding.
2. List the advantages and limitations of manual metallic arc welding compared with oxy-fuel gas welding.
3. Describe the principles of MIG and TIG welding and compare the advantages and limitations of these processes.
4. The principal factors to be assessed when judging the quality of a weld are:
   (a) shape of the profile
   (b) uniformity of surface and freedom from surface defects
   (c) degree of undercutting
   (d) smoothness of join where weld is recommenced
   (e) penetration of bead and degree of root penetration
   (f) degree of fusion
   (g) non-metallic inclusions and cavities

   Describe the effect of **five** of these factors on the strength of the joint, suggest possible causes of defects in the factors chosen, and suggest suitable remedies in technique to prevent the defects occurring in the first place.

3.9 Resistance welding

The welding processes discussed so far in this chapter have been fusion processes – that is, the filler metal and the edges of the parent metal have been melted and allowed to run together. In *resistance welding* processes the metal is raised to just below its melting point and the weld is completed by the application of pressure, as in forge welding.

3.9.1 *Spot welding*

This is the most common of the resistance-welding processes. It is much quicker than riveting as a technique for joining sheet metal components and, since no holes are drilled in the components being joined, they are not weakened. The components are joined together by making a series of spot welds side by side at regular intervals. Such joints are not 'fluid tight' and have to be sealed to prevent leakage or corrosion. Apart from ensuring that the joint faces are clean and free from corrosion, no special joint preparation is required.

The temperature of the components to be joined is raised locally by the passage of a heavy electric current, at low voltage, through the components, as shown in Figs. 3.13(a) and 3.13(b). When an electric current flows through a resistor the electrical energy is partially converted into heat energy and the temperature of the resistor is raised. Resistance welding uses this principle. Resistance to the flow of current occurs between the two surfaces of the components being joined, over the cross-section of the electrodes. Sufficient heat is generated to raise the components to the welding temperature at this spot. The current is then switched off and the weld is completed by squeezing the components together tightly between the electrodes. The electrodes are made from copper as this metal has high thermal and electrical conductivity. The electrodes are water cooled to prevent them from overheating. The complete cycle of events is controlled automatically. No additional material filler metal is required to make a resistance-welded joint. Since the process is akin to forge welding, the grain structure of the metal in the weld zone is in the wrought condition, rather than in the 'as cast' condition associated with fusion welding. This makes resistance welds stronger and more ductile than fusion welds of similar cross-sectional area.

Fig. 3.13 *Principles of resistance welding: (a) electric spot-welding machine (schematic diagram); (b) spot welding; (c) seam welding; (d) projection welding*

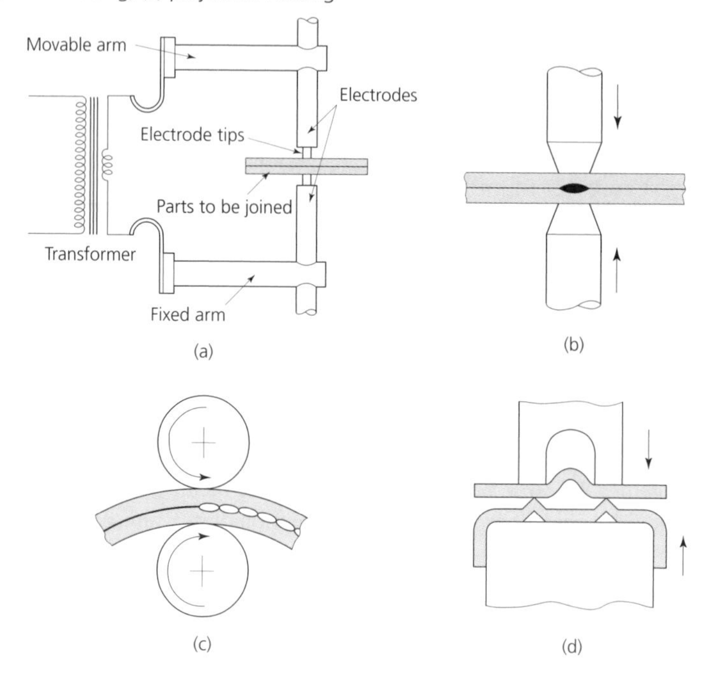

3.9.2 *Seam welding*

The components to be joined are gripped between revolving, circular electrodes, as shown in Fig. 3.13(c). The welding current is applied in pulses resulting in a series of overlapping spot welds being made along the seam. This method of resistance welding is used for the manufacture of containers and fuel tanks.

3.9.3 *Projection welding*

With this process the electrodes act as locations for holding the parts to be joined. The joint is so designed that *projections* are preformed on one of the parts, as shown in Fig. 3.13(d).

Projection welding enables the welding pressure and the heated welding zone to be localised at predetermined points. This technique is largely used for small components that need to be located accurately.

3.9.4 *Butt welding*

The joints described so far in this section have been lap joints connecting sheet metal components. Butt welding is used for connecting more solid components such as carbon steel shanks on to the high-speed steel bodies of large twist drills. The principle of resistance butt welding is shown in Fig. 3.14. The two ends of the rods are brought together with just sufficient force to ensure that the current can flow without arcing. The resistance of the joint face ensures that local heating will then take place on the passage of a heavy, low-voltage, electric current. When the metal in the joint zone has reached its welding temperature, the current is switched off and the axial force on the joint is increased to complete the weld.

Fig. 3.14 *Resistance butt welding*

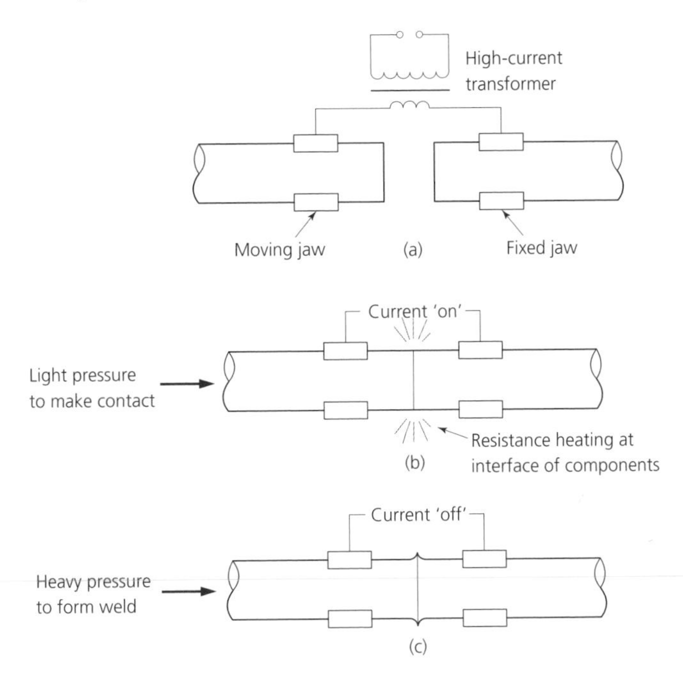

As for all resistance-welding processes, a sound weld is achieved by raising the temperature of the metal to just above the pressure-welding temperature. Should the surfaces of the components at the joint interface commence to melt, this molten metal is displaced by the force on the components until metal at the correct pressure-welding temperature is reached in the joint substrate and the weld is completed.

3.10 Braze welding

Braze welding, also known as bronze welding, differs from fusion welding in three ways.

1. Braze welding can be used to join dissimilar metals as well as similar metals.
2. The melting point of the filler metal is lower than the metals being joined.
3. In braze welding the parent metal is never melted.

Figure 3.15 shows a typical braze-welded butt joint. Its similarity with a welded joint is immediately apparent. However, the joint faces of the parent metal are not melted, which they are when fusion welding. Instead, the copper-rich filler alloy reacts with the surfaces of the parent metal to form intermetallic compounds. This *interdiffusion* of the parent metal by the filler alloy is referred to as 'tinning'. For tinning to occur successfully, the joint faces must be thoroughly clean both physically and chemically. No scale, oil, grease, paint or other contaminant can be present if a successful joint is to be made. A borax-based flux is required and suitable fluxes are supplied in powder form by the manufacturers of the brazing rods.

Fig. 3.15 *Braze welding: the metal in the region of the joint is first painted with flux*

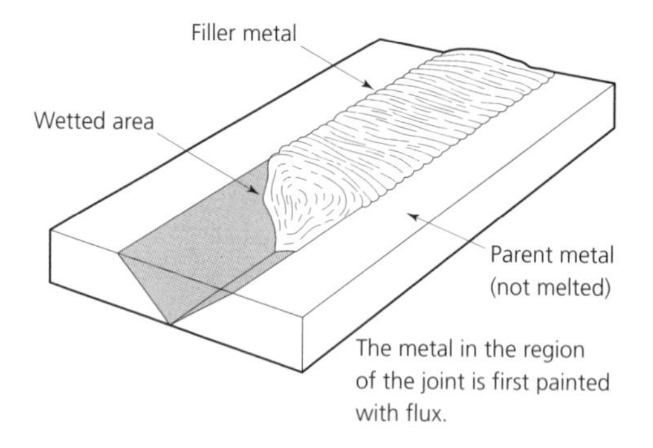

Apart from the ability to join dissimilar metals, one of the main advantages of braze welding over fusion welding is the lower melting temperature of the filler material. This reduces such problems as distortion and oxidation of the parent metal. The lower process temperature reduces the extent of the heat-affected zone each side of the joint. The lower process temperature also results in less grain growth in the parent metal with correspondingly improved properties. Table 3.4 lists some typical B 1453 copper alloy filler rods for braze welding, together with their compositions and melting temperatures.

3.11 Brazing (hard soldering)

Brazing is also known as hard soldering. The difference between hard soldering and soft soldering lies in the melting temperature range of the filler materials (solders and

Table 3.4 ***Copper alloy filler rods for bronze welding***

Compositions and melting points

BS 1453 Type	*Composition (%)*						*Melting point (°C)*
	Copper min.–max.	*Zinc min.–max.*	*Silicon min.–max.*	*Tin max.*	*Manganese min.–max.*	*Nickel min.–max.*	
C2	57–63	36–42	0.2–0.5	0.5	—	—	875
C4	57–63	36–42	0.15–0.3	0.5	0.05–0.25	—	895
C5	45–53	34–42	0.15–0.5	0.5	0.5	8–11	910
C6	41–45	37–41	0.2–0.5	1.0	0.2	14–16	

Note: Type C2 is termed a *silicon–bronze* filler rod and is specially recommended for the braze welding of *brass* and *copper* sheet and tubes as are used for sanitary and hot water installations. It is also suitable for the braze welding of *mild steel* and *galvanised steel*.

Type C4 is termed a *manganese–bronze* filler rod, it has a higher melting point and is especially suitable for braze welding *cast* or *malleable iron*, and also for building up worn parts such as gear teeth.

Types C5 and C6 are *nickel–bronze* rods which are recommended for braze welding *steel* or *malleable iron* where the *highest mechanical strength* is required. These are the high-melting point welding rods and have a valued application in the reclaiming and building up of wearing surfaces.

spelters). Brazing spelters are alloys of copper and zinc that have a melting temperature range of approximately 850–900 °C. Silver solders are alloys of copper, zinc and silver that have a melting temperature range of approximately 620–775 °C. Soft solders are alloys of tin and lead that have a melting temperature range of approximately 180–255 °C.

The essential differences between welding, braze welding and brazing is that, when brazing, only the filler metal becomes molten and that it *flows into the joint by capillary attraction*. In both fusion welding and braze welding a fillet of molten filler metal is built up in the joint and there is no capillary flow. To ensure that the molten filler material flows into the joint, scrupulous cleanliness of the joint surfaces is required and this is ensured by the use of an appropriate flux. The term brazing (hard soldering) also applies to silver soldering. The success of the brazing process depends upon the following general conditions:

- The selection of a suitable filler material that has a melting range appreciably lower than that of the parent metals being joined.
- Thorough cleanliness of the surfaces being joined.
- Complete removal of any oxide film from the joint surfaces of the parent metal and from the filler material by an appropriate flux.
- Complete 'wetting' of the joint surfaces by the filler material. The flux should assist this 'wetting' to take place.

When a surface is 'wetted' by a liquid, a continuous film of the liquid remains after any surplus liquid has been drained off. Similarly, it is essential in any brazing process that the molten filler material forms a continuous film over the entire joint surfaces and completely fills the joint into which it is drawn by capillary attraction.

Unlike fusion welding where only similar metal and alloys can be successfully joined, metals and alloys of a dissimilar composition can be joined together by brazing – for example, steel to cast iron.

3.12 Filler materials (brazing alloys)

The oldest filler material, from which the 'brazing' process gets its name, is a *brass alloy* containing equal amounts of copper and zinc. This was originally called 'brazing spelter' (*spelter* being the old name for zinc). The high zinc content reduces the melting temperature range, thus enabling a joint to be made with components manufactured from ductile brass alloys that have a higher copper content. It might be thought that the use of a high zinc content brazing spelter would result in a relatively weak joint. However, during the brazing process some of the volatile zinc content is lost and the final composition of the filler material has a lower zinc content and a higher strength. When brazing with a copper–zinc alloy spelter, a paste of borax in water provides a suitable flux. The composition of some BS 1845 brazing spelters are listed in Table 3.5 together with their approximate melting ranges.

Table 3.5 *Brazing spelters*

Compositions and melting points

BS 1845 Type	*Composition (%)* Copper min.–max.	Zinc	*Approximate melting range (°C)*
8	49–51	Balance	860–870
9	53–55	Balance	870–880
10	59–61	Balance	885–890

Note: This group of copper alloys tends to lose zinc by vaporisation and oxidation when the parent metal is heated above 400 °C. This loss of zinc produces relatively higher tensile strength. The brazing alloys containing a high percentage of zinc, therefore, produce joints of the lowest strength. Type 8 is used for medium-strength joints, whilst the strongest joints can be produced by using type 10.

3.13 Filler materials (brazing alloys containing phosphorus)

Filler materials that contain phosphorous are usually referred to as self-fluxing. These alloys contain silver, copper and phosphorus or simply copper and phosphorus, the former possessing a lower melting temperature range. Examples of both these alloy types are listed in Table 3.6. The phosphorus content lowers the melting range sufficiently for copper and copper-based alloys to be brazed without the parent metal melting. Copper and steel components can be brazed in air since the oxidation products form a liquid compound that acts as an effective flux. Copper-based alloys require a separate flux. Self-fluxing alloys are widely used for furnace brazing and other manufacturing techniques where the filler material is applied as 'preforms'.

Table 3.6 *Brazing alloys containing phosphorus*

Compositions and melting points

BS 1845 Type	*Composition (%)* Silver min.–max.	Phosphorus min.–max.	Copper	*Approximate melting range (°C)*
6	13–15	4–6	Balance	625–780
7	—	7–7.5	Balance	705–800

3.14 Filler materials (silver solders)

These are more expensive than the common brazing alloys because they contain a high percentage of the precious metal silver. Typical silver solders are listed in Table 3.7 together with their compositions and approximate melting ranges. However, they offer the advantage of producing strong and ductile joints at much lower temperatures than those associated with brazing spelters and, consequently, have little heat-effect on the parent metals being joined. Silver solders flow very freely, which speeds up the process, results in a neat joint on fine and intricate work, and requires little finishing. Proprietary fluxes are required and these can be obtained from the manufacturers of the silver solder. After all brazing and silver soldering operations, it is important to remove any residual flux deposits. Suitable cleaning solutions are listed in Table 3.8.

Table 3.7 ***Silver solders***

Compositions and melting points

BS 1845 Type	*Composition (%)* Silver min.–max.	Copper min.–max.	Zinc min.–max.	Cadmium min.–max.	*Approximate melting range (°C)*
3	49–51	14–16	15–17	18–20	620–640
4	60–62	27.5–28.5	9–11	—	690–735
5	42–44	36–38	18.5–20.5	—	700–775

Note: Type 4 possesses a high conductivity and is, therefore, very suitable for making electrical joints. It is the most expensive because of its high silver content.

Type 3 is extremely fluid at brazing temperatures which makes it ideal when brazing dissimilar metals. A *low melting point alloy*.

Type 5 is a general-purpose silver solder which can be employed at much higher brazing temperatures.

Table 3.8 ***Methods of flux removal after brazing and silver soldering***

Stage	*Method A* Solution	*Temp.*	*Time (min.)*	*Method B* Solution	*Temp.*	*Time (min.)*
1	Concentrated nitric acid	Room	5–15	10% nitric acid 0.23% hydrofluoric acid	Room	5–10
2	Water rinse	Room	—	Water rinse	Room	—
3	10% nitric acid 5–10% sodium dichromate	Room	5–10	—	—	—
4	Water rinse	Hot	—	Water rinse	Hot	—
5	Dry using warm air	—	—	Dry using warm air	—	—

Note: Method A is the more common, and method B gives a good uniform appearance to the joint surfaces, as it counteracts the darkish appearance in the joint caused by the silicon content of the filler metal.

3.15 Brazing techniques

There are various ways in which metals may be joined by brazing. In every case, however, the joint must have a *uniform capillary gap* between the components so that the molten filler metal will flow into the joint. For any one combination of liquid and solid, the smaller the gap the deeper the capillary penetration. The principle of capillary flow is illustrated in Fig. 3.16.

Fig. 3.16 *Capillary flow*

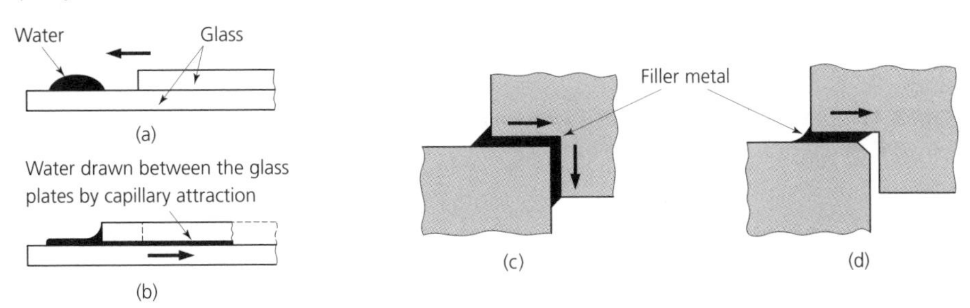

Consider a glass plate containing a small droplet of water, as shown in Fig. 3.16(a). Another piece of glass is then laid on the first piece and made to slide until its edge is brought into contact with the water droplet. Some of the water will immediately flow by *capillary attraction* into the very narrow gap between the mating glass surfaces for a considerable distance, as shown in Fig. 3.16(b). Exactly the same effect occurs when brazing and hard soldering, except that instead of glass plates there are the mating surfaces of the metals being joined and, in place of water, there is the molten filler material.

The effectiveness of capillary attraction and the soundness of the joint are governed by the maintenance of appropriate joint clearances. Under normal conditions the joint clearance (gap) should lie between 0.01 and 0.02 mm. The mating surfaces should be parallel and there should be no break in the uniformity of the clearance. If a break occurs due to the widening or closing of the clearance, then capillary flow will stop in that vicinity and may not go beyond it. Figure 3.16(c) shows that metal can be drawn round a corner by capillary attraction when the gap is uniform. However, the chamfered corner shown in Fig. 3.16(d) changes the gap size (increases the clearance) and no capillary flow takes place beyond the chamfer. The joint is therefore weakened.

3.15.1 *Flame brazing*

This technique may be used to fabricate almost any assembly on a small quantity basis. An oxy-propane gas torch is commonly used, and a skilled operative can produce neat joints that require no finishing. Flame brazing is also used where the assembly is too large to braze in a furnace. To ensure a sound joint, the correct flux must be applied, and the parent metal must be sufficiently hot to melt the spelter on contact after the flame of the torch has been momentarily withdrawn.

3.15.2 *Furnace brazing*

This technique is used when:

- The parts to be brazed can be pre-assembled or jigged to hold them in position.
- The brazing filler material can be preformed and preplaced, as shown in Fig. 3.17.
- A controlled atmosphere is required.

The method of heating varies according to the application. The work to be brazed can be loaded into a muffle furnace so that the atmosphere can be controlled and the products of combustion will not affect the joint, or the work can be packed into sealed containers. Alternatively, the work can be passed continuously through the furnace on a conveyor.

Fig. 3.17 *Use of prepared filler metal spelter preforms*

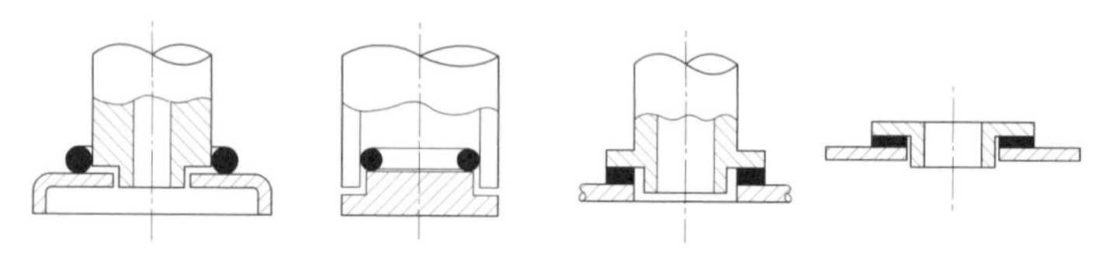

3.15.3 *Dip brazing (molten spelter)*

The parts to be brazed are assembled together and submerged in a bath of molten filler material that is drawn into the fluxed joint by capillary attraction. It does not adhere to the surface of the work. The filler material is melted in a graphite crucible and a layer of flux is floated on the surface of the molten metal. Large assemblies have to be preheated before being lowered into the molten filler material.

3.15.4 *Dip brazing (salt bath)*

The molten salts, in a salt bath furnace, provide uniform heating of the work and prevent atmospheric contamination. They do not take part in the fluxing of the joint or the intermetallic reactions between the filler metal and the metal components being joined. Again, the work has to be pre-assembled, fluxed and the filler material applied as preforms.

3.15.5 *Electric induction brazing*

The component to be brazed is placed in the magnetic field of an induction coil through which is passed a high-frequency electric current, as shown in Fig. 3.18. This induces eddy currents in the component that causes it to heat up, thus generating heat within the component itself. Preformed and prepositioned filler material is used together with a suitable flux if required.

Fig. 3.18 *Electric induction brazing: (a) external coil; (b) internal coil – it is usual for induction coils to be designed to surround the joint, but internal coils can be used for certain applications*

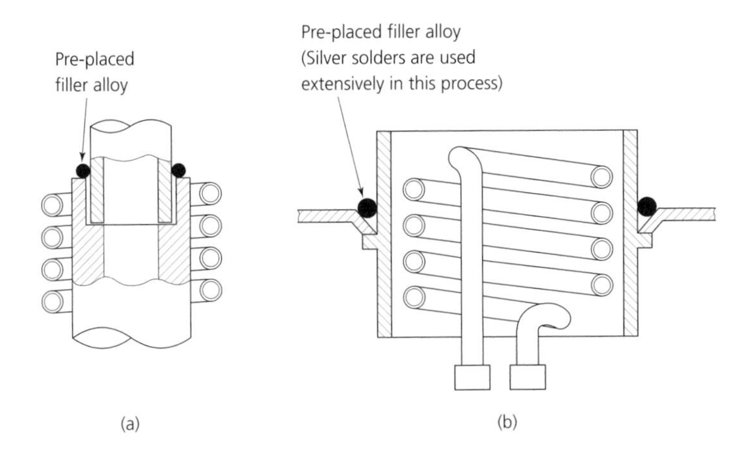

3.15.6 *Electric resistance brazing*

In this process the heat required to melt the filler material is developed by:

- Resistance at the joint interface (direct heating).
- Resistance between the work and the electrodes (indirect heating).

The basic principle of resistance heating is that a heavy electric current at a low electrical potential (voltage) is passed through the assembly in such a way that a hot spot is generated at the joint interface, as shown in Fig. 3.19. Heating can be localised and this ensures that the parent metal suffers no general loss of mechanical properties.

Fig. 3.19 *Electric resistance brazing: (a) direct heating; (b) indirect heating; both techniques apply pressure at the brazing temperature; special machines are used which are very similar in operation to spot welders, except the electrodes are usually made of carbon, molybdenum, tungsten or steel*

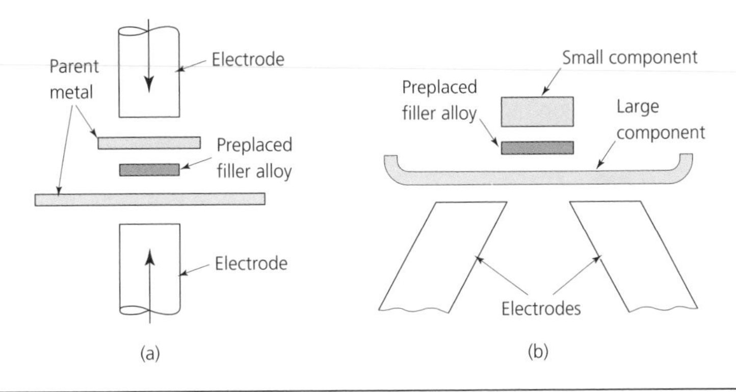

3.16 Aluminium brazing

In aluminium brazing the filler material is an aluminium alloy having a melting point below that of the parent metal. The various grades of commercially pure aluminium alloys containing 1.25 per cent manganese – and certain aluminium–magnesium–silicon alloys – can be successfully joined by brazing. Aluminium–magnesium alloys containing more than 2 per cent magnesium are difficult to braze, as their oxide films are tenacious and difficult to dissolve with ordinary brazing fluxes. Table 3.9 lists some typical aluminium filler alloys. Proprietary fluxes should be used, and these are basically mixtures of alkali metal chlorides and fluorides. Care should be taken when handling and using the fluxes as they can cause skin irritation and the fumes given off when heated can be toxic.

Table 3.9 ***Common filler alloys for aluminium brazing***

Compositions and properties

	Composition (%)				
BS 1942 Type	*Aluminium*	*Copper min.–max.*	*Silicon min.–max.*	*Approximate melting range (°C)*	*Brazing range (°C)*
1	Balance	2–5	10–13	550–570	570–640
2	Balance	—	10–13	565–575	585–640
3	Balance	—	7–8	565–600	605–615
4	Balance	—	4.5–6.5	565–625	620–640

Note: It is impossible to braze certain aluminium alloys whose melting points are below those of the available brazing alloys.

SELF-ASSESSMENT TASK 3.3

1. Discuss the essential differences between soft soldering, silver soldering and brazing and, with the aid of typical examples, explain where each process could be used to the greatest advantage.
2. List the requirements for making a successful joint by flame brazing.
3. Discuss the essential differences between braze (bronze) welding and brazing.
4. Research the manufacturers' literature and list the composition, properties and applications of typical fluxes used when silver soldering and brazing.

3.17 The effect of welding on material properties

Let's now consider the effects of welding on the properties of the materials being joined. When two pieces of metal are joined by fusion welding, the weld pool and the edges of the parent metal are molten. As the joint cools down, the molten metal solidifies and becomes a miniature casting. The fact that a cast metal structure is weaker than a wrought metal structure results in a joint that is usually weaker than the surrounding metal. Figure 3.20 shows a typical cross-section through a solidified weld pool when welding mild steel and indicates the grain structure that may usually be found. Fortunately the chilling effect of the parent metal inhibits grain growth and minimises the loss of strength. However, the condition of the parent metal each side of the joint must also be considered because many metals and alloys rely upon cold working and/or heat treatment to enhance their mechanical properties. The temperatures reached in the parent metal, each side of the weld zone, are sufficiently high to modify the grain structure resulting from such cold working and/or heat treatment. The areas each side of the weld zone, where structural modifications occur as a result of welding, are referred to as the *heat-affected zones*. The upper limit of the temperature gradient across the heat-affected zone is the temperature of the molten weld pool, and the lower limit is the temperature of the parent metal remote from the weld zone (room temperature). Figure 3.21 shows, schematically, the heat-affected zone for a welded joint in mild steel, whilst Table 3.10 summarises the effect of the temperature gradient on the crystal structure of the metal.

Fig. 3.20 *Structure of weld metal (mild steel): (a) large single-run weld; (b) multi-run weld*

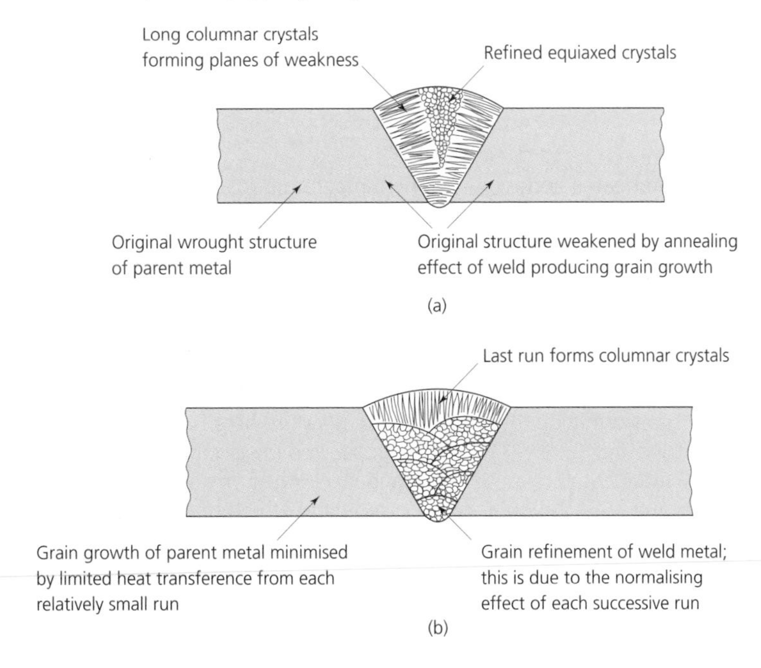

Identification of the various heat-affected zones discussed above can be achieved by examining the variation in hardness across the welded joint. The rapid fall off in hardness associated with the heat-affected zones is due to annealing or over-ageing (depending upon the metal or alloy). The partial recovery of hardness associated with the weld zone is a result of chilling of the cast structure and the addition of alloying elements from the filler

Fig. 3.21 *Structure of a welded joint (schematic)*

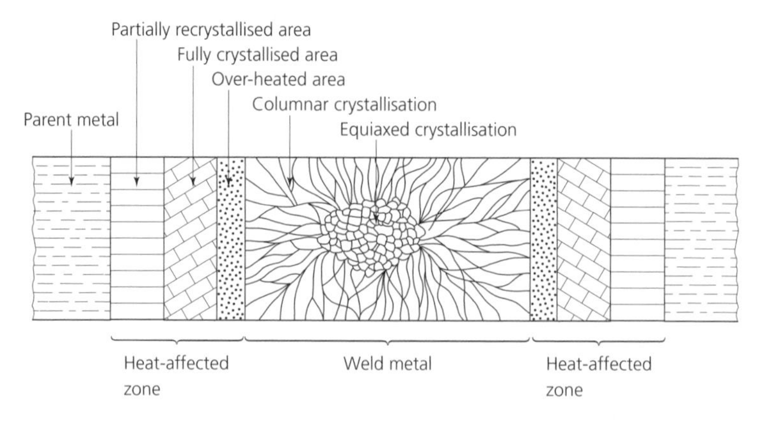

Table 3.10 *Effect of temperature gradient on crystal structure of metal*

Temperature zones	*Remarks*
Fusion zone	Temperature reaches melting point. The cooling rate is in the order of 350–400 °C/min, which is the maximum quenching range. The weld is less hard than the adjacent area of the parent metal because of loss of useful elements (carbon, silicon and manganese)
Overheated zone	The temperature reaches 1100–1500 °C. Cooling is extremely rapid in the order of 200–300 °C/min. Some grain coarsening occurs
Annealed zone	Here the temperature reaches slightly higher than 900 °C. The parent metal has a refined normalised grain structure. The change is not complete because the cooling rate is still high, in the order of 170–200 °C/min.
Transformation zone	The temperature here is between 720 and 910 °C. These are the upper and lower critical temperatures between which the iron in steel transforms from a body-centred-cubic to a face-centred-cubic structure. The parent metal tends to recrystallise

material. Changes in tensile strength across the joint are similar to the changes in hardness. Cracking, resulting from brittleness, depends upon the cooling rates of the various zones. It is, therefore, sometimes necessary to apply heat treatment to welded assemblies (e.g. anneal or normalise) to stabilise and enhance their mechanical properties. The adverse effects of welding on certain alloy steels (e.g. stainless steels and maraging steels), and the precautions that must be taken when welding such steels, were introduced in Section 1.6 and are discussed further in Section 3.15.

3.18 The effect of welding on mild steel

Mild steel is one of the easiest metals that can be joined by fusion welding. With care, the joint that is formed is almost as strong as the parent metal. When arc welding mild steel, the heat-affected zones will be much narrower than in gas welding, because the heat of the arc is more localised and the temperature is raised more quickly. However, when the flame of a gas-welding torch is applied to the edges of the components being joined, the time taken to raise them to the welding temperature causes appreciable heating of the parent metal and appreciable spread of the heat-affected zone. This not only softens the parent metal each side of the joint, but reduces the 'chill casting' effect on the weld deposit. Further, since the heat from the welding flame is applied for a longer period, and cooling of the joint is slower, grain growth is also more pronounced with a corresponding loss of strength. Thus, for thick plates, arc welding should be used. Figure 3.22 compares the macrostructures of a single run weld in mild steel plate: 3.22(a) when gas welding and 3.22(b) when arc welding.

Fig. 3.22 *Macrostructure of single-run welds: (a) oxyacetylene weld in mild steel; (b) metallic arc weld in mild steel*

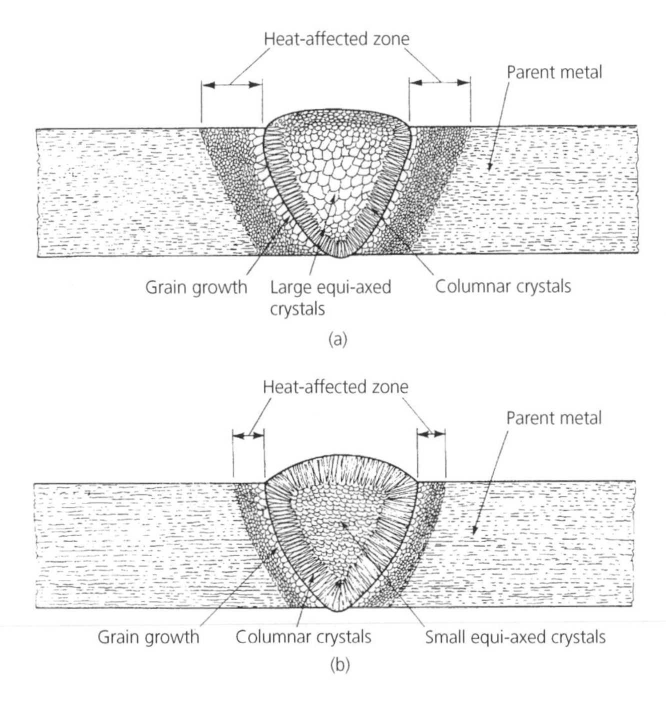

When a fusion weld is made in low-carbon steel with the addition of filler material, the following structures will result:

- Metal that has been molten will make up a cast structure of deposited metal alloyed with parent metal.
- There will be a *fusion line* at the junction between the metal that has been melted and the parent metal that has not been melted.

- There wil be a *heat-affected zone* that extends from the fusion line to that part of the parent metal that has not been heated sufficiently for structural changes to occur.
- In the heat-affected zone there is an area, adjacent to the 'fusion line', that has cooled slowly and thus has a coarse grain structure. This results in loss of hardness and strength.
- Progressing away from the weld zone through the coarse grain structure just described, the grains become smaller, and the zone where they become very small is called the *refined zone*. This metal has been heated to the transformation temperature just long enough for the metal to recrystallise, but the metal has then cooled before grain growth can occur.
- Progressing away from the refined zone described above, there is a zone of *mixed structure*. In this zone some of the grains have recrystallised and some have remained unaffected.
- Finally, the last zone, remote from the weld zone, is the *unaffected zone*. This is where the parent metal has not been sufficiently heated to cause any structural change.

3.19 The effect of welding on alloy steels

The problems associated with the welding of alloy steels, such as stainless steels, have already been introduced in Section 1.6. Weld decay occurs in 18/8 austenitic stainless steels unless such steels have been 'proofed' by the addition of molybdenum, niobium or titanium (1.6 per cent). The reason for the onset of weld decay is shown in Fig. 3.23. The temperature curve across the heat-affected zone results in bands each side of the joint in which the metal will be at temperatures ranging from 500 to 900 °C. This results in carbide particle precipitation that, in turn, leads to *intergranular corrosion* and failure through *weld decay*. This

Fig. 3.23 *Carbide precipitation in a non-stabilised (unproofed) 18/8 stainless steel*

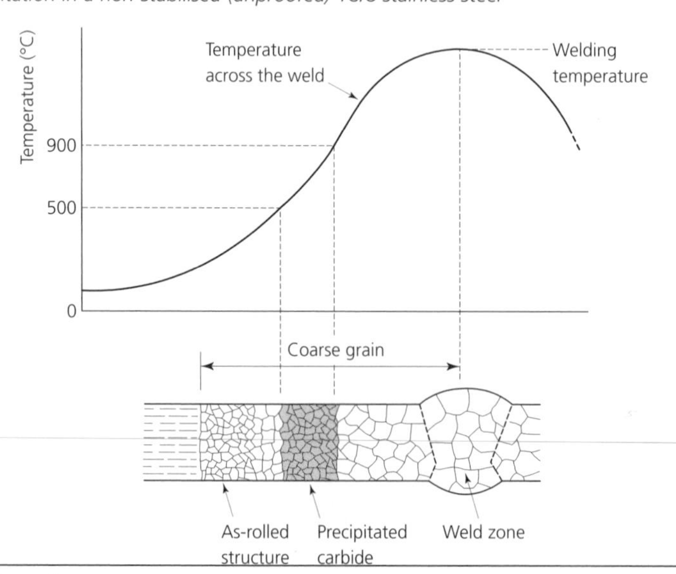

phenomenon will be discussed further in Section 4.8. As an alternative to using 'proofed' steel, 18/8 stainless steels can be heat treated by quenching them from 1050 °C immediately after welding. Ferritic stainless steels suffer from 'sigma-phase' formation, and this tendency can also be reduced by suitable heat treatment (see Section 1.6).

Figure 3.24 shows what happens when a 13 per cent chromium air-hardening 'cutlery grade' steel is welded. Remember that in the previous chapter it was stated that chromium used as an alloying element in steels was a carbide stabiliser but promoted grain growth. Therefore, with 13 per cent chromium present in the steel under discussion, not only is there considerable grain growth in the heat-affected zone leading to weakness, but the formation of martensite in a band each side of the joint leads to brittleness and fracture.

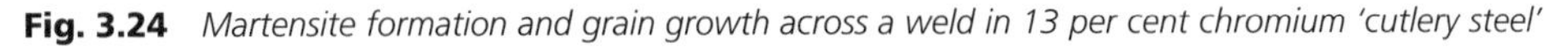

Fig. 3.24 *Martensite formation and grain growth across a weld in 13 per cent chromium 'cutlery steel'*

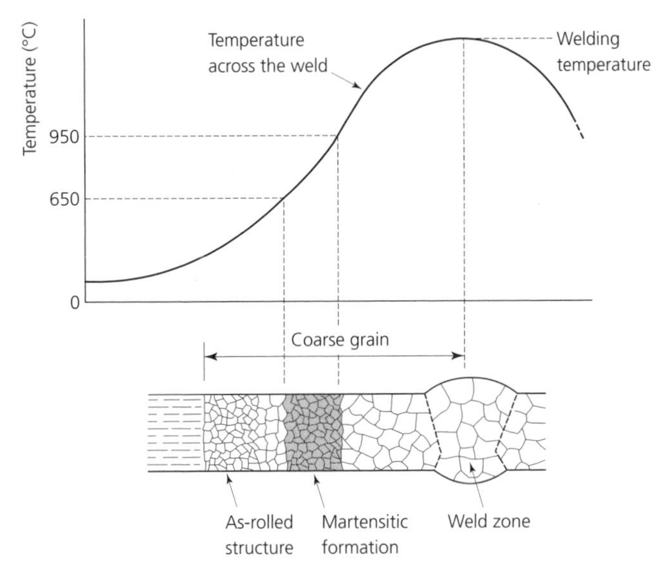

Maraging steels are relatively soft when cooled from the austenising temperature. Therefore, the heat-affected zones are softened by the heat of welding, with the result that the residual stresses are lowered and there is less tendency for hydrogen cold cracking than with conventional alloy steels. Post-weld ageing treatment at 480 °C results in the precipitation of hard intermetallic particles (dispersion hardening) and this precipitation raises the strength of the joint close to the original plate strength, and the toughness of the joint is equal to the plate toughness. Most welding processes will produce a joint efficiency exceeding 90 per cent in terms of strength alone. However, the choice of welding process can have a significant effect upon the toughness of the joint. The TIG welding process produces the most satisfactory joints in maraging steels providing the following precautions are taken.

- The time the heat-affected zone is maintained at elevated temperatures should be minimised.
- Preheating should be avoided and interpass temperatures should be kept below 120 °C.
- Weld input energy should be kept to a minimum.

- Conditions causing slow cooling rates should be avoided.
- The joint should be kept as clean as possible since impurities cause a fall off in toughness by creating barriers to the movement of dislocations (Section 8.7).

3.20 The effect of welding on copper

The various grades of copper available have already been discussed in *Engineering Materials*, Volume 1. Ordinary tough pitch copper contains oxygen in the form of copper oxides, which give the metal its increased strength. Unfortunately the welding flame reacts with the oxide particles to produce steam, resulting in gas porosity and weakness of the joint. This effect can be overcome by using a slightly oxidising flame and a filler rod containing phosphorus. Where it is known that copper components are going to be assembled by welding, one of the 'deoxidised' grades of copper and a neutral welding flame should be used.

The high thermal conductivity of copper (seven times that of steel) is a disadvantage when welding. As heat energy is conducted away rapidly from the weld zone, it is not only difficult to achieve complete fusion of the edges of the parent metal but grain growth in the parent metal is excessive. To achieve fusion a large jet has to be used in the torch. This is not only wasteful in the use of welding gases, but exacerbates the problem of softening and grain growth in the parent metal. Too small a jet causes the filler material to melt before the parent metal edges melt, resulting in lack of penetration and planes of weakness at the joint edges.

Like most non-ferrous metal, copper depends upon cold working to increase its strength and hardness. The heat conducted back from the weld pool anneals the parent metal, resulting in general weakness in the vicinity of the joint. A typical cross-section through a welded joint in copper is shown in Fig. 3.25.

Fig. 3.25 *Structure of the weld zone (copper)*

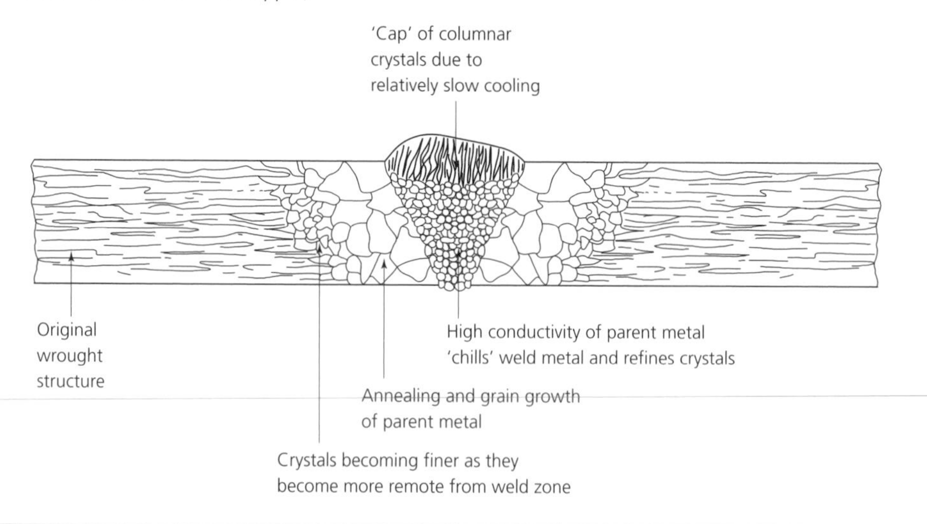

3.21 The effect of welding on aluminium and its alloys

Like copper, aluminium also has a high thermal conductivity and depends upon cold working to improve its strength. Therefore the conditions discussed in Section 3.20 also apply to aluminium. Furthermore, aluminium oxidises very easily and has to be protected from atmospheric oxygen by the use of fluxes and a reducing flame setting. Unfortunately, aluminium and its alloys absorb hydrogen more readily than any other metal in the molten state, and the hydrogen comes from various sources, such as:

- Incomplete combustion in the welding flame.
- The fluxes.
- Atmospheric moisture.

As the weld cools, the dissolved hydrogen is expelled, resulting in 'gas porosity'. The fact that the conditions for preventing gas porosity and preventing oxidation conflict with each other is the main reason why aluminium and its alloys are such difficult metals to weld. There are usually three zones in welded joints in pure aluminium and non-heat-treatable aluminium alloys. These are shown in Fig. 3.26.

Fig. 3.26 *Structure of the weld zone (aluminium)*

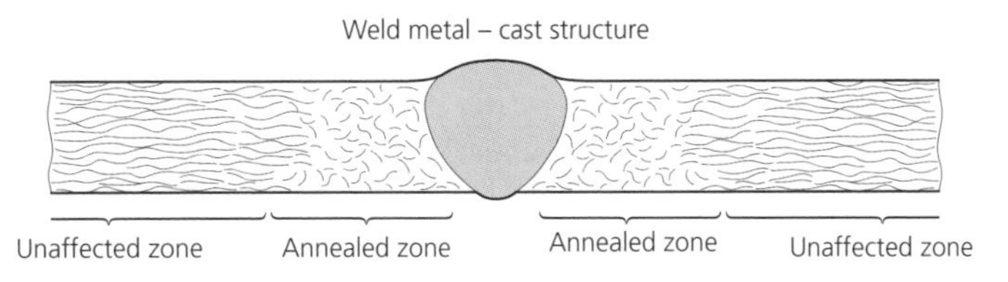

- *Weld metal* The weld bead with its 'as cast' structure, where the parent metal is alloyed with the deposited filler material.
- *Annealed zone* The region where heat conducted from the weld zone has caused annealing, resulting in loss of hardness and strength.
- *Unaffected zone* The region where heating from the weld zone has not affected the wrought structure of the metal.

The heat-treatable aluminium alloys contain alloying elements that exhibit a marked change in solubility with change in temperature. The solution and precipitation heat treatment of these alloys was described in *Engineering Materials*, Volume 1, Section 7.3. The uncontrolled solution of the micro constituents in the weld pool during welding, and their uncontrolled precipitation during cooling, results in undesirable effects on the microstructure and mechanical properties of the alloy. Welds in heat-treatable alloys generally exhibit five zones, as shown in Fig. 3.27.

- *Weld metal* The weld bead with its 'as-cast' structure where the parent metal is alloyed with the deposited filler material.
- *Fusion zone* The region where partial melting of the parent metal occurs primarily at the grain boundaries.

Fig. 3.27 *Structure of the weld zone (heat-treatable aluminium alloys). Key: A = weld metal; B = fusion zone; C = solid solution; D = partially annealed or over-aged; E = unaffected*

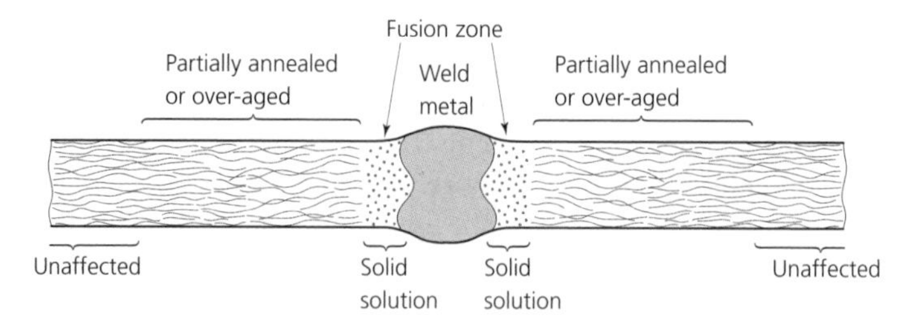

- *Solid solution zone* The region where heat conducted from the weld pool raises the temperature of the parent metal sufficiently to promote solution treatment.
- *Over-aged zone* Continued heating at the rather lower temperature of the next zone results in over-ageing and partial annealing to occur. The heat-treatable aluminium alloys rely upon precipitation age hardening to increase their hardness and strength. However over-ageing results in softening, grain growth and weakness.
- *Unaffected zone* The region where heating from the weld zone has not affected the structure.

The five zones are generally quite evident in welds made in heat-treatable alloys in which copper and zinc are the major alloying elements. Alloys of the magnesium–silicon type exhibit structural changes in the heat-affected zone that are somewhat different, with the principal heating effect being over-ageing. In these alloys the over-aged and partially annealed zones are of much greater widths.

The speed of welding has a marked affect upon the properties of welds in heat-treatable alloys. High welding speeds not only decrease the width of the heat-affected zones, but they minimise effects such as grain boundary precipitation, over-ageing and grain growth. Heat-treatable aluminium alloys may, in some cases, be heat treated after welding to bring the heat-affected zones back to their original strength and structure. Where this is not possible, the 'as cast' strength must be accepted. For example, an aluminium alloy containing 4 per cent copper has a permissible axial stress after heat treatment of 130 N/mm^2 (before welding), whereas the permissible axial stress in the 'as cast' condition is only 50 N/mm^2 (after welding).

3.22 The effect of brazing on cold-worked, low-carbon steels

As explained in Section 3.10, brazing processes do not involve such high temperatures as welding and the joint edges of the parent metal are not melted. Nevertheless, the process temperature is above that required to stress relief anneal cold-worked low-carbon steels. Further, the parent metal is often heated long enough for grain growth to occur in the vicinity of the joint. Cold-worked steel will have gained considerably in strength and hardness through becoming work hardened. Therefore, the effect of brazing will be to anneal the steel in the vicinity of the joint, resulting in local loss of strength and hardness.

3.23 Comparison of joining techniques

3.23.1 *Welding*

Welding has largely superseded riveting for making permanent joints in fabricated steel work and pressure vessels. It has a number of advantages over riveting, apart from being quicker and less labour intensive:

- The joint is continuous and the stresses between the individual members are transmitted more uniformly.
- The joints are fluid tight and therefore ideal for pressure vessels.
- The joints surfaces are smooth, more pleasing in appearance and easier to maintain on site (painting).

Welding does have some disadvantages compared with riveting: the equipment is more expensive, more energy intensive, and relatively dangerous (particularly compressed gases). The labour required for welding is more skilled and joint defects are more difficult to detect. Riveting does not affect the properties of the metal being joined whereas the heat-affected zone of a welded joint can lead to weakness, to weld decay and eventually to fatigue failure. Further, cracks in welded plates can run throughout the structure, crossing joint lines, and causing catastrophic failure, whereas cracks in riveted structures will only run to the nearest rivet hole where the strain energy will be reduced to a value too low to sustain crack propagation (Section 8.10).

3.23.2 *Brazing*

Considering the relatively low strength of brazing spelter compared with welding filler material, brazed joints are remarkably strong. In fact they can even be stronger than welded joints under some conditions. This is because of the greater joint area of properly designed brazed joints, as shown in Fig. 3.28(a), compared with the joint area of a corresponding welded joint, as shown in Fig. 3.28(b), results in reduced stress per unit area in the filler material. Further, as previously explained (Section 3.18), although the process temperature for brazing is sufficient to cause some modifications of the material structure, these are far less than the structural modifications associated with the heat-affected zone of a welded joint. Again, the lower process temperature for brazed joints results, and the

Fig. 3.28 *Comparison of (a) brazed and (b) welded joints*

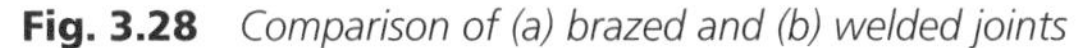

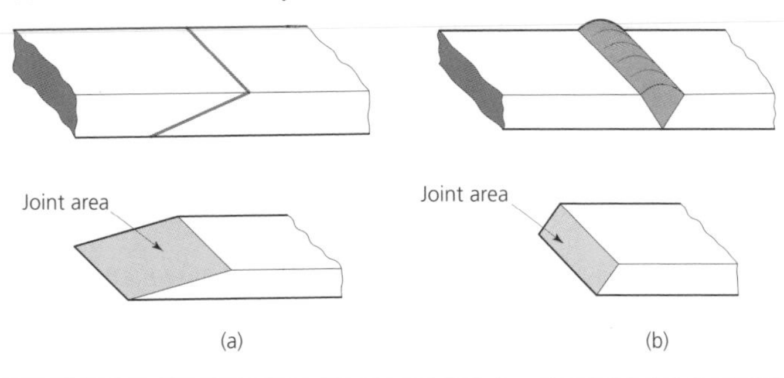

corresponding reduction in structural modification of the materials being joined enhances both the strength of brazed joints and their fatigue resistance when compared with welded joints.

SELF-ASSESSMENT TASK 3.4

1. Explain what is meant by the heat-affected zone of a welded joint, why structural changes take place in the heat-affected zone, and how these changes can affect the strength of the joint.
2. Compare the effects of welding and brazing on the structural changes that occur in low-carbon steels joined by these processes, and how this can result in brazed joints that are as strong as or, in some cases, stronger than welded joints.

EXERCISES

3.1 Compare and contrast the differences between welding and brazing processes and show, with the aid of sketches, the essential difference in the design of joints for these processes.

3.2 Describe *the heat-affected zone* of a welded joint and explain how it affects the properties of the metals being joined and the strength of the joint.

3.3 Describe the chemical reactions that take place during oxy-acetylene welding and how these reactions can affect the joint.

3.4 Describe the composition, need for, and function of the flux coating of an arc-welding electrode.

3.5 Describe the principles of the spot- and butt-welding processes.

3.6 Compare the effects of welding and brazing on the properties of the materials being joined.

3.7 List the advantages and limitations of brazing compared to welding.

3.8 Compare the advantages and limitations of welding to the advantages and limitations of riveting as a means of making strong, permanent joints between metal components.

3.9 Explain what is meant by *weld decay* in alloy steels, describe its affect on the strength of a joint, and explain how this effect can be avoided.

3.10 Explain how maraging steels may be welded and describe any precautions that must be taken to ensure a sound joint.

4 Corrosion prevention

The topic areas covered in this chapter are:

- The principles of dry and wet corrosion.
- Uniform and local corrosion of various types.
- Corrosion prevention including cathodic and anodic protection.
- Protection coatings and their application.

4.1 Dry corrosion

The corrosion of metals and preventative treatments were introduced in *Engineering Materials*, Volume 1. Let's now consider the basic principles of a number of corrosion mechanisms in greater detail.

Dry corrosion occurs as the result of a metal–gas reaction either at ambient temperature in an indoor environment, or during processing at elevated temperatures as when hot-forging or hot-rolling metals. The corrosion film produced is the result of an oxidation reaction. Chemical oxidation reactions occur when a metal is converted from its elemental atomic form to the ionic form of its compound (M to M^{n+}). This loss of electrons in the metal occurs whenever it forms a compound by a chemical oxidation, for example:

$$4Al + 3O_2 \rightarrow 2Al_2O_3 \quad \text{and} \quad Fe + S \rightarrow FeS$$

Thus oxidation reactions can occur with gases other than oxygen. However, in the practical world of engineering, the term 'oxidation' invariably refers to reactions between metals and atmospheric oxygen and reactions with other gases are named accordingly, e.g. sulphidation, nitridation.

A typical example of dry corrosion at room temperature is the formation of the transparent, protective film of alumina that forms on aluminium products in dry air indoors. After polishing, such films usually achieve a maximum thickness of 1–5 nm in about 2–4 weeks, after which any further growth becomes negligible. The corrosion products are found not only on the surface of the metal but also just below the surface. This is due to the reaction gases diffusing into the atomic structure of the metal. The higher the surface temperature of the metal the greater will be the agitation of the metal atoms and the easier it is for diffusion to occur. This results in a more rapid reaction and a thicker oxidation layer.

At higher temperatures the tempering colours on quench-hardened and tempered carbon steel tools are an example of dry oxidation. As the tempering temperature is increased, the oxide film becomes thicker and the colour of the film becomes darker.

At processing temperatures above 570 °C, a scale commences to form on ferrous metals – that is, the oxide film exceeds 1 mm in thickness and the film becomes more complex as shown in Fig. 4.1. The ability of an oxide film to protect the metal beneath it from further corrosion depends upon many factors. If the film is porous or subject to flaking then it will offer little protection as new metal is constantly being exposed to attack. The oxide film can often be improved by the addition of alloying elements. For example, when chromium is added to carbon steel it forms a continuous barrier layer of chrome oxide (Cr_2O_3) which opposes the migration and diffusion of the metal and oxygen atoms at the surface of the metal. Other alloying elements that help to improve the protection offered by the oxide layer when added to steels and cast irons are aluminium, silicon and nickel.

Fig. 4.1 *Oxide film on plain carbon steels above 570 °C*

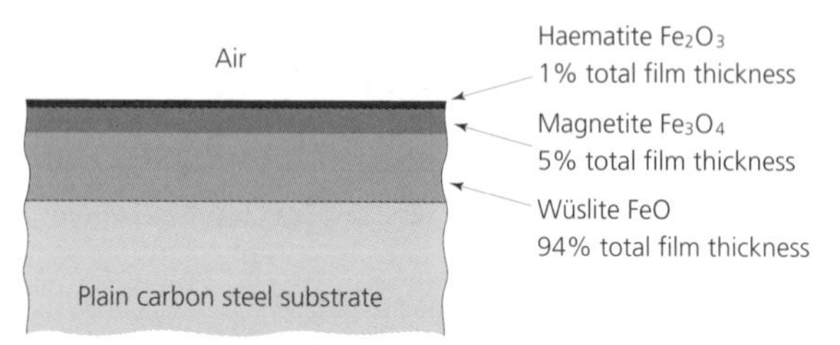

Table 4.1 shows the effect of adding some of these alloying elements to iron. It also shows that small additions of a third or fourth element to the alloy can be more effective than simply increasing the quantity of the second element, (e.g. compare Fe + 1% Si with Fe + 3% Si and Fe + 2% Si + 2% Cr). When alloying reduces the oxidation rate it is often possible to use the material at higher temperatures. For example, the temperature at which a low-carbon steel commences to form a scale is only 480 °C compared with an 18/8 stainless steel, which does not commence to scale until it has been heated to 900 °C.

Table 4.1 ***Corrosion (oxidation) rate of iron***

Effect of alloying elements

Composition (%)				*Average mass gain per unit area (mg/cm²)*
Fe	*Si*	*Al*	*Cr*	
100	—	—	—	15.50
99	1	—	—	1.00
99	—	1	—	1.50
99	—	—	1	4.00
97	3	—	—	0.50
95	—	5	—	0.20
86	—	—	14	0.01
98	1	1	—	0.03
98	1	—	1	0.50
96	2	—	2	0.02

Note: Oxidation of iron in dry air at 700 °C.

Further effects of oxidation in metals and alloys will now be considered. 'Growth' in grey cast irons occurs if the castings are raised to the temperature at which they become austenitic. The volume increase associated with this transformation allows the infusion of atmospheric oxygen that reacts with the flake graphite to form carbon monoxide. Some of this gas becomes trapped in the casting as it cools and causes a permanent increase in volume or 'growth' to occur, resulting in warping and loss of strength. This effect can be overcome by using alloy cast irons such as *silal* or *nicrosilal* for castings that have to be used at elevated temperatures.

Intergranular corrosion of nickel–chromium alloys can occur at temperatures of 1000 °C or above in the presence of hydrocarbon gases or even residual mineral oils or greases on the surface of the metal. Some of the chromium in the alloy reacts to form non-protective compounds such as chromium carbide (Cr_2C_6). This reduces the amount of chromium present in the alloy and oxidation of the alloy becomes easier. This results in the strength of the alloy being reduced so that it becomes more brittle. If the metal fractures *green chromic oxide* (Cr_2O_3) colours the broken surfaces and the fracture is said to be the result of 'green rot'.

Flue gases resulting from the burning of fossil fuels contain a mixture of corrosive gases. These gases prevent the formation of protective oxide films and the corrosion rate is increased. Notably amongst the flue gases, sulphur and sulphur dioxide are the most reactive. Nickel and nickel-based alloys are particularly susceptible to *intergranular attack* by sulphur leading to *embrittlement*. Iron-based alloys are less susceptible to sulphur attack than nickel-based alloys.

The combustion of fossil fuels also produces complex ashes that contain highly reactive components at high temperatures. Generally, high-chromium-content alloys are required to resist such corrosion and whether the base of the alloy is nickel or iron will depend upon whether or not sulphur compounds are present.

Catastrophic oxidation occurs when no protective oxide film is formed and the reaction continues until no metal is left. For example, a steel containing 1 per cent chromium and 1 per cent molybdenum is completely destroyed when heated to 650 °C in air. The molybdenum forms a volatile oxide that continually breaks down any protective oxides formed by the chromium. The chromium content has to be increased to at least 9 per cent to prevent this happening, and the presence of nickel in the alloy also helps.

4.2 Wet corrosion

Let's now consider wet corrosion. This type of corrosion occurs where a liquid, usually water, is present. For corrosion to occur, solid, liquid or gaseous contaminants must be dissolved in the water to form an *electrolyte* because wet corrosion is an electrochemical reaction. Molten metals, molten salts and organic solutions can also cause wet corrosion.

A common example of wet corrosion is the *rusting* of ferrous metals. Iron will not rust in dry air nor will it rust in pure water from which any dissolved oxygen has been removed. However, in a moist atmosphere rusting will occur as the distinctive, reddish-brown rust film of *ferric hydroxide* builds up on the surface of the metal:

$$4Fe + 6H_2O + 3O_2 \rightarrow 4Fe(OH)_3$$

It is the presence of the dissolved oxygen in the water that makes this reaction possible. A simple experiment will prove this fact. Take three test tubes and place some iron filings or turnings in the bottom of each. Add boiled water to one test tube so as to completely fill it and seal the top. Boiling the water will remove the dissolved oxygen, and sealing the test tube will prevent the ingress of atmospheric oxygen. To the second test tube add silica gel and seal the top. The silica gel will remove any residual water vapour in the air in this test tube, and sealing the test tube will prevent the ingress of atmospheric moisture. The third test tube is left open to the atmosphere (preferably damp). Only the iron filings or turnings in the open-topped test tube will eventually rust. This proves that *both water and oxygen* are necessary for rusting to take place. Water alone does not cause rusting. Oxygen alone does not cause rusting. The reaction described above is achieved as follows and is typical of the electrolytic processes of wet corrosion generally.

Elemental atomic iron becomes ionised by the loss of two electrons (this is the oxidation mechanism described in Section 4.1) and enters the solution as ferrous ions:

$$Fe \rightarrow Fe^{2+} + 2e^- \text{ (2 electrons)}$$

These ferrous ions are further oxidised to ferric ions:

$$Fe^{2+} \rightarrow Fe^{3+} + e^- \text{ (1 electron)}$$

Fig. 4.2 *Rusting on steel reinforcing rods (reproduced courtesy of the University of Luton)*

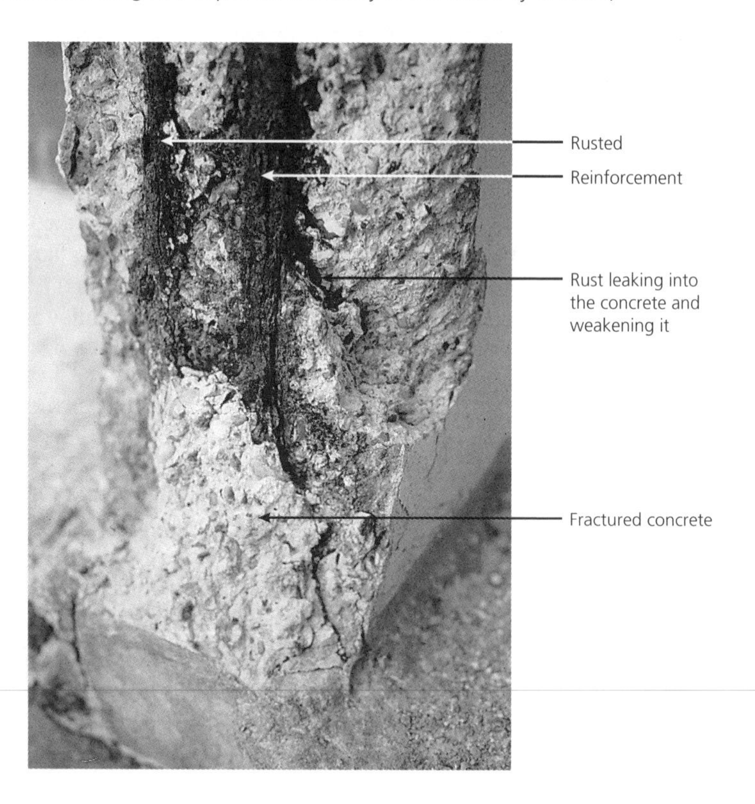

At the cathodic area of the reaction, the electrons that flow from the anodic area, as the result of the ionisation of the iron, are intercepted by the oxygen atoms present:

$$4e^- + O_2 + 2H_2O \rightarrow 4(OH)^-$$

The ferric ions then combine with the hydroxil groups so that ferric hydroxide (rust) is formed and the overall electronic equilibrium is maintained. Figure 4.2 shows the effect of wet corrosion (rusting) on steel reinforcing rods at the base of a concrete column.

Wet corrosion also commonly occurs when two dissimilar metals come into contact in the presence of an electrolyte and an electrical cell is formed. This type of corrosion is also known as *galvanic corrosion* or *bimetallic corrosion* and results in one or other of the metals being eaten away. Metals can be arranged in a special order called the *electrochemical series*. Some of the metals used in engineering are listed in the order of the electrochemical series in Table 4.2 and it should be noted that, in this context, hydrogen gas behaves like a metal. If any two metals in the table come into contact in the presence of an electrolyte, the more electronegative metal will be attacked and eaten away whilst the more electro-positive metal will be protected, as shown in Fig. 4.3.

Table 4.2 *Electrochemical potentials of common metals*

Metal	*Electrode potential (volts)*	
Sodium	−2.71	*Corroded (anodic)*
Magnesium	−2.40	
Aluminium	−1.70	
Zinc	−0.76	
Chromium	−0.56	
Iron	−0.44	
Cadmium	−0.40	
Nickel	−0.23	
Tin	−0.14	
Lead	−0.12	
Hydrogen (reference potential)	0.00	
Copper	+0.35	
Silver	+0.80	
Platinum	+1.20	
Gold	+1.50	*Protected (cathodic)*

In the case of galvanised iron (zinc-coated low-carbon steel), any porosity in the coating or damage to the coating results in the zinc corroding away whilst protecting the steel from rusting. Thus the zinc is said to be *sacrificial*. The iron is only protected as long as some zinc remains in the vicinity of the surface discontinuity. Once the zinc is destroyed, rusting will commence. Figure 4.2(a) shows the conventional electric current between the two metals as flowing from the iron to the zinc. Thus the *electron current* flows from the anodic zinc to the cathodic iron and, as a result, the Zn^{2+} ions enter solution in the electrolyte. The current flow will cause hydroxil ions to form at the surface of the iron as previously described. The

Fig. 4.3 *Galvanic corrosion: (a) protection by a sacrificial coating (coating is eaten away whilst protecting the base); (b) protection by a purely mechanical coating (coating only protects the base if intact; if coating is damaged, base is eaten away quicker than if coating were not present)*

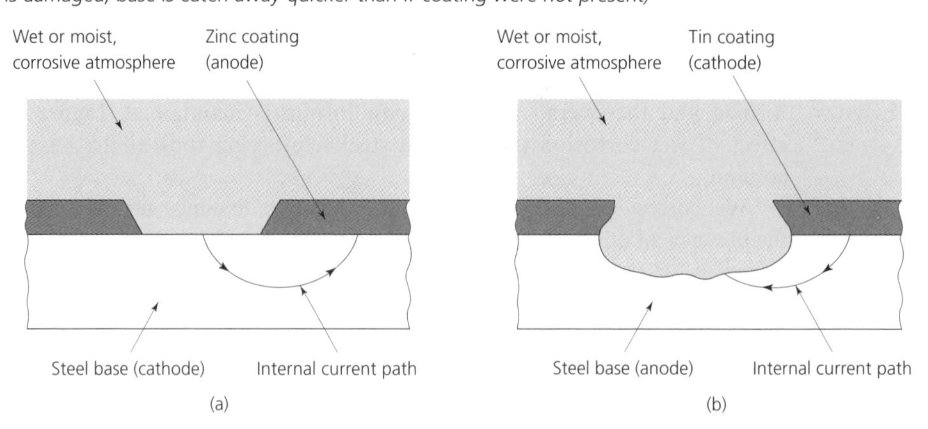

Zn^{2+} ions and the OH^- ions will react to form a white deposit of zinc hydroxide on the iron:

$$Zn^{2+} + 2(OH)^- \rightarrow Zn(OH)_2 \downarrow$$

In the case of tin plate, the mild steel substrate is eaten away if the tin film is broken. Hence cut edges should always be tinned and bend lines should be marked with a soft pencil and not cut with a scriber. Figure 4.3(b) shows the conventional electric current flow from the tin to the iron. Thus the *electron current flow* is from the anodic iron to the cathodic tin and, as a result, the Fe^{3+} ions will enter solution in the electrolyte. Hydroxil groups will form at the surface of the tin cathode as previously described. The Fe^{3+} ions and the OH^- ions will react to form ferric hydroxide (rust):

$$Fe^{3+} + 3(OH)^- \rightarrow Fe(OH)_3 \downarrow$$

It should be noted that the behaviour of metals is not solely dictated by their position in the electrochemical series. For instance tin is cathodic to steel (iron) in most aqueous solutions in the open air. However, when formed into sealed cans, the presence of food acids and the lack of oxygen causes the tin to behave as though it were anodic with respect to steel. For this reason many food tins are often lacquered on the inside to protect the tin plating. Let's now consider some other factors that affect the wet corrosion mechanisms.

SELF-ASSESSMENT TASK 4.1

1. Distinguish between wet and dry corrosion and give two examples of each.
2. A coating of zinc is often used to protect steel (galvanising). Explain why the steel is still protected from corrosion even if the coating becomes damaged.
3. Explain why tin plate should be marked out with a pencil where bending is to take place and why cut edges must be sealed with solder if the steel substrate is to remain protected.

4.3 Uniform corrosion

Some of the most important factors in the behaviour of electrolytic corrosion are the relative areas of the electrodes. As its name implies, uniform corrosion occurs where the anodic and cathodic regions have approximately equal areas. Metal loss is greatest with uniform corrosion and it may be specified as the corrosion rate from which the service life of the material may be predicted. The SI unit for the rate of corrosion is the penetration per year in millimetres (mm/y). Other units such as milligrams per square decimetre per day (mdd) and, in the USA, mils per year (mpy or m/y) are also used. Note that one mil equals one thousandth of an inch.

4.4 Preferential corrosion

Although local corrosion destroys less metal than uniform corrosion, it is more dangerous since it leads to unpredictable and local failure. This is because the pitting caused by local corrosion penetrates more deeply into the metal than uniform corrosion, which is a surface effect. This is because when the cathode area is large compared to the anode area, the corrosion mechanism is intensified. Thus, if the paint film over the steel body of a car is damaged or becomes porous, the wet *galvanic corrosion* that takes place will be locally intense, leading to deep pitting as shown in Fig. 4.4.

Fig. 4.4 *Preferential corrosion (wet contaminated atmosphere (electrolyte))*

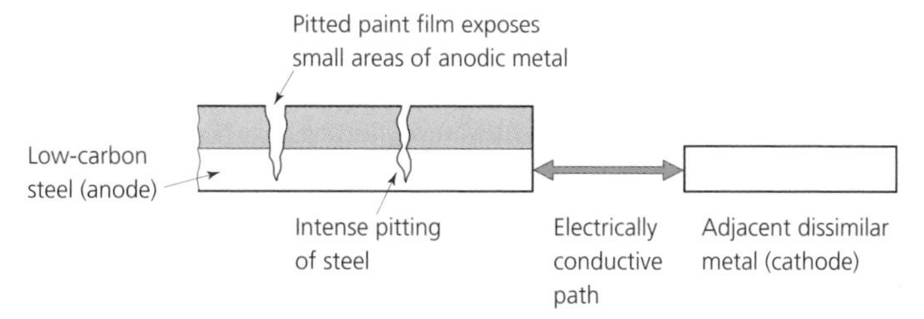

Where bolts or rivets are used to join metal components together, the following rules should apply. The rivets or bolts should be of the same material as the components being joined. If the bolts or rivets have to be of a different metal they should be more 'noble' in the electrochemical series so that they are cathodic relative to the metal of the components being joined. That is, steel bolts or rivets of small surface area will soon dissolve away if used to join brass, bronze or copper components, whereas brass, bronze or copper bolts or rivets, of small surface area, will be unaffected when joining steel components.

4.5 Crevice corrosion

Figure 4.5 shows two plates of the same material riveted together (the effect would be the same if they were bolted together) and immersed in oxygenated water. It might be thought that where the plates are in contact with each other little corrosion could take place. However, variations in the amount of oxygen present from one region of the joint to another results in corrosion occurring where it is least expected. This is because the water drawn into the joint between the plates by capillary attraction becomes starved of oxygen as the reactions proceed whilst those regions of the plates exposed to free water have an adequate supply of dissolved oxygen. Further, it is much easier for electrons to migrate out of the joint than it is for replacement oxygen atoms to move into the joint. Hence the exposed surfaces of the metal become cathodic as electrons migrate to these regions and react with the oxygenated water present to form hydroxil ions as previously described.

Fig. 4.5 *Crevice corrosion*

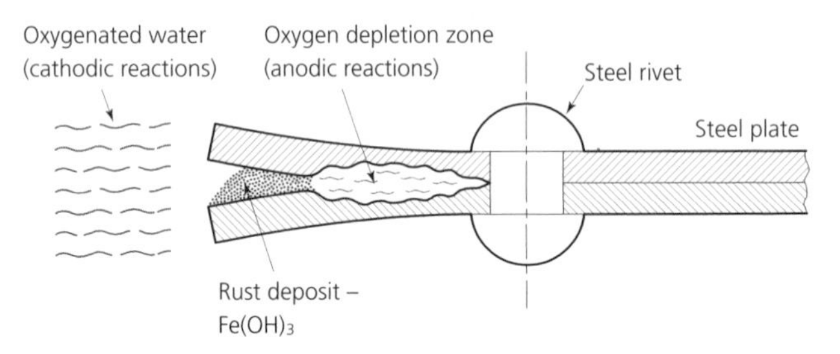

Therefore, relatively, those regions of the plates exposed to water containing less oxygen will become anodic and go into solution. That is, the metal surfaces in the joint become corroded away as shown:

$$Fe^{3+} + 3(OH)^{-} \rightarrow Fe(OH)_3 \downarrow \text{(rust)}$$

Since the Fe^{3+} ions can migrate through the gap between the plates more freely than the OH^{-} ions, the rust so formed builds up at the entrance to the gap between the plates and still further hinders the oxygenation of the trapped moisture and the migration of OH^{-} ions. This corrosion mechanism is referred to as *crevice corrosion*.

4.6 Galvanic corrosion

Galvanic corrosion between two dissimilar metals was shown in Fig. 4.3. In Fig. 4.3(b) it was the iron that was corroded away rather than the tin coating because the iron was anodic in that instance. Similarly, hot-rolled or forged steel corrodes (rusts) more quickly than bright steel because the scale is not only porous and offers no protection, but is cathodic to the metal beneath it.

Galvanic corrosion on a microscopic scale can also occur due to impurities. These impurity particles may be anodic or cathodic relative to the metal in which they are found. Where the impurity is cathodic the metal surrounding it will be eaten away, and where the impurity is anodic the impurity is eaten away. Again, galvanic action can take place within the metal structure. For example, lamellar pearlite consists of alternate laminations of ferrite and cementite. In the presence of oxygenated water the ferrite is anodic and eaten away whilst the cementite (iron carbide) is cathodic and unaffected. Thus metals that are very pure, or metals that form homogeneous solid solutions are the least likely to suffer from galvanic corrosion.

4.7 Pitting

Pitting caused by preferential corrosion has already been considered. Pitting is caused by the presence of:

- Local anodic impurities.
- Differential aeration.
- Surface debris such as mill scale.
- Aggressive ions in the electrolyte.

Residual stresses can accelerate pitting and, as pitting reduces the effective cross-sectional area under stress, premature and unpredictable failure will occur.

Pitting in steelwork can cause problems when the surface is repainted. Rust blisters will appear over any pits where residual electrolyte has been trapped and corrosion recommences. Before repainting, previous pitting must be removed by grinding or any residual electrolyte and products of corrosion in the pits must be neutralised chemically. For example, resprayed car body shells frequently blister very quickly if they are not properly prepared before painting. Figure 4.6 shows a reinforcing beam from a bridge deck in which pitting is evident.

Fig. 4.6 *Reinforcement beam from bridge decking showing pitting (reproduced courtesy of the University of Luton)*

4.8 Intergranular corrosion

The effects of impurities on local corrosion have already been discussed in Section 4.7. Since impurities in metals frequently migrate to the grain boundaries as the metal solidifies, they cause intergranular corrosion resulting in embrittlement and weakness.

However, intergranular corrosion can occur without impurities being present, possibly due to the irregular arrangement of ions at the grain boundaries and the fact that the grain boundaries are regions of higher energy levels. This is particularly so when the material has been severely cold worked. Under suitable wet corrosion conditions the grain boundaries become anodic relative to the body of the grains that exhibit cathodic characteristics. Under such conditions the anodic boundaries are attacked and this attack is intensified by the fact that there is a small anode area compared to the large cathode area (Section 4.4).

An example of intergranular corrosion is the 'season cracking' of α-brass after severe cold working. Corrosion follows the grain boundaries, reducing the effective cross-sectional area of the metal, until it can no longer sustain the applied load and failure occurs. Stress relief by a low-temperature annealing process following the cold working allows the atoms and ions to realign themselves sufficiently to reduce the energy levels at the grain boundaries, and corrosion is eliminated for all practical purposes.

The intergranular corrosion of high-alloy steels occurs not only when such alloys are welded but also when castings of such alloys are cooled slowly or when they have been held too long at the 500–800 °C temperature range during heat treatment (temper brittleness). Slow cooling through this temperature range results in the formation of chromium carbide at the grain boundaries, leaving the surrounding areas deficient in chromium and unable to form the chromium oxide necessary to prevent corrosive attack, as shown in Fig. 4.7.

Fig. 4.7 *Intergranular corrosion of high-chromium alloy steels*

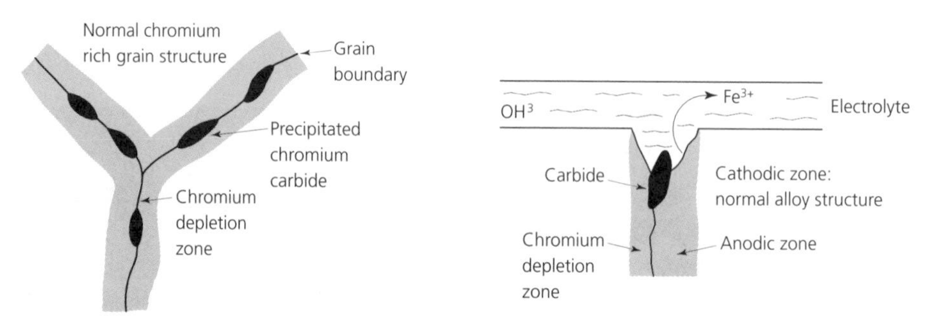

Intergranular corrosion can be largely overcome in various ways depending upon the limitations of service conditions. The metal can be reheated to 1100 °C after processing to dissolve the carbides. It is then quenched to prevent the carbides precipitating out again. Alternatively, a low-carbon stainless alloy ('stainless iron') such as BS 304S12 could be used to lessen the risk of carbide precipitation. Alloys such as BS 321S20 containing titanium or BS 347S17 containing niobium are said to be 'stabilised' or 'proofed'. Titanium and niobium both form carbides more readily than chromium and leave the chromium content undepleted. Further, the carbides formed do not congregate at the grain boundaries.

Unfortunately the temperatures achieved during arc welding may exceed 1100 °C, resulting in the titanium or niobium carbides being dissolved in the heat-affected zone. As the weld zone cools, chromium carbide is precipitated out rendering the joint liable to 'weld decay'. Under these conditions precipitation occurs in a narrow band each side of the weld itself and is referred to as 'knife-line attack', as shown in Fig. 4.8. This can be overcome by reheating the weld zone to just above 1000 °C and allowing the assembly to cool naturally. This avoids the distortion and stresses associated with quenching.

Fig. 4.8 *Intergranular corrosion zones for welded high-chromium alloy steels*

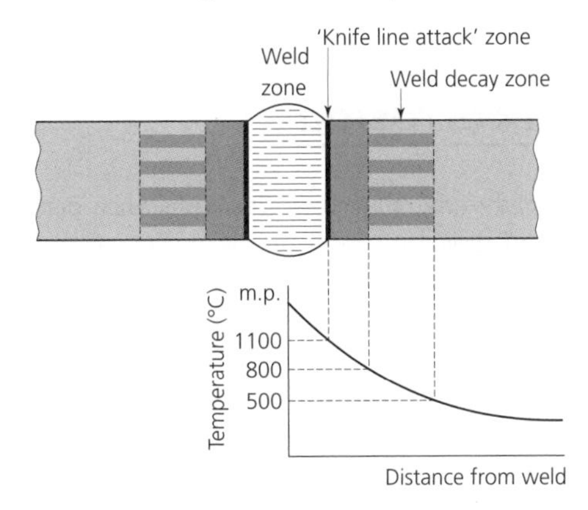

Although stainless steels are normally unreactive and corrosion resistant in the presence of oxygenated water and many other aqueous solutions because of their natural oxide film, they corrode rapidly in the presence of sea water. This is because the chlorine ions in the sea water are particularly aggressive towards stainless steels. They cause the breakdown of the protective oxide film at localised regions, causing pitting and activation of the alloy. This is aggravated where welding has occurred, resulting in catastrophic failure even after the precautions described previously have been taken.

4.9 Selective leaching

This is the electrolytic attack of one element in an alloy by another in the presence of an aqueous electrolyte. For example, the de-aluminiumification of aluminium–copper alloys where the aluminium content is corroded out to leave a spongy mass of copper. Again, there is the de-zincification of brass where the zinc content is corroded out leaving a spongy mass of copper. In both these examples the anodic aluminium and the anodic zinc are destroyed in the presence of the cathodic copper in what is a simple electrolytic reaction. This reaction is particularly rapid when chloride ions are present, as in brine and sea-water. The residual copper has no mechanical strength and, because of its spongy nature, is itself increasingly susceptible to further corrosion.

Selective leaching can also occur between metals and non-metals in an alloy – for example, the graphitisation of grey cast iron. Here, the uncombined flake graphite is cathodic and the anodic iron surrounding the flakes is corroded away as rust to leave a spongy mass with no mechanical strength.

The only way selective leaching can be overcome is to replace the materials under attack with materials that are resistant to the aqueous solutions present. For example, replacing the simple, binary brass alloy with Admiralty brass that contains tin and arsenic in addition to zinc. Alternatively, the brass alloy may be replaced by a cupronickel alloy that would be even more resistant to salt solutions and sea water, but would be more expensive in first cost. Again, 'white' cast irons that contain no free graphite are not susceptible to selective leaching; however, they may be too hard and brittle for many purposes.

4.10 Erosion corrosion

We have already seen that corrosion resistance depends largely upon the formation of impervious and resistant oxides on the surface of the metal. During erosion corrosion the flow of electrolyte over the metal surface destroys the protective film by a process of abrasion. This abrasion is exacerbated if solids are present in suspension in the liquid. Erosion is usually localised and caused by turbulence due to partial obstructions or bends in pipelines, as shown in Fig. 4.9. The erosion corrosion attack that occurs at pipe bends is often referred to as impingement attack. Impingement attack also affects turbine blades although, in this instance, cavitation attack, which will be discussed later in this section, may also be present.

Fig. 4.9 *Erosion corrosion caused by turbulence*

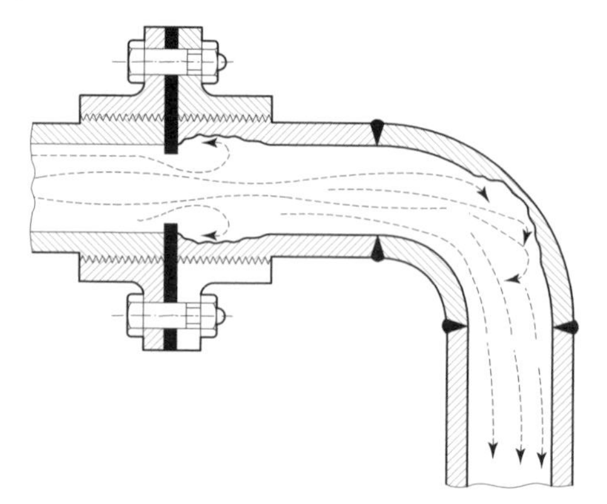

Erosion corrosion may be limited by careful design to reduce turbulence – for example, using bends instead of elbows where pipework changes direction, or increasing the bore of the pipe so that the velocity of the fluid is reduced below the critical velocity. Some materials, such as nickel-based alloys, have a greater resistance to erosion corrosion than iron-based alloys. If iron-based alloys have to be used, then stainless steels and cast irons such as 'ni-resist' and high silicon alloys are reasonably resistant to erosion corrosion providing chloride ions are not present. In closed systems, such as central-heating systems, erosion corrosion can be reduced by the use of an 'inhibitor' in the circulating fluid. This is a chemical that encourages the formation of protective films on the metal. However, the use of an inhibitor does not excuse bad design and the system should be as free from turbulence as possible.

Another form of erosion corrosion is *cavitation attack*. This is often present in the vicinity of ships' propellers, turbine blades and pump impellers. The pressure drop at the trailing surface of modern high-speed blades can fall below the vapour pressure of the fluid in which it is operating so that the liquid actually 'boils' on the surface of the metal. Vapour bubbles formed in this way burst rapidly. Each time a bubble bursts it produces a pressure wave, and the intensity of such a pressure wave hammering on adjacent metal surfaces can reach intensities of 1.5 GN/m. Since more than one million bubbles can form and collapse in one second over a relatively small area, the cumulative effect damages the surface of the blades and any other adjacent metal. Careful choice of materials that can resist such attack, accompanied by careful design to reduce cavitation to a minimum, is the only way to overcome this problem.

SELF-ASSESSMENT TASK 4.2

1. Explain why a strip of magnesium is sometimes connected to galvanised steel tanks below the water line.
2. Discuss whether or not it is a sensible idea to attach copper roofing sheets using steel nails.
3. Explain what is meant by erosion corrosion and how the design of the pipework can reduce this type of corrosion.
4. Compare and contrast the advantages and limitations of riveting and welding as means of construction of a steel boiler barrel in terms of crevice corrosion and intergranular corrosion.

4.11 Corrosion fatigue

Mechanical fatigue failure is exacerbated by corrosion, particularly in the case of ferrous metals. The effect of aerated sea water on the *S–N* diagram for a low-carbon steel was shown in Fig. 13.10 in *Engineering Materials*, Volume 1.

The damage ratio compares the fatigue limit for a particular material in dry air with the fatigue limit for the same material in a corrosive environment.

$$\text{Damage ratio} = \frac{\text{fatigue limit in dry air}}{\text{fatigue limit in corrosive environment}}$$

For a low-carbon steel in aerated sea water the damage ratio is about 0.2 compared to a damage ratio of 1.0 for a material that is not corroded in such an environment – for example, pure copper. The level and effects of corrosion fatigue can be reduced by control of the following factors:

- Solution composition.
- Aeration.
- pH value.
- Temperature.

However, the only successful resolution of the problem lies in the selection of a material which is resistant to corrosion in its working environment. Note that, unlike fatigue failure in dry air where the component under test is subjected to cyclical stressing, corrosion fatigue can occur even when the component is subjected to constant load conditions below the yield point of the material from which it is made. This latter phenomenon is referred to as *stress-corrosion cracking*. Some examples of metals and the environments that can lead to stress-corrosion cracking are listed in Table 4.3.

Table 4.3 ***Stress corrosion cracking***

Common material–environment combinations

Material	*Environment*
Aluminium alloys	Aqueous chloride solutions
Copper alloys	Ammonia
Low-carbon steels	Hot concentrated alkaline solutions
Magnesium alloys	Aqueous chloride solutions
Nickel alloys	Molten caustic soda
Stainless steels	Aqueous chloride solutions
Titanium alloys	Aqueous chloride solutions

4.12 Hydrogen damage

Atomic (nascent) hydrogen produced by such processes as electroplating, electrolytic cleaning and polishing, pickling and by some micro-organisms can cause 'blistering' and embrittlement of a metal. Particularly susceptible are high-strength steels. Although the hydrogen is generated at the surface of the metal, some of the relatively small hydrogen atoms diffuse readily between the larger metal atoms and recombine as molecular hydrogen to form pockets of hydrogen gas. These gas pockets cause internal stress, surface blistering and embrittlement of the material.

4.13 Biological corrosion

Micro-organisms do not attack metals directly, but often exude and thrive upon substances that do attack metals. For example, the anaerobic organisms that may be found in seawater and soils thrive on nitrogen, hydocarbons (e.g. methane caused by vegetable decay) and sulphur compounds; in turn, they exude hydrogen and corrosive compounds.

Organic substances can also attack metals, and careful selection of materials is particularly important in food-processing plant. (*Note*: some metals that are corrosion resistant are also toxic and cannot be used in such situations, e.g. lead.) An example of organic attack is the dark-green verdigris that builds up on copper and which must not be confused with the pale-green patina that is formed by inorganic chemical action and is a protective coating. Copper utensils used for food preparation should, therefore, be heavily tin plated.

4.14 Factors affecting aqueous corrosion

Having discussed the mechanisms of aqueous corrosion, Let's now consider the factors affecting the rate of such corrosion. We can summarise these as follows:

- Metal composition and structure.
- Environment surface defects.
- Structural design.
- Applied or internal stresses.
- Temperature.
- Aeration.
- Chemistry of the electrolyte.

4.14.1 *Metal composition and structure.*

The position of a metal in the electrochemical series is a good indication of its relative rate of corrosion. Metals at the anodic end of the table will corrode more rapidly than metals at the cathodic end of the table. Metals with a high degree of purity tend to corrode less rapidly than the same metals when impurities are present since electrolytic reactions take place between the anodic metal and the cathodic impurities in which the metal is eaten away. Metals with coarse-grained structures generally corrode more rapidly than metals with fine-grained structures. This is because a coarse-grained structure is more susceptible to ionic diffusion. Alloys can be developed to be resistant to a wide range of corrosive environments. However, it should be remembered that ferrous-based alloys, such as stainless steels, that are resistant to most rural, urban and industrial environments, are attacked in coastal and marine environments due to the presence of chloride ions. The metallurgical structure of the material can also affect its corrosion. For example, the electrolytic attack on the iron content in lamellar pearlite and the de-zincification of brass were discussed in Sections 4.6 and 4.9 respectively.

4.14.2 *Environmental surface defects*

Air consists of oxygen (20 per cent), nitrogen (nearly 80 per cent) together with other gases such as argon, helium and neon in very small amounts. Atmospheric oxygen is responsible for both the corrosion of ferrous metals when it is dissolved in water (rain) and the formation of protective oxides on metals that are corrosion resistant. When the oxides of nitrogen and the carbon dioxide, produced by the burning of fossil fuels, become dissolved in rain water to form acid solutions, the range of metals attacked and the rate of attack is increased. Atmospheric corrosion depends upon a number of local factors. In rural communities corrosion is generally due to oxygenated rain water alone, although prevailing winds may carry urban and industrial pollutants from neighbouring areas. In urban areas additional pollutants such as carbon dioxides and oxides of nitrogen are present from the burning of fossil fuels. In industrial areas, sulphur compounds and ammonia compounds may also be present. In coastal and marine environments highly reactive chlorides will also be present. The greater the concentration and range of pollutants present, the greater will be the rate and range of corrosive damage.

Damage to protective coatings allows the aqueous solutions responsible for corrosion to come into contact with the structural metal and corrosion can occur. As has been previously explained, the rate of corrosion is increased because of the small exposed anode/cathode area ratio that is present under such conditions.

4.14.3 *Structural design*

The following factors should be observed during the design stage of a component or fabrication to reduce corrosion to a minimum:

- Materials that are inherently corrosion resistant should be chosen or, if this is too costly, then an anticorrosion treatment should be specified.
- The design should avoid crevices and corners where moisture and silts may become trapped. Adequate drainage and ventilation should be provided.
- The design should allow for easy washing down and cleaning.
- Joints that are not continuously welded should be sealed, for example, by the use of mastic compounds.
- Where dissimilar metal have to be joined their electrolytic compatibility should be established at the design stage. Further, they should be insulated from each other either by the use of a suitable adhesive in the case of a permanent joint, or by suitable bushings and washers made from insulating materials, or by the use of non-metal connections.

4.14.4 *Applied or internal stresses*

Chemical or electrochemical corrosion is intensified when a metal is in a stressed condition. Internal stresses are usually caused by cold working and, if not removed by stress relief annealing, results in corrosive attack along the crystal boundaries. This is because distortion of the grain structure facilitates the diffusion and migration of ions between heavily cold-worked zones of the component that have anodic characteristics and less heavily cold-worked zones that have cathodic characteristics. Hence the more severely cold worked and stressed zones of the component will be subjected to corrosive attack and destruction. The 'season cracking' of an α-brass after severe cold working is an example of corrosion that can be overcome by stress relief annealing after cold working.

4.14.5 *Temperature*

All chemical and electrochemical reactions have a critical temperature below which the reaction cannot take place. Hence, metal is preserved from corrosion in the very cold climates met with at the North and South polar regions. Also corrosion is negligible in desert areas for, although the ambient temperature is above the critical reaction temperature, the atmosphere is too dry for 'wet corrosion' to occur. The worst conditions for corrosion are found in the tropics where high temperatures combine with high humidity and engineering products need to be 'tropicalised' if they are to have a reasonable service life. Even in temperate zones corrosion is a constant problem.

4.14.6 Aeration

It has already been shown that dissolved oxygen in water is essential for many forms of corrosion. Hence the rate of corrosion will be greater in aerated water than in stagnant water. Consider the legs of the piers found at many coastal resorts. It can be seen that corrosion is heaviest and fouling by marine organisms is most prevalent from the low-tide watermark through the high-tide watermark into the splash and wave zone. This is shown in Fig. 4.10. Below the low-tide watermark little aeration takes place, but the agitation of the surface of the sea increases aeration and this is locally increased still further by the turbulence and splashing that occurs round any obstruction such as the legs of the pier. This accounts for the increase in the rate of corrosion in this zone.

Fig. 4.10 *Effect of partial immersion on rate of corrosion*

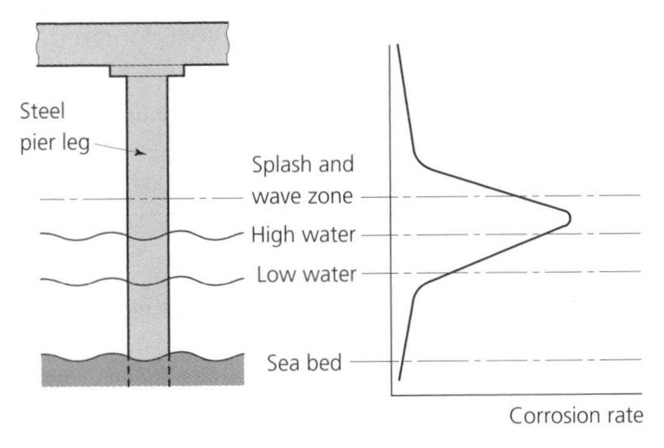

4.14.7 Chemistry of the electrolyte

The moisture necessary to form solutions which can act as electrolytes can be classified as:

- Water vapour.
- Rain water.
- Fresh (natural water).
- Brackish water.
- Sea water.

Water vapour and rain water will contain dissolved atmospheric pollutants (gases and dusts) and will generally be acidic. Fresh water from underground springs and rivers will have a variety of contaminants depending upon such factors. For example, the impurities in the rain from which such water originates, the rock layers through which it has passed, the soils over which it has passed, and any debris or effluents that have been allowed into it. Deep-sea water is fairly constant in composition, but coastal waters can vary widely in composition depending upon changing concentration due to evaporation where it is shallow, effluent discharge from industry and sewage treatment plants, and fresh water dilution at the estuaries of large rivers. Thus the chemical composition and corrosive effects of aqueous electrolytes can vary extensively and anticorrosion treatment suitable in one wet environment can be quite unsuitable or even counter-productive in another wet environment. For example, dissolved oxygen can result in passivity under some conditions, yet cause differential aeration cells resulting in corrosion under other conditions.

Dissolved salts affect the electrical conductivity and pH value of the electrolyte by changing its alkalinity or its acidity. The concentration of the solution is also important. Concentrated sulphuric acid tends to passivate steels and render them less susceptible to corrosion, yet dilute sulphuric acid tends to attack steels. Lead, which is considered a very corrosion-resistant metal, is affected in exactly the reverse manner. It is resistant to attack from dilute acids but is corroded by concentrated sulphuric acid.

SELF-ASSESSMENT TASK 4.3

1. (a) Explain how corrosion effects can exacerbate mechanical fatigue.
   (b) Explain what is meant by 'damage ratio' and its importance in the design of equipment operating in hostile environments.
2. Explain the difference between 'patina' and 'verdigris' and state where examples of each may be found.
3. Discuss the importance of prior knowledge of the environmental conditions under which a product will be used when selecting a suitable material for its manufacture.
4. Research some case studies for an industry with which you are familiar and discuss the cost of corrosion to that industry.

4.15 The cost of corrosion and corrosion prevention

It is estimated that annual loss and damage due to corrosion in the United Kingdom costs about £5000 million, and that approximately one tonne of steel is lost through corrosion every 90 seconds. Further, it is estimated that 25 per cent of this loss could be avoided by correct design, correct material selection, and proper preventative processes. The cost of corrosion is not only the replacement of damaged or destroyed equipment, but also such factors as preventative maintenance, loss of production due to unexpected failure and compensation when plant failure leads to destruction, environmental contamination, injury and death. The prevention of corrosion and its effect on safety, performance and cost are therefore of prime importance to engineers.

The importance of choice of material and the effect of alloying elements and heat treatment (e.g. composition, change of phase and impurities) on corrosion problems have already been considered in this chapter. Metals that resist corrosion and corrosion-resistant coatings were introduced in *Engineering Materials*, Volume 1. Let's now consider some further techniques for the prevention of corrosion.

4.15.1 *Anodic protection*

There are two fundamental techniques of anodic protection.

1. The use of *sacrificial anodes*. For example, the proximity of a manganese bronze propeller (cathodic) to the low-carbon steel hull of a ship (anodic) in highly agitated and aerated sea water should result in the rapid corrosion and destruction of the hull. This tendency is largely eliminated by bolting large slabs of zinc onto the hull near the propellers. Since the zinc is anodic to both the manganese bronze propellers and the steel hull of the ship, the zinc will corrode sacrificially whilst protecting the hull. The zinc anodes are replaced from time to time.
2. *Anodic passivation* may be achieved by two techniques:
   (a) *Galvanically* (but not sacrificially – see above). An electrolytic cell is created between the metal to be protected and a more noble metal by plating (e.g. platinum on stainless steel) or by alloying. The alloying additions need only be small and the addition of only 0.5 per cent of such metals as platinum or rhodium to titanium, or chromium to carbon steels, is sufficient. This technique is only satisfactory in oxygen-free conditions. The corrosion rate for titanium when boiled in dilute sulphuric acid is 100 mm/y compared with 1 mm/y for a titanium–platinum alloy.
   (b) *Impressed electromotive force* (e.m.f.) Previously it has been stated that when a material is anodic in an electrolytic cell it is corroded away, whilst the cathodic material is protected. Therefore it appears strange that making a material increasingly anodic can protect it. It should be noted that this technique can only be applied to a very few metals capable of forming passive surface oxides; that is, surface oxides that are unreactive and protective. For example, low-carbon steel and stainless steel are increasingly corroded as they are made more anodic until, at a critical e.m.f., a surface oxide forms which is passive and the corrosion rate drops significantly. For example, BS 304 stainless steel requires the impressed e.m.f. to be increased until a current density of 50 A/m^2 is achieved in order to form a passive oxide film. Once passivation has been achieved the impressed e.m.f. is lowered so that the current density is reduced to 0.04 A/m^2, which is sufficient to maintain passivation.

4.15.2 *Cathodic protection*

Unlike anodic protection where corrosion is not eliminated but reduced to an acceptable level, cathodic protection can prevent corrosion completely. Further, whilst anodic protection can only be applied to a limited range of materials, cathodic protection can be applied to most metals. A typical example of cathodic protection is shown in Fig. 4.11. Buried iron pipes would normally be anodic compared with the surrounding moist soil and would quickly corrode. However, when subjected to an impressed e.m.f. – that is, electro-negative relative to the surrounding soil, the iron pipe behaves as though it was cathodic relative to the soil and does not corrode. A low-voltage direct current (d.c.) generator is used to provide the impressed current.

Fig. 4.11 *Cathodic protection: impressed current*

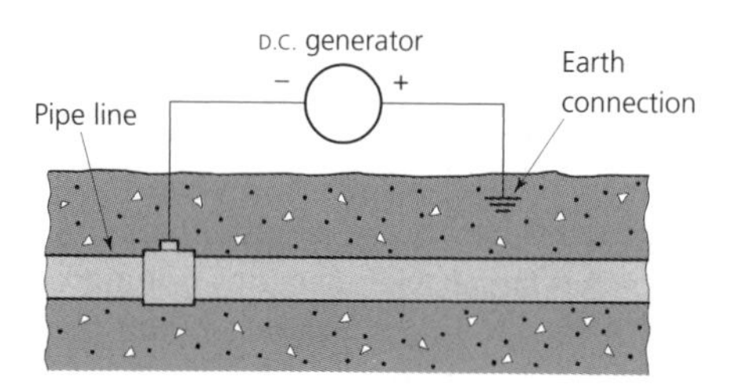

4.15.3 *Inhibitors*

Corrosion inhibitors are chemicals that reduce or prevent anodic or cathodic reactions. For example, chromates (CrO_4^{2-}) and nitrites (NO_2^-) are oxidising anodic inhibitors which are used in anticorrosive primers in paint systems. They promote the formation of passive oxide films on the painted metal. Nitrite inhibitors are also incorporated in protective oils and greases and in cutting fluids. Care must be taken when using anodic inhibitors. If the concentration falls below the minimum for a given system, severe pitting can occur because of the high cathode/anode area ratio of the unprotected zones.

Vapour phase inhibitors (VPI) and volatile corrosion inhibitors (VCI) are used in confined places and in packing materials (e.g. VPI-impregnated papers). Silica gel may also be used to absorb any moisture present and so prevent wet corrosion.

Cathodic inhibitors reduce the rate of electron production at the anode of an electrolytic cell indirectly by forming a protective barrier at any cathodic sites. Unlike anodic inhibitors, there are no minimum concentration levels for cathodic inhibitors and they do not require such skilled monitoring for safety. For example, if the concentration level of a cathodic inhibitor falls below the minimum critical level, only mild uniform corrosion will occur and there will be no aggressive pitting. However, cathodic inhibitors are not so efficient as the anodic types because the deposit formed is more soluble and less adherent.

Mixtures of inhibitors tend to reinforce each other's advantages whilst reducing their individual disadvantages. Since the primary inhibitor in the mixture becomes more effective at low concentrations there is less likelihood of pitting when the more efficient anodic inhibitors are used.

4.16 Protective coatings (preparatory treatments)

Corrosion-resistant metals and corrosion-resistant metallic coatings were discussed in *Engineering Materials*, Volume 1, and some inorganic and organic coatings were briefly introduced. Let's now consider these inorganic and organic coatings in greater depth. First, however, it is necessary to examine the various preparatory techniques that are essential for the satisfactory protection of the substrate. These techniques are equally applicable both for first time treatment and for subsequent, remedial treatments.

Most component surfaces are contaminated with one or more of the following:

- *Oxide and hydroxide films* resulting from the reaction of the base material with atmospheric oxygen and moisture.
- *Metal salt deposits*, such as sulphates and carbonates of the base or its cladding, as the reaction products when attacked by the dissolved impurities in rain or surface water (e.g. 'acid rain' resulting from the burning of fossil fuels);
- *Soils* in the form of grease, dust and dirt together with swarf and grinding wheel and polishing wheel dross from machining and finishing processes;
- *Previous protective coatings* (e.g. paint films which need to be stripped to provide a sound base for replacement).

Failure to prepare the surface of the base material correctly results in either lack of adhesion so that the protective coating flakes away, or self-perpetuating corrosion resulting in destruction of the base material under the protective film. For example, if all traces of rust are not removed from steelwork before painting, rusting will continue under the paint film and will show itself by the paint film blistering and lifting off the base metal (e.g. evidence of the onset of body-rot in motor vehicles). Let's now consider some common preparatory processes.

4.17 The removal of existing coatings

- *Organic coatings*, such as paint films, can be removed by use of a chemical solvent or a propane torch to soften the coating followed by mechanical scuffing of the surface using a manual or power scraper.
- *Anodic films* can be removed by mechanical means or, more usually, by etching and chemical polishing after solvent or chemical cleaning.
- *Chromate films* can be removed by chemical stripping and cleaning.
- *Phosphate films* can be removed by chemical stripping and cleaning.

4.18 The removal of corrosion products

In addition to the removal of deteriorated or obsolete protective coatings it is also important to remove corrosion products and scale before applying any decorative or protective coating.

- *Acid pickling* in hydrochloric or sulphuric acid is used to remove rust and scale. The acid cannot distinguish between the oxide and the metal and will often attack the metal if the oxide film is thick. This results in uneven pickling and pitting of the metal surface. This attack of the metal surface can be prevented by the addition of an *inhibitor* chemical to the pickling bath. It is essential to wash and neutralise the pickled metal and treat it with a temporary protective film such as lanolin or oil to prevent the freshly exposed surface from corroding again.

- *Conversion coatings* Where it is not possible to remove the corrosion products completely, then, prior to painting, proprietary chemical compounds (in liquid form) may be applied to convert and passivate the residual corrosion products – for example, the conversion of rust to a phosphate coating that acts as a protective film as well as a key for painting.

4.19 The removal of miscellaneous debris

Miscellaneous loose debris, soils and corrosion products are usually removed by physical rather than chemical means. Let's now consider some typical techniques.

4.19.1 *Wire brushing*

Wire brushing with a rapidly rotating coarse wire brush is used to dislodge loose debris and soils from structural steelwork before painting or repainting. The slightly roughened surface left by brushing provides a key for the first paint coat and helps it to adhere to the metal. Fine wire brushing can also be applied to aluminium and its alloys both as a decorative finish and as a key prior to painting.

4.19.2 *Shot and vapour blasting*

In shot blasting fine steel, shot or grit particles are blasted against the metal surface at high velocity using compressed air. This is used to remove soils, previous coatings and scale from structural steelwork on site as well as smaller components under factory conditions. It is also used for cleaning and descaling forgings and sand castings. Vapour blasting uses high-pressure water vapour. The water droplets travelling with high velocity have much the same effect but are less stringent and can be used on softer materials such as stonework and aluminium and copper alloy components.

4.19.3 *Flame descaling*

Flame descaling depends upon the difference in expansion between the scale, soils and other surface debris and the base metal when locally heated. This process is used for cleaning heavily rusted and soiled steelwork before initial and maintenance painting. The surface is heated with an oxy-fuel gas torch fitted with a specially designed nozzle that gives a broad fan-shaped flame. The rapid expansion of the scale compared with the cooler metal substrate causes the scale, debris and rust to flake away. Any entrapped moisture is driven off as steam and helps in the stripping process.

4.19.4 *Abrasive finishing*

Abrasive finishing can range from the use of portable grinding machines, where heavy corrosion layers need to be removed on site, to polishing and decorative finishing in the factory.

4.20 The removal of oil and grease (degreasing)

Greases and oils prevent wetting of the surface to be treated and must be removed before any pretreatment or finishing process can be applied. Let's consider some of the techniques available.

4.20.1 *Solvent degreasing*

Trichloroethylene and perchloroethylene are still used wildely in vapour-degreasing plants despite the highly toxic nature of these chemicals. Oil and grease removal is effective, but inorganic soils are only removed by the washing action of the condensed liquid. This type of a washing action is not particularly effective.

Kerosene (paraffin) will dissolve most oils and greases. It is usually blended with oil-soluble surface-active agents (surfactants) and becomes emulsifiable. Such systems have the advantage over vapour degreasing in as much that soils and residues can be rinsed away from the metal surfaces by the detergent and flushing action of the liquid. The objects to be cleaned can be either immersed in a tank of paraffin and scoured with a brush, or supported over the tank on a grill and hosed down with paraffin under pressure.

4.20.2 *Alkali cleaning*

Alkali cleaning is used where degreasing is to be followed by electroplating, because any residual solvent film will lead to poor adhesion of the plating. Further, alkalis do not have the toxicity of chlorinated hydrocarbons or the flammability of kerosene. Alkali detergents range from washing soda and caustic soda to sophisticated blends of silicates, phosphates, carbonates and surface-active agents. Alkali solutions are used at temperatures of 80–90 °C. It is important that work so treated is thoroughly rinsed so as to avoid 'carry over' into the plating baths, where the presence of alkalis would be highly undesirable. Alkali cleaning must **NOT** be used with aluminium-based or zinc-based alloys, unless suitable buffered solutions are used, as these metals suffer from alkali attack.

4.21 Protective coatings (inorganic)

These consist of ceramic materials (to be discussed in Section 5.12) applied over metallic components. The ceramic coating acts as a barrier to corrosive and erosive agents. Such coatings are susceptible to thermal and mechanical shock and, therefore, can only be applied to rigid components and structures for a limited range of applications. Such applications could be the lining of chemical and water storage tanks, the protection of pipework and the insulation of rigid electrical conductors.

Vitreous finishes are often used for the protection of low-carbon steel cooking utensils and domestic appliances as an alternative to using more expensive stainless steel. Unfortunately such coatings are very easily chipped. The vitreous finish consists of applying an opaque, powdered-glass slurry, called a *slip*, to the metal surface that is to be protected. The coated component is then dried and *fired* in a kiln so that the glass matrix melts and flows evenly over the metal surface. This finish is often referred to as 'vitreous enamelling and must not be confused with organic enamelling which is a painting process.

4.22 Protective coatings (organic)

Organic coatings can be divided into three general categories:

- Bitumastic coatings.
- Plastic and elastomer coatings.
- Paint films (Section 4.23).

4.22.1 *Bitumastic coatings*

Coatings based upon bitumen, pitch and tar are used to form barriers against the absorption of moisture. Such materials are used to protect underground pipes either as a direct coating or by wrapping the pipes in impregnated woven material such as hessian. Underground electric mains armoured cables are also protected in this way. Bitumastic paints are also available for protecting underground steel structures and the steelwork on ships.

4.22.2 *Plastic and elastomer coatings*

Plastics and rubbers, when used as protective coatings, can be functional as well as corrosion resistant and decorative. The wide range of these materials available for coating purposes provides the designer with means of achieving:

- Abrasion resistance.
- Cushion coating (up to 6 mm thick).
- Electrical and thermal insulation.
- Flexibility over a wide range of temperatures.
- Non-stick properties.
- Permanent protection against weathering and atmospheric pollution, subject to the inclusion of anti-oxidants and ultraviolet filter dyes.
- Reduction in maintenance costs.
- An impervious, protective barrier to a wide range of chemicals that would otherwise be potentially corrosive to the metal substrate.
- The covering and sealing of mechanical joints, welds and porous castings.

It must be remembered that neither the rubber nor the plastic coating in any way passivates the metal component that it is covering. Therefore, it is important that the component is treated before coating so that corrosion does not occur *under* the coating due to residual impurities. It is important that the plastic or rubber coating is not broken at any point, otherwise moisture will seep in between the coating and the component by capillary attraction. Further, unless the coating is securely bonded to the component any moisture that may also seep in due to damage will spread under the coating.

There are many processes by which plastic coatings may be applied. For example, fluidised bed dipping, as shown in Fig. 4.12. The plastic powder is kept in a state of agitation by compressed air passing through the porous bed from the plenum chamber below it. In its state of agitation, the powder offers little resistance to the immersion of the preheated workpiece to which it adheres to form a homogeneous skin.

Liquid plastisol dipping is used for coating the workpiece with PVC. The plastisol consists of the resin powder held in suspension in a plasticiser and no dangerous solvent is used. The heated workpiece is dipped into the thixotropic (non-drip) liquid plastic and a plastic film is formed on the heated surface. For larger work, spraying can be used, the plastic powder being sprayed onto the heated surface of the workpiece in a similar manner to paint spraying.

Fig. 4.12 *Fluidised-bed dipping*

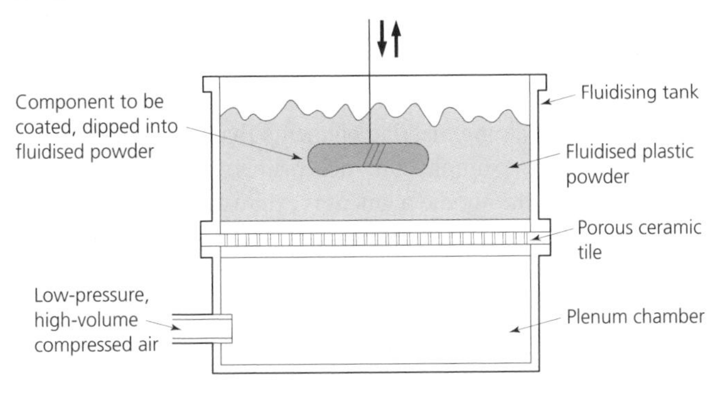

4.23 Paint films

Painting is widely used for the protection and decoration of metallic components and structures. It is the easiest and cheapest coating that can be applied with any degree of permanence and, by careful choice, painting can provide a wide range of protective properties. Paint films can be used as sealants over such finishes as galvanising, sherardising and phosphating. This is particularly useful in urban and industrial environments where the sulphur products in the atmosphere destroy the sacrificial zinc coating of galvanised or sherardised steelwork.

Paints may generally be described as consisting of finely divided solids (pigments) in suspension in a liquid (binder or 'vehicle') which dries or sets to provide a coherent film over the metal surface. Usually a paint is made up of three main constituents:

- *Binder* This contains the film-forming component in a volatile solvent. The binder is a natural or synthetic resinous material and reflects the four essential properties of the paint: durability, protective ability, flexibility and adhesion.
- *Pigment* This provides the paint with its opacity and colour. Further, some pigments have special properties and act as corrosion inhibitors, fungicides, insecticides, etc.
- *Solvent or thinner* This controls the consistency of the paint and its application. Since the solvent is volatile and evaporates once the paint has been spread, it forms no part in

the final film. In addition, a paint may contain small quantities of a catalyst or accelerator to speed up the drying and setting reactions, together with anti-skinning, and thioxotropic (anti-drip) agents.

Let's now consider a complete paint *system*, which normally consists of the following components detailed below.

4.23.1 *Primer*

This is the first paint film to be applied to the component. It must adhere strongly to the surface to which it is applied and form a 'key' to which the subsequent coats can adhere. Primers used on some metals, such as aluminium and brass, contain an etching agent to attack the surface of the metal and produce a suitable 'key' so that the primer will adhere strongly to the metal substrate. Since the primer may also contain an anodic pigment or a corrosion inhibitor, it must be matched to the material being painted. There are, basically, three types of primer:

- Primers containing metallic pigments that are anodic and sacrificial to the metal being painted – for example, zinc-rich primers for steel. This prevents under-rusting of the primer until the anodic pigment is exhausted.
- Primers containing pigments that are not only inhibitors but also dissolve in any moisture, permeating the paint film to form solutions that stifle the corrosion process in the presence of aerated water – for example, the use of chromates, phosphates and red lead when painting steelwork. The red lead oxide reacts with the oil binder of the paint to form lead azelate which is a strong corrosion inhibitor. Therefore, red lead oxide can only be used in oil-based paints. Note that chromates and red lead oxide are toxic.
- Primers that have high adhesion- and chemical-resistant properties – for example, two-part epoxy primers. These offer no anodic protection, nor do they offer any inhibition to corrosion reactions but merely act as a barrier. Therefore, the metal being protected must be free from impurities and must be carefully prepared before application of the primer to ensure that corrosion does not occur beneath the primer.

4.23.2 *Putties or fillers*

These are applied using a putty knife or a spatula to fill surface defects in castings or dents and defects in sheet metal. After setting, the putty is sanded down smooth ready for undercoating.

4.23.3 *Undercoat*

One or more undercoats are used to build up the thickness of the paint film, to give opacity to the colour and to provide a smooth surface for the finishing coat. To this end, undercoats should be thoroughly 'flatted down' between each coat. Highly pigmented undercoats decrease the permeability of the paint to oxygen, and the use of laminar pigments reduces and delays the penetration of moisture.

4.23.4 *Finish or top coat*

This is not only decorative because of its high gloss, but provides most of the corrosion resistance of the system. This is because the finish or top coat contains a varnish which seals the undercoats and prevents the absorption of moisture. The varnish content is usually tough and abrasion resistant, being based upon acrylic or polyurethane rubbers.

SELF-ASSESSMENT TASK 4.4

1. Explain precisely how a paint coating acts as a protection against corrosion.
2. Describe what preparation must be undertaken before applying a protective paint coating. State the consequences if preparation is not properly carried out.
3. Describe the essential differences between *inorganic* and *organic* protective coatings.

4.24 Types of paint

Paints can be broadly classified, by the manner in which they dry, into four main groups.

4.24.1 *Group 1*

In this group, atmospheric oxygen reacts with the binder causing it to polymerise into a solid film. This reaction is speeded up by forced drying at 70 °C. Paints that dry by oxidation include the traditional linseed oil-based paints, the oleoresinous paints, and the modern general-purpose air-drying paints based on oil-modified alkyd resins. Note, paints based upon drying-oil type binders must not be used in alkaline environments or the binder will soften and dissolve by *saponification*. Therefore, such paints cannot be used in the presence of cathodic protection systems – for example, over galvanised steelwork.

4.24.2 *Group 2*

These paints are based upon amino-alkyd resins that do not cure (set) at room temperature but have to be 'stoved' at 110–150 °C to promote the polymerisation reaction. When set, such paints are tougher and more resistant to abrasion than air-drying paints. Further, the drying cycle is much quicker than for air-drying paints. Paints in this group are often used for motor car bodies.

4.23.3 *Group 3*

In this group, polymerisation is achieved by the addition of an activator or hardener. Since this is stored separately and only added to the paint immediately before use, such paints are referred to as 'two-pack' paints. Polymerisation (hardening) commences as soon as the hardener is added to the paint. At first this will be slow and give ample time for application of the paint but, as soon as the paint is spread, a solvent (thinner) commences to evaporate increasing the relative concentration of the hardener. When this increase in concentration reaches a critical level, rapid polymerisation occurs and the paint is soon 'touch-dry'. However, it does not attain its full mechanical properties and resistance to damage for a few days. Paints in this category are based upon polyester, polyurethane and epoxy resins. The tendency nowadays is to use 'one-can' paints. The hardener is added at the time of manufacture but below the critical concentration level. Polymerisation cannot, therefore, occur until the paint has been spread and the volatile solvent has evaporated, increasing the concentration of hardener to above the critical level.

4.23.4 *Group 4*

These are the laquers – that is, paints that dry by simple evaporation of the volatile solvent with no hardening or polymerisation reactions taking place; for example, cellulose nitrate dissolved in acetone with a pigment in suspension. As soon as the acetone volatilises (evaporates) a dry film of coloured cellulose nitrate covers the surface to which the lacquer has been applied. Since both the base and the solvent are highly flammable and the fumes given off during volatilisation are toxic, great care must be taken in their use.

4.25 Application of paints

The success of a paint system depends upon the satisfactory preparation of the surface being treated and the correct application of the paint. Brush painting is labour intensive and the skill required depends upon the quality of finish required. Except for maintenance and remedial treatment it is seldom used in the engineering industry.

4.25.1 *Spraying*

This is one of the most versatile methods of coating surfaces with paint. Originally introduced for finishing mass-produced products such as cars, it is now used for large panels and small components alike. There are three basic techniques:

- Conventional spraying (compressed air).
- Airless spraying.
- Electrostatic spraying (conventional and airless).

Conventional spraying

This method employs compressed air to atomise the paint in a spray gun and project it onto the component being coated. It is a quick and relatively simple process requiring relatively low-cost equipment. Further, it is versatile and can accommodate frequent colour changes. It gives consistently high standards of finish but paint and solvent wastage is high due to overspray and bounce.

Airless spraying

In this process the paint is not atomised by compressed air but is pumped under pressure through a fine jet. Airless spraying gives less overspray and bounce, resulting in less hazardous spray-dust than with conventional spraying. As a result the process is more suitable for use outside a booth, for example, when used on large structures and for maintenance on-site. Airless spraying can give better penetration into corners of awkwardly shaped components and the rate of covering is greater. However, the film thickness is difficult to control.

Electrostatic spraying

This method applies opposing electrical charges to the paint particle as they leave the spray gun and to the component being sprayed (up to 150 kV). Since like charges repel, the spray droplets disperse and form a cloud. This gives uniformity to the film thickness and helps to dispel runs and prevents the formation of 'tear drops'. The ionised paint droplets are strongly attracted to the work that is earthed. This not only removes wastage due to overspray and bounce, but also produces a wrap-around effect – that is, spray droplets are deposited on the back as well as the front surfaces of components and edge coverage is greatly improved.

4.25.2 *Dipping*

In this process the work to be coated is suspended in a bath of paint, then lifted out and allowed to drain off. The surplus paint drains back into the bath and little is wasted. The drying process is usually accelerated by 'stoving' at elevated temperatures. In order to maintain consistency of the paint bath it is neither usual nor desirable to use air-drying paints and paints specially formulated for dipping should be used exclusively. Dipping plants usually operate on a continuous conveyor system where the components pass through the dipping bath, then over a drainage area, and finally through a stoving-tower before cooling and being off-loaded from the conveyor with the paint film set and ready for handling. Dip painting is highly productive and the labour costs are low; however, the capital cost is high, especially for large components, and close control is required if consistent results are to be achieved.

4.26 Hazards of paint application

The hazards of industrial paint application fall into two main categories:

- Explosion and fire hazards resulting from the use of flammable solvents and the formation of flammable dust particles as the spray mist dries in the atmosphere.
- Toxic and irritant effects due to the inhalation of paint mist (wholly or partially solidified) and solvent fumes.

These hazards are particularly related to spraying and stoving processes but the storage of paints also presents special problems. The local Health and Safety Inspector and the fire authorities should be consulted before painting on an industrial scale is undertaken.

4.26.1 *Spray painting*

It is essential when spray painting to provide an efficient means of extraction to remove excess spray mist and solvent fumes. Spray booths serve the double purpose of removing spray mist and fumes from the working area and then treating the exhausted air so that it is cleansed before being released back into the atmosphere. Spray booths can be classified under three headings:

- *Dry back spray booths* These consist of three-sided roofed canopies made from sheet steel and provided with electrically driven exhaust fans. The air drawn from the booth must be passed through disposable filters before passing back to the atmosphere.
- *Down-draught water-washed spray booths* These have floor gratings through which the air from the booths is exhausted. This system has the advantage that large components can be accommodated more easily than any other type. The air drawn down from the booth passes through water sprays and is washed clean before being returned to the atmosphere. The dross removed from the exhausted air forms a floating spongy mass that can be easily removed. The water is recirculated and must not be released into the environment through the public drainage system until it has been treated to render it environmentally safe.
- *Wet screen water-washed spray booths* These have a vitreous enamelled back screen over which water cascades in a continuous waterfall whilst spraying is taking place. Spraying takes place towards the screen, and excess spray mist droplets are carried down into a tank where separation takes place. The water is recirculated.

The lighting, fans and other electrical equipment associated with spray painting booth and equipment have to be to *Buxton Approved Standards* for flame and explosion proof fittings.

4.26.2 *Stoving*

The main hazards associated with stoving ovens results from:

- The use of unsuitable paints having volatile and flammable solvents that are liable to ignite at the stoving temperature.
- The accumulations of explosive dusts and gases in the fume extraction ducts. Stoving ovens and their extractors should have pressure relief vents so that any explosion is carried upwards and away from the working area.

4.27 Testing protective coatings

As for any other manufacturing process, the quality of any decorative and/or protective coatings and finishes can only be maintained by a rigorous programme of inspection and testing. To detail such tests is beyond the scope of this chapter since the tests vary for each type of coating and its application. Full information on the causes of corrosion, its prevention, and the testing and inspection of protective and/or decorative finishes can be obtained from various sources, for example:

- Manufacturers of the chemicals and materials used.
- Institution of Corrosion Science and Technology.
- Department of Industry Committee on Corrosion.
- National Physical Laboratory.
- British Standards Institution.

The efficiency of a protective coating is dependent upon a number of factors. Let's now look briefly at these factors.

4.27.1 *Type*

Permanent protective coatings must be selected to satisfy the service requirements of the component – that is, they must be economic to apply yet satisfy such design criteria as appearance and level of corrosion prevention (passivation). If corrosion resistance is the primary objective, the process selected will depend upon the material from which the component has been made and the environment in which it is to operate.

4.27.2 *Adhesion*

The satisfactory adhesion of a protective finish to the metal substrate is extremely important. Any lack of adhesion will result in the protective film flaking or peeling away, resulting in exposure of the substrate and its corrosion. Satisfactory adhesion is largely dependent upon the correct preparation of the surface to be protected.

4.27.3 *Thickness*

An unnecessarily thick protective coating is a waste of relatively expensive corrosion-resistant material and a waste of processing time. However, a coating that is too thin will not present an adequate barrier to the corrosive environment and corrosion will occur. Uneven coating can result in both waste and inadequate protection.

4.27.4 *Uniformity*

The composition of any coating must be uniform otherwise, even if the thickness is constant and correct, the efficiency of the protective coating may vary and allow local corrosion such as pitting.

4.27.5 *Chemical stability*

Except where coatings are deliberately sacrificial, they must be inert to, or become passivated, by the reactive agents in the environment against which they are to provide protection. Further, the process by which the protection is applied must not, itself, affect the substrate to which it is applied. For example, the processing temperature must not cause temper brittleness, neither should any hydrogen released during electro-deposition be allowed to cause embrittlement.

4.28 Remedial measures

These consist of the replacement of a deteriorated or obsolete protective coating with a replacement coating to the original or an improved specification. Where corrosion and/or mechanical damage has occurred, a combination of the preparatory treatments previously discussed will be required together with the replacement or repair of the structure itself, after which the replacement coating can be applied by a method appropriate to the size and type of the plant or structure, its situation and its environment.

4.29 Effects of finishing processes on material properties

All remedial and preparatory processes affect the mechanical properties of the metal being processed. For example, shot blasting enhances the mechanical properties of metals by putting the surface into a state of compression which improves the fatigue performance of the metal. On the other hand, machining, grinding, or scratch-brushing the surface reduces the fatigue performance of the metal since they leave the surface of the metal in tension. However, negative rake machining and polishing can enhance the properties of the metal since these processes tend to leave the surface of the metal in compression.

Chemical processing, such as pickling in acid, can lead to hydrogen embrittlement particularly in the case of high-strength alloy steels. Chemical etching and chemical polishing can also lead to hydrogen embrittlement. Hence, particle blasting and mechanical polishing is preferable for highly stressed components. Even a light vapour blast after chemical treatment is all that is required to restore the fatigue properties of the metal.

Many finishing processes are carried out above ambient temperatures and such processes can have the effect of impairing the mechanical properties of the metal from which the workpiece is made. For example, low-carbon steels become brittle when heated to 200 °C for any prolonged period of time, yet many finishing processes are carried out around this temperature. Similarly, aluminium alloys are particularly susceptible to processing at temperatures between 100 and 150 °C, yet this is the temperature for force drying paint. Grinding and polishing and machining processes can also raise the temperature of the metal surface to a level which can adversely affect the mechanical properties of the metal.

Finally, it should be noted that there is inevitably some chemical interaction at the interface between the cladding and the substrate. This usually results in some loss of fatigue strength. However, the protection from corrosive fatigue often far outweighs the slight lowering of the mechanical fatigue performance.

EXERCISES

4.1 With the aid of sketches, show how component design can influence the onset of corrosion.

4.2 Describe the essential reactions of dry corrosion, the causes of such reactions, and how they may be resisted.

4.3 Describe the essential reactions of galvanic or bimetallic corrosion, the causes of such reactions, and how they may be resisted.

4.4 With the aid of sketches, compare the mechanisms of uniform corrosion with preferential corrosion, and give examples of where these are likely to occur.

4.5 With the aid of sketches, describe the mechanism of crevice corrosion, together with an example of the circumstances in which it is liable to occur, and show how this form of corrosion can be prevented.

4.6 Describe the causes of, and the methods of preventing:

(a) pitting
(b) intergranular corrosion

4.7 Discuss the effect of intergranular corrosion on welded joints, and describe how such corrosion can be avoided.

4.8 Describe the mechanism of leaching, together with **two** examples of this form of corrosion, and suggest ways in which it can be avoided.

4.9 With the aid of sketches, show what is meant by erosion corrosion and suggest ways in which it may be avoided.

4.10 Discuss the effect of corrosion fatigue on the properties of metals, quoting case studies of structural failures resulting from this form of metal fatigue.

4.11 With the aid of examples, explain what is meant by:

(a) hydrogen damage
(b) biological corrosion

4.12 Discuss the factors affecting aqueous corrosion. Give examples of how this form of corrosion is affected by good and bad design and the choice of materials.

4.13 Describe, with examples, how corrosion can be prevented by the use of:

(a) sacrificial anodes
(b) anodic passivation
(c) cathodic protection

4.14 (a) Explain why the successful application of a corrosion-resistant coating is dependent upon the correct preparatory treatment of the substrate surface.
(b) Discuss the effects of finishing processes on material properties.

4.15 Discuss how the following surfaces could be prepared ready for the application of a paint system:

(a) bright low-carbon steel pressings protected by an oil film
(b) new structural steelwork
(c) rusted structural steelwork which has previously been painted and is in need of refurbishment
(d) aluminium alloy components

4.16 With the aid of typical examples, discuss the use and relative merits of bitumastic, plastic and elastomer protective coatings.

4.17 With the aid of typical examples, discuss the use and relative merits of paint systems as decorative and protective coatings.

Part B
Non-metallic materials

5 Ceramics and composites

The topic areas to be covered in this chapter are:

- Ceramic structures.
- Refractory products.
- Properties and manufacture of crystalline ceramics.
- Glass.
- Composite materials.
- Concrete and its reinforcement.
- Fibres for reinforcement.
- Particulate and dispersion hardening.

5.1 Introduction to ceramics

Ceramics are inorganic, non-metallic materials that are processed and/or may be used at high temperatures. They consist mainly of silicon and/or metallic elements chemically combined with non-metallic elements such as oxygen, carbon and nitrogen. Metallic compounds are also frequently present. The internal structure and bonding of these materials accounts for their unique properties. Ceramics are used in engineering for a wide range of products including cutting tool tips, abrasives, piezo-electric transducers, insulators, magnets, refractories, and fibres for reinforcement and optical data transmission.

Ceramics can be divided conveniently into the following three main groups that are related to their composition and structure.

Crystalline ceramics

These are widely used for cutting tools and abrasives. They may be single-phase materials such as aluminimum oxide (corundum), or mixtures of such compounds. Some of the carbides and nitrides also belong to this group.

Amorphous ceramics

Although exhibiting the characteristics of solids inasmuch that they are of definite and permanent shape at ambient temperatures, amorphous ceramic materials do not have a crystalline structure and their molecules are not arranged in regular geometric patterns. They are usually regarded as *super-cooled liquids*. This group of ceramic materials includes the 'glasses' used for such applications as glazing, mirrors, optical lenses, reinforcement fibres for glass-reinforced plastic (GRP) products and optical fibres for data transmission.

Bonded ceramics

This group includes the 'clay' products. These are complex materials containing both crystalline and amorphous constituents in which individual crystals are bonded together by a glassy (vitreous) matrix after 'firing'. The uses of ceramic products from this group include electrical insulators and refractory bricks for furnace linings. Ceramic insulators are suitable for use out of doors (electricity grid system) as their hard glazed surface renders them weather resistant. They can also be used at high temperatures – for example, sparking plug bodies and radiant heater formers.

5.2 The basic structure of crystalline ceramics

The crystalline group of ceramics may be subdivided into:

- Those with simple crystal structures.
- Those with complex crystal structures.

Examples of those ceramic materials with simple crystal structures are magnesium oxide and silicon carbide. Magnesium oxide is widely used in refractory furnace linings for steel making, and for refractory electrical insulation in mineral-insulated, copper-sheathed cables. Silicon carbide is the 'green grit' abrasive used for sharpening carbide-tipped cutting tools.

The element silicon is present in most ceramic materials in the form of complex silicates. (Note that *silica* is silicon oxide, SiO_2, whilst *silicates* are compounds of metal ions with silicon and oxygen – e.g. $MgSiO_3$.) Silicon and carbon have many similar chemical properties and, just as carbon compounds can be built up into long polymer chains to produce 'plastic' materials, silicon–oxygen groups can also be built up into long-chain, sheet or three-dimensional framework structures.

5.2.1 *Chain structures of ceramics*

Having considered some examples of ceramic materials with simple crystalline structures, let's now consider some ceramic materials with more complex crystal (chain) structures. Figure 5.1 shows a single SiO_4 group tetrahedron that forms a 'building block' for many different ceramic structures. For example, Fig. 5.2 shows a typical double-chain structure based on SiO_4 group tetrahedra cross-linked by ionically bonded metal ions. Unlike organic polymer materials (plastics), where adjacent chain molecules are only held together by relatively weak van der Waals forces, the stronger ionic bonds in ceramic materials makes them harder and stronger than organic polymers. However, the ionic bonds cross-linking adjacent silicate chains are weaker than the covalent bonds linking the silicate groups. Thus ceramic materials tend to fracture along the ionic bonds parallel to the silicate chain. Note that single chains based on SiO_4 groups may also be formed but these are less important.

Fig. 5.1 *The SiO_4 group: (a) silicon–oxygen bond; (b) three-dimensional representation of the SiO_4 group tetrahedron*

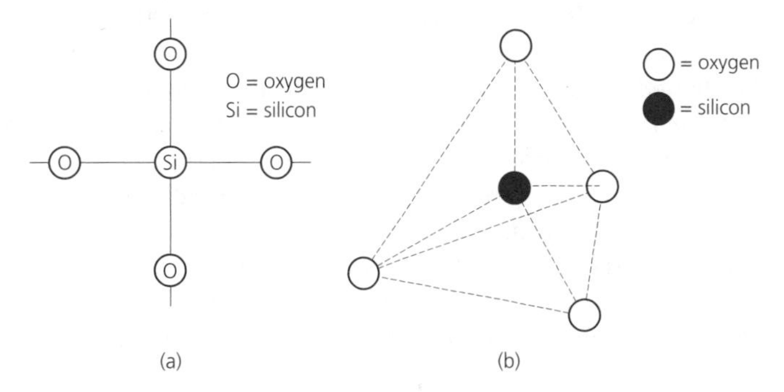

Fig. 5.2 *Double chain structure of silicate tetrahedra with metallic ion cross-linking*

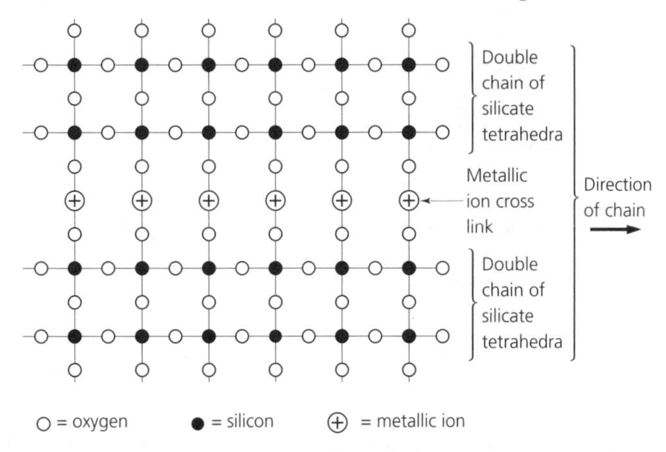

The differences in behaviour between metallic crystals and ionically bonded ceramic crystals can be explained by reference to Fig. 5.3. In a metallic crystal the positively charged metal ions are prevented from repelling each other by the negatively charged electron 'cloud' surrounding each metal ion. The mutual attraction of the positive metal ions and the negatively charged electron 'clouds' results in a state of equilibrium that holds the metal ions in position in the crystal, providing no external disturbing force is applied. It can be seen in Fig. 5.3(a) that when such an external disturbing force (F) is applied to a plane of metal ions they tend to move along a slip plane, so that the ions become nearer to those of an adjacent plane. This results in an increase in the electrostatic repulsion forces between adjacent planes of metal ions and they try to move back to their original positions (spring-back of elastic materials). The repulsion forces reach a maximum at the elastic limit for the material but, if overcome by the external disturbing force, the ions move to the next incremental position. Equilibrium between the positive metal ions and the negative electron 'cloud' is once more achieved and slip occurs. The plane along which slip occurs by dislocation is called the *slip plane* and the theory of slip by dislocation, was introduced in *Engineering Materials*, Volume 1, will be developed further in Section 12.2 of this book.

In an ionically bonded ceramic crystal, as shown in Fig. 5.3(b), it can be seen that positively charged metal ions and the negatively charged silicate ions are arranged so that each ion is surrounded by ions of opposite charge, and this equilibrium system holds the ions rigidly in place. However, if an external disturbing force of sufficient magnitude is applied, the equilibrium position of the ions will be disturbed. This will result in ions with like charges being brought closer together; the like charges will repel each other and the crystal will split apart – that is, the crystal will shatter along the cleavage plane.

Although the ionic bonding and covalent bonding present in ceramic materials results in relatively high compressive strengths and moduli, the lack of slip systems in ceramic materials is reflected in their lack of plastic properties. The low tensile strengths of most ceramic materials is due to the surface micro-cracks which act as stress raisers (see Sections 12.13 to 12.15 inclusive). It is the high integrity of the surface finish of ceramic fibres (glass fibres and carbon fibres) that results in the high tensile strengths of these materials.

Fig. 5.3 *Deformation of crystals: (a) metallic crystal; (b) ionic crystal*

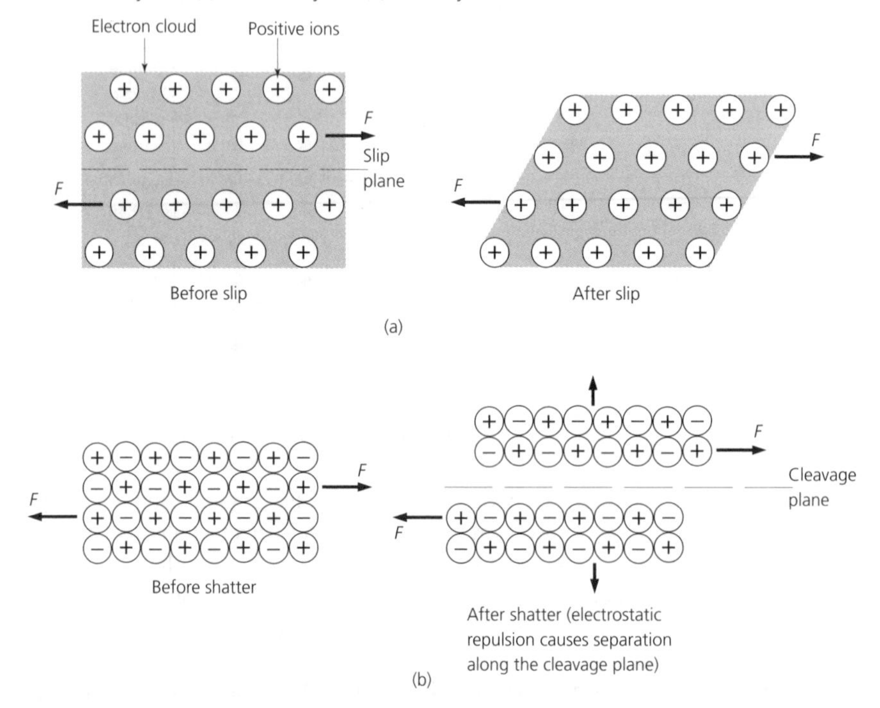

SELF-ASSESSMENT TASK 5.1

1. Review the various types of atomic bonds. Refer also to Chapter 2 in *Engineering Materials*, Volume 1.
2. State the major disadvantage of ceramic materials. Describe three techniques used to overcome this disadvantage.
3. During the next week take note and make a list of the products that you come across that exploit the special properties of ceramic materials. Identify the properties that make ceramic materials most suitable for the products that you list.

5.2.2 *Sheet structure of ceramics*

Ceramic materials that consist of flakes or plates, as found in clays and micas, have their silicate groups arranged in sheet structures rather than in chain structures as previously described. All the valencies within the sheet structure are satisfied and adjacent silicate sheets are only attracted to each other by relatively weak van der Waals forces.

5.2.3 *Framework structure of ceramics*

If the basic silca (SiO_2) tetrahedra develop into three-dimensional framework structures so as to produce a giant molecule, a hard, rigid framework structure is produced. This structure fractures in an irregular manner as the bonding forces are equal in all directions and no preferential cleavage planes exist.

Two important examples of this type of ceramic structure are *cristobalite*, which is used in refractory bricks for high-temperature furnace linings, and *quartz*, which occurs naturally in sandstones and in beach sands. Quartz is used as an abrasive (sand paper) and, industrially made quartz crystals are used for resonant frequency applications (e.g. in electronic oscillator circuits). Quartz is extremely resistant to solution in water, hence its long life in beach sands. Unfortunately quartz is unsuitable for high-temperature applications since the Si–O–Si bonds change their angles at 573 °C. Since silica is a rigid compound, this change of angles causes an abrupt change of volume resulting in cracking, unless the rate of heating or cooling is very slow.

5.3 Refractory products (clays)

These are ceramic materials whose silicate tetrahedra form sheet structures as described earlier. The flaky particles of dry clay are transformed into a plastic mass (dough) by the addition of water. The water molecules are attracted to the surface layers of the 'sheets' by polarisation forces and provide boundary layer lubrication between adjacent sheets. Clays are frequently 'blended' by mixing two or more ingredients to control the composition and plasticity. This is particularly important when producing high-grade articles such as porcelain insulators, chemical ware and whiteware. The blend consists of finely ground clay mixed with such non-plastic materials as ground quartz, flints and fluxes. The ground quartz and flints are non-absorbent and reduce the amount of water required to make the blend plastic. This reduces shrinkage, distortion and cracking during drying. Fluxes are added to reduce the fusion point of vitreous (glass) content of the blend and thus reduces the firing temperature required for complete vitrification. Clay products may be shaped by hand or by machine moulding.

Drying is used to remove the water from the moulded mass before firing. Excessive drying shrinkage can lead to distortion and even cracking; therefore, drying ovens should operate at between 85 and 96 °C and the atmosphere of the oven should have a high humidity. This prevents the surface of the mould clay product drying out before the core of the product and ensures uniform drying throughout the mass of the product.

Firing vitrifies the moulded and dried clay product so as to give it strength, rigidity and durability. Firing temperatures depend upon the composition of the clay and the physical characteristics of the product. The process of firing takes place in a series of stages as follows:

110–260 °C The final shrinkage water and pore water (hygroscopic water) is driven off.

430–650 °C The clay minerals break down into silica and alumina and any chemically combined water molecules are driven off as water vapour. At this point the clay loses its ability to absorb water and can never again become a plastic dough. However, there is, as yet, very little change in the strength and porosity of the moulding.

800–900 °C At this temperature an oxidising atmosphere should be maintained in the furnace to burn off any residual organic material and to oxidise any iron pyrites that may be present as an impurity. All gas-forming reactions should also be completed at this stage whilst the moulding is still porous, and any gases that are produced can easily escape.

900–1000 °C At these temperatures vitrification and firing shrinkage begins to take place. Vitrification is the gradual fusion of some of the compounds present to form a liquid that fills up the pores in and between the particles forming the moulding. Full vitrification is reached at about 1400 °C. Further heating is unnecessary and could lead to the softening and collapse of the moulding. When cooled to ambient temperatures this vitreous liquid solidifies to provide a glass-like matrix that cements together the inert particles. This decreases the porosity and increases the strength of the product.

5.4 Refractory products (common)

These are heat-resistant products such as firebricks and fireclays that are used for furnace linings. Refractoriness is defined as *the ability of a material to withstand high temperatures without appreciable deformation under service conditions*. It is assessed generally by the softening or melting point of the material. Since refractory materials in service are subjected to loads of varying intensity, it is important to assess their refractoriness under similar loads. Some refractories, such as high-alumina (aluminium oxide) firebricks and fireclays, soften gradually over a wide range of temperatures and collapse under load well below their true fusion temperatures. Others, such as silica (silicon oxide), soften over a narrow range of temperatures and continue to exhibit good load-bearing properties close to their fusion temperatures. However, the melting temperature of silica is very much lower than that of alumina, and this must be taken into account when determining their relative usefulness for load-bearing refractories.

The majority of commercial refractory materials used in high-temperature processes and environments are represented by the 'common' refractories because of their availability and low price. Such refractories consist of crystalline and some amorphous constituents cemented together by a vitreous (glassy) matrix. Some of the most 'common' refractories are based upon alumina and silica compositions and vary from almost pure silica through a wide range of alumina silicates, to almost pure alumina.

The silica–alumina phase equilibrium diagram is shown in Fig. 5.4. Note that *cristobalite* and *tridymite* are allotropes of silica, and *mullite* is the name given to the composition formed when excess alumina reacts with residual silica ($3Al_2O_3.2SiO_2$). This composition corresponds, by weight, to 71.8 per cent alumina and 28.2 per cent silica. Corundum is the name given to crystalline alpha alumina that forms when excess mullite dissociates at 1840 °C.

Fig. 5.4 *Silica–alumina phase equilibrium diagram (adapted from Higgins, The Properties of Engineering Materials, Arnold)*

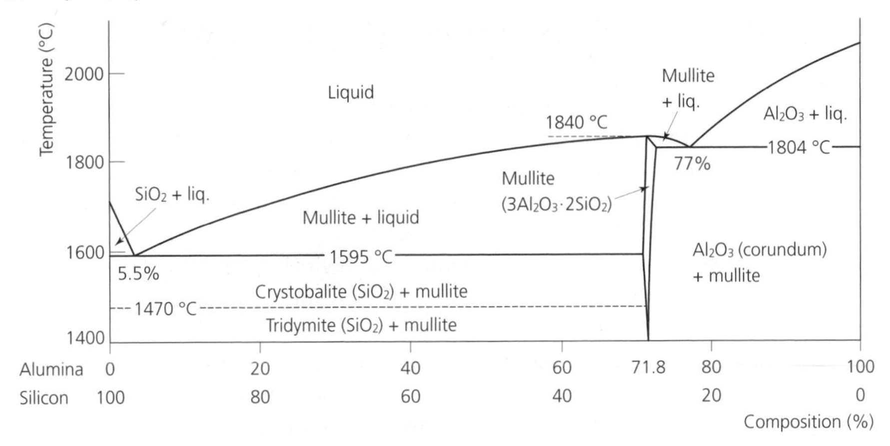

A detailed explanation of the silica–alumina phase equilibrium diagram is beyond the scope of this book; however, some practical points should be noted:

- Figure 5.4 shows that refractories containing between 4 per cent and 8 per cent alumina should be avoided since they melt completely at or a little above the eutectic temperature of 1595 °C.
- Refractoriness increases with an increasing alumina content and the load-bearing properties of the material also increases. This may appear to contradict an earlier statement that alumina has a wide softening range and tends to collapse before it melts. However, it must be remembered that silica has a relatively low melting point compared with alumina and that refractoriness depends upon the amount of liquid present and the viscosity of that liquid.
- Silica–alumina refractories soften over a wide range of temperatures. Softening commences at the temperature at which the silica starts to melt and liquid starts to form, and extends to the temperature at which the refractory is totally molten. As the alumina content increases above the eutectoid point of 5.5 per cent alumina, the quantity of liquid produced at 1595 °C will decrease and the refractory will become stronger at that temperature.
- For severe conditions, where the temperature exceeds 1800 °C, the alumina content must exceed 71.8 per cent to yield a solid phase containing only corundum and mullite in a vitreous matrix.

Apart from silica and alumina, other 'common' refractories are magnesite, forsterite, dolomite, silicon carbide and zircon. Refractories can be classified chemically as: acid, basic or neutral. Magnesite is a *basic* refractory found in the linings of iron- and steel-making plant where high phosphorus content ores are used. The phosphorus content in the metal is reduced to a safe level because the phosphorus reacts with the basic furnace lining to form a *basic slag*. This slag provides a cheap source of high phosphorus content fertilisers. Although the furnace lining is gradually eaten away, this process allows low-cost, low-grade ores to be used. Silca brick is an *acid* refractory, and slica–aluminium brick is a *neutral* refractory, providing the alumina content is high enough.

5.5 Refractory products (high grade)

The 'common' refractories discussed so far depend upon a ceramic bond when fired. The term 'ceramic bond', as used in the ceramic industry, is not a true chemical bond but a cementing or binding together of refractory particles in a vitreous (glassy) matrix. This ceramic bond is responsible for the cold strength of ceramic materials but its presence reduces the refractoriness of the material at high temperatures. Thus for 'high-grade' refractories, where strength is required at high temperatures, the vitreous matrix should be reduced to the minimum necessary to provide adequate strength at room temperature. When the temperature during firing is sufficiently high and the duration of the firing process sufficiently long, the glassy matrix may be gradually replaced by crystals. This results from the dissolution of some of the compounds and a change in the crystal structure of mullite. The change results in the formation of an interlocking crystal structure that greatly improves the bonding and, therefore, the strength of the material under load at high temperatures (refractoriness).

Phase equilibrium diagrams for refractories, such as the example shown in Fig. 5.4, indicate that generally higher refractoriness can be obtained by using pure, high-melting-point oxides since no eutectoid compositions will be present. The presence of even small quantities of impurities not only lowers the melting point considerably but also reduces the refractoriness under load to a greater extent than is indicated by the phase equilibrium diagram. Thus 'high-grade' refractory materials consist of very pure basic ingredients and the elimination of the ceramic bond by special processing techniques. Let's now consider some individual 'high-grade' refractories in greater detail.

5.6 Oxides

Pure oxide refractories consist essentially of such oxides as alumina, beryllia, magnesia, thoria and zirconia. These are the names given to the oxides of aluminium, beryllium, magnesium, thorium and zirconium respectively. This is not an exhaustive list but represents the more readily available refractory oxides. Products from these materials are made by slip-casting, pressing, and extrusion. Firing takes place at about 1800 °C at which temperature the fine oxide particles sinter rapidly and solid surface reactions occur between the individual particles. This results in a crystalline bond that produces a coherent mass, and the fired article so produced is said to be a *self-bonded refractory*. Since the ceramic bond is eliminated there is no glassy matrix with its low melting point to reduce the refractoriness of the product. The crystalline bond is composed of crystals of the same materials as the particles being bonded and this results in self-bonded products having refractory properties approaching that of the particle material itself.

5.7 Borides

Borides are essentially interstitial compounds of transition metals with boron. The properties of these materials are:

- Very high hardness.
- Good resistance to chemical attack.
- Melting points ranging from 1800–2500 °C depending upon the compound.

Although borides are more resistant than carbides to oxidation at high temperatures, they do start to oxidise at about 1400 °C. The more commonly available industrial borides are those of chromium, molybdenum, titanium, tungsten and zirconium.

5.8 Nitrides

Nitrides are compounds of metals and nitrogen. Although nitrides have high melting points, they have a low resistance to oxidation and chemical attack generally. However, 'Borazon', which is a synthetic boron nitride produced under high temperatures and pressures, has a hardness approaching that of diamond and can withstand temperatures up to 1925 °C without appreciable oxidation.

5.9 Carbides

Although carbides have very high melting points, they lack resistance to oxidation at high temperatures. The more important refractory materials are the carbides of boron, silicon, titanium and zirconium. Carbides for cutting tools will be considered in Chapter 10. Silicon carbide (carborundum) refractories are the oldest and the most widely used, their properties depending upon the type and quantity of bond.

The most widely used binding agent is refractory clay that forms a ceramic bond on firing. However, as with all ceramic bonds, the refractory formed starts to soften at the relatively low temperatures of 1200–1500 °C. Other binding materials used are silicon nitride and silicon. When silicon nitride is used as the binding material the refractory produced has high strength and a high resistance to thermal shock. Silicon has only limited application as a binding material since its melting point is only 1426 °C.

Self-bonded silicon carbides are formed by mixing the carbide compound with temporary organic binding agents. They are then pressed to shape and fired at 1700 °C. The temporary binder is burnt off and a crystalline bond develops between the silicon carbide particles. This results in a product of high refractoriness, high strength, high density, high abrasion resistance and high resistance to chemical attack. All silicon carbide refractories have high thermal conductivity and low coefficients of thermal expansion but, unfortunately, they tend to oxidise slowly to silica in the temperature range 900–1300 °C.

Boron carbide is very hard and abrasion resistant. Components are produced by hot pressing or by bonding with sodium silicate and firing. Although the melting point of boron carbide is 2450 °C, its maximum usable temperature is restricted to about 980 °C as boron carbide oxidises rapidly at higher temperatures and also reacts with hot or molten ferrous metals.

Cerium, molybdenum, niobium, tantalum, tungsten and zirconium carbides can be used at temperatures above 2000 °C in neutral or reducing atmospheres, whilst niobium, titanium and vanadium carbides can be used above 2500 °C in an atmosphere of nitrogen. Hafnium carbide has the highest melting point of any known substance at 3900 °C.

5.10 General properties of crystalline ceramics

The ceramic materials considered so far have been crystalline or largely crystalline with some amorphous content. All these materials share the following general properties to a greater or lesser degree.

5.10.1 *Refractoriness*

The refractoriness of the various groups of ceramics has already been dealt with in some detail. All ceramic materials have a greater refractoriness than most metals. The melting temperatures of some typical high-purity ceramic materials are listed in Table 5.1.

Table 5.1 *Melting points of common refractory ceramics*

Refractory material	*Melting point (°C)*
Halfnium carbide	3900
Tantalum carbide	3890
Thorium oxide	3315
Magnesium oxide	2800
Zirconium oxide	2600
Beryllium oxide	2550
Aluminium oxide	2050
Silicon nitride	1900

5.10.2 *Strength*

Ceramic materials have high compressive strengths compared with their tensile strengths and the ability to retain this strength at high temperatures is one of their most important properties. Titanium diboride, for example, has a compressive strength of 250 MPa at 2000 °C which makes it one of the strongest materials known at such a high temperature. The lack of tensile strength in ceramic materials is due to the presence of micro-cracks. These act as potential stress raisers and the lack of ductility in ceramic materials prevents any stress concentrations at the micro-cracks from being relieved by the onset of plastic flow. The reason why ceramic materials lack ductility and suffer from sudden cleavage was explained in Section 5.3.

5.10.3 *Hardness*

Ceramic materials are harder than any pure or alloyed metallic materials even after heat treatment. This hardness makes ceramics useful as abrasives and as cutting-tool tips. The Knoop hardness number for some hard ceramic materials is listed in Table 5.2.

Table 5.2 ***Hardnesses of common ceramic materials***

(At room temperature)

Ceramic material	*Knoop hardness number*
Cubic boron nitride*	7000
Boron carbide	2900
Silicon carbide	2600
Aluminium oxide	2000
Beryllium oxide†	1220

*Compare with *diamond*, Knoop hardness 7000.
†Compare with quench-hardened, high-carbon *steel*, Knoop hardness 700.

5.10.4 *Electrical properties*

Ceramic materials have been used for electrical insulation purposes for a long time. Glazed porcelain insulators are used for such purposes as supporting high- and medium-voltage overhead electric cables and also telephone and telegraph cables. The hard glazing prevents the insulators from 'weathering'. Softer plastic insulators would be unsuitable for such purposes since they would quickly 'weather' and become roughened. They would then become covered in dirt from atmospheric pollution, and this surface layer of dirt would provide a conducting path over the insulator leading to a 'flashover'. Any dirt deposited on hard glazed porcelain is quickly washed away by rain as it cannot adhere to the smooth surface. Unglazed ceramics are used for formers for wire-wound resistors and heating elements.

Magesium oxide powder is used in mineral-insulated, copper-sheathed cables and in sheathed heating elements. This material has the advantage that it is able to retain its insulating properties at very high temperatures that would destroy more conventional insulating materials. Ceramic materials are also used for low-loss high-frequency insulators, ferromagnets and semiconductor devices.

5.11 Shaping methods

Traditional shaping methods for ceramics (some of which have already been mentioned) consist of hand and machine moulding, powder pressing, the extrusion and rolling of a plastic, clay–water, mixture and 'slip–casting'.

5.11.1 *Slip casting*

This process consists of filling a porous mould with a liquid suspension of the powdered clay (slip). On standing, the powder forms a deposit on the walls of the mould. After sufficient time has elapsed for the deposit to have acquired the specified thickness, the surplus slip is poured off leaving the 'casting' behind in the mould. This technique is used for the production of thin-walled products. The casting is dried to give it sufficient strength to withstand handling, after which it is removed from the mould and fired.

5.11.2 *Isostatic pressing*

Mass-produced components such as sparking plug bodies can be made by the compaction of ceramic powder in dies, followed by sintering. 'Firing' converts the amorphous ceramic particles in the mix into a glassy vitreous 'bond' or matrix, whereas sintering is the heating of the compact of high-purity crystalline ceramic material, in a controlled atmosphere to prevent oxidation, to a temperature that is sufficient to cause diffusion and recrystallisation across the particle boundaries causing them to bond together.

Simple pressing in metal dies leads to non-uniform density distribution in the compact, resulting in the distortion and cracking of the final product. This can be overcome by the use of isostatic pressing. In this technique the powder to be compacted is placed in a strong rubber bag mould and subjected to external fluid pressure, as shown in Fig. 5.5, thus ensuring uniform compaction in all directions. Unfortunately this technique does not allow

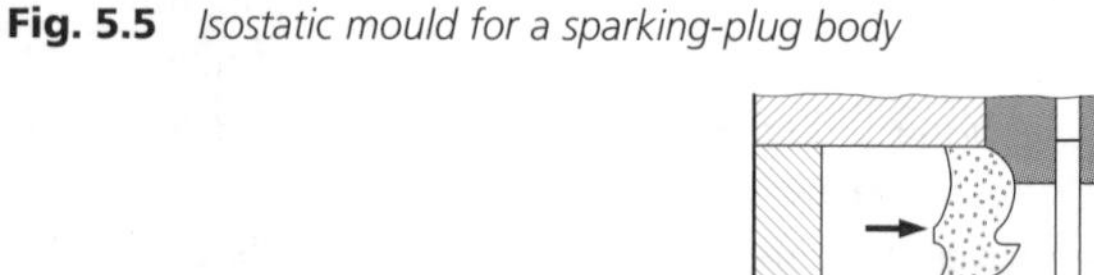

Fig. 5.5 *Isostatic mould for a sparking-plug body*

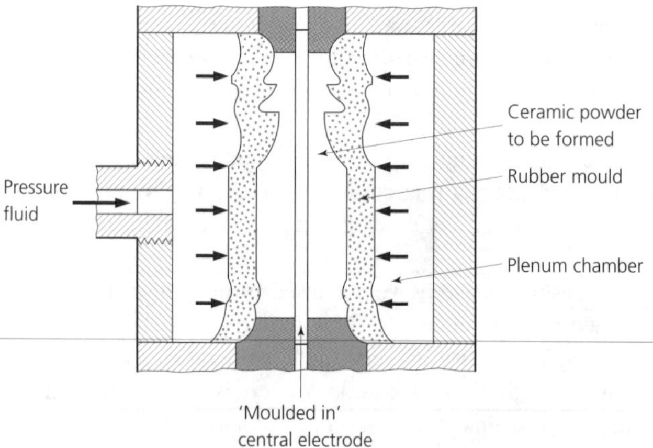

the forming of complex shapes of high accuracy. However, prior to sintering machining of a compact which has been formed by isostatic pressing can be used to improve the accuracy of the finished product.

5.11.3 *Hot pressing*

Unfortunately, ordinary sintering processes do not increase the density of many materials. This applies particularly to materials that dissociate at the high temperatures involved in sintering. For example, silicon carbide dissociates to silicon and carbon at temperatures exceeding 2300 °C. However, the density of many of these materials can be increased at lower temperatures by the simultaneous application of pressure and heat. Hot pressing results in densities approaching the theoretical maximum and also results in a finer grain structure. Unfortunately, this process is not only expensive but it can only be applied to simple shapes such as equiaxial cylinders. The uniform increase in density (densification) becomes difficult when the length/area ratio becomes large.

5.11.4 *Gas pressure bonding*

The Battelle process of gas pressure bonding is, in fact, hot isostatic pressing. A flexible, thin-walled, metal pressure vessel is used in place of the rubber bag of a conventional isostatic moulding. The liquid is replaced by hot, pressurised gas. Alumina can be densified to 99.3 per cent of its theoretical maximum using gas at a temperature of 1290 °C and a pressure of 69 MPa.

5.12 Ceramic coatings

Let's now consider the coating of metal components with ceramic materials to improve their corrosion and erosion properties and their high temperature resistance

5.12.1 *Flame spraying*

There are two basic variants of the flame-spraying technique for depositing ceramic coatings.

- Ceramic powder is melted as it is passed through an oxyacetylene flame, an oxy-hydrogen flame, or a plasma arc. The molten ceramic material is sprayed onto the surface of the component in a similar way to metal spraying.
- A prefabricated ceramic rod is used in place of the powder described above. The heat source both melts and 'atomises' the ceramic material and the molten particles are sprayed onto the surface of the component. Although this is a more expensive process because of the cost of prefabricating the ceramic rods, it ensures that only molten particles are sprayed onto the component surface. In the process previously described there is a danger that some unmolten powder is also carried over in the spray, reducing the integrity of the coating.

Since there is no chemical reaction between the coating and the component being coated, the surface of the component must be roughened to provide a mechanical 'key'. In fact better adhesion is achieved if the surface to be protected is sprayed with metal prior to spraying with a ceramic material. Sprayed ceramic coatings are not impermeable and, therefore, only offer limited protection against oxidation or chemical corrosion of the substrate. Although the substrate is cold in most spraying processes, higher densities and lower permeabilities can be achieved if the substrate is heated before and during the spraying process.

5.12.2 *Chemical vapour deposition*

Chemical vapour deposition (CVD) consists of heating the substrate and passing a vapour or mixture of vapours over it. Reactions take place at the surface with the deposition of a dense impermeable coating. For example, passing methyl trichlorosilane (CH_3SiCl_3) over heated graphite results in the deposition of a layer of silicon carbide on the surface of the graphite component. The coefficients of thermal expansion of the coating and the substrate must be closely matched to prevent cracking or peeling of the coatings. Deposits of this type (eg. titanium carbide on steel) are useful for increasing the resistance of the component to abrasive wear and to chemical attack.

Thin-walled components have been manufactured experimentally by building up the thickness of the deposit and then removing the substrate. For example, thin-walled impermeable silicon-carbide tubes have been manufactured by depositing the silicon carbide on a graphite mandrel and burning away the graphite.

5.13 Finishing processes

Many ceramic products used in engineering applications require some finishing process. Since fired ceramics are very hard they can only be shaped or finished by grinding using diamond-impregnated grinding wheels or machining using diamond-tipped tools. Therefore, it is economically important to ensure that the fired component only requires the minimum of machining. Although finishing improves the dimensional accuracy of a fired component, the surface finish produced by grinding or machining tends to be unsatisfactory for engineering applications due to the porosity of the unglazed material. Pore-free ceramics are now available which can be finished to high accuracies and very low values of surface roughness, however such materials and processes tend to be costly.

5.14 Joining ceramics to metals

To obtain a satisfactory joint between ceramic materials and metals, it is essential to ensure that the materials being joined have closely matching coefficients of thermal expansion. Also, if there are any residual stresses in the assembly, these must be such that the ceramic component is maintained in compression.

Metals and ceramic materials can be joined using intermediate layers of glass as a bond. Alternatively, the ceramic component in the assembly can be metallised with silver or nickel and soft soldered to the metal component. Some of the strongest joints produced by the latter process are used in the manufacture of high-power thermionic valves used for such purposes as power amplifiers in radio transmitters. For this application, high-temperature metallisation of the ceramic component is carried out using a molybdenum–manganese alloy or titanium. The layer so formed is built up by nickel plating and then brazed to the metal component of the assembly.

5.15 Glass

Unlike the ceramic materials considered so far, glass can be considered as a product of the fusion of inorganic materials that have cooled to a rigid condition without crystallisation – that is, glass behaves as a supercooled liquid. At high temperatures, when molten, glasses form normal liquids. Their atoms are free to move and they respond readily to shear stresses. In fact molten glass can be used as a lubricant for some metal-forming processes where extreme pressures are involved. As supercooling takes place without crystallisation, a sudden change in the coefficient of thermal contraction occurs, as shown in Fig. 5.6, due to the rearrangement and more efficient packing of the atoms. This is the glass transition (T_g) or *fictive* temperature for glass. Below this temperature no further rearrangement of the atoms takes place and any further contraction is due only to the reduced thermal agitation of the atoms. Other physical properties that change at the glass transition temperature are:

- Density.
- Viscosity.
- Refractive index.
- Electrical resistivity.

Although glass is an excellent insulator at ambient temperatures, it becomes a conductor of electricity at red heat.

Fig. 5.6 *Glass transition temperature (T_g) for silicon glass*

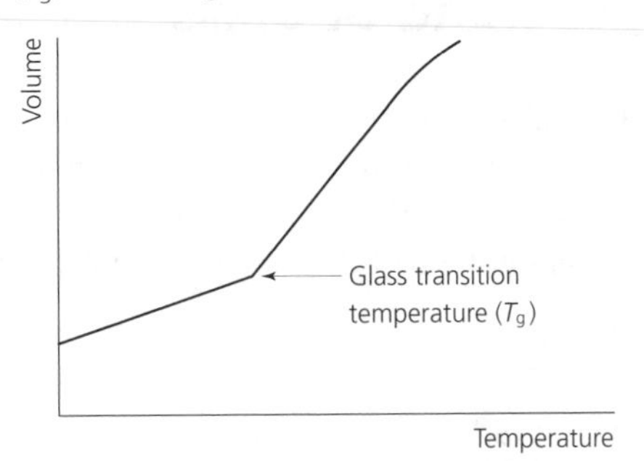

5.16 Types of glass

Although silica is, theoretically, an excellent glass-making material, its high melting temperature makes it uneconomical to use alone and basic metal oxides are added to lower the fusion temperature and viscosity of the melt. A eutectic mixture of 75 per cent silica with 25 per cent sodium oxide reduces the fusion temperature to 793 °C and results in the formation of sodium disilicate ($Na_2O.2SiO_2$). Unfortunately such a glass would be water soluble and unsuitable for all practical purposes. The addition of calcium oxide to the mixture produces soda-lime glass that is insoluble in water and widely used for window glass and bottles. Glass will sometimes crystallise over a period of several hundred years. This process is called devitrification and results in increased brittleness and reduced transparency. For this reason calcium oxide, which tends to promote devitrification, is deliberately kept below 20 per cent. The most commonly available commercial glasses are listed in Table 5.3 together with some typical applications. Lead glasses are also known as 'flint' glasses and, where extra density is required for special optical purposes and for protective shields to absorb X-ray radiation, the lead oxide content can be increased to as much as 80 per cent.

5.17 Manufacture of glass

The raw materials from which glasses are manufactured are mixed together and heated in 'tank' furnaces where they react together and melt to form the complex ceramic substances called 'glass'. For example, the main raw materials for common soda-lime glass are silica sand, soda ash (crude sodium carbonate) and lime obtained from limestone. 'Cullet' (broken scrap glass) may also be added.

The furnace temperature to ensure the charge is completely molten is about 1500 °C, depending upon the viscosity required for the forming process. The glass is taken from the furnace as required and blown and/or moulded to various shapes or, in the case of sheet glass, manufactured by a flotation process.

5.18 Strengthening of glass

The cooling of glass from its processing temperature to room temperature is relatively rapid and leads to thermal stresses being set up in the glass. Such cooling stresses adversely affect the strength and other physical properties of the glass, but these stresses can be relieved by the application of suitable heat-treatment processes, such as:

- Thermal tempering.
- Chemical tempering.
- Annealing;

Table 5.3 Compositions and properties of common types of glass

Type of glass		*Composition*								*Properties and uses*
		SiO_2	*Na_2O*	*CaO*	*MgO*	*K_2O*	*PbO*	*Al_2O_3*	*B_2O_3*	
Soda-lime glass		70–75	12–18	5–14	0–4	0–1	—	0.5–2.5	—	Window glass, bottles and general usage
Leaded glasses	(a)	53–68	5–10	0–6	—	1–10	15–40	0–2	—	High electrical resistance – lamp and valve envelopes
	(b)	40	2.5–5	—	—	2.5–5	45–50	0–5	—	High refractive index and dispersive powers. Used for lenses, prisms, and other optical devices. Used for cut 'crystal' glass for tableware
Borosilicate glass		73–82	3–10	0–1	—	0.4–1	0–10	2–3	5–20	Low thermal expansion and resistant to chemical attack. Heat-resistant cooking and tableware (Pyrex) and laboratory apparatus. Will perform a seal with some metals
Aluminosilicate glass		70	2–4	—	—	2–4	—	3	20	High softening temperature, $T_g = 800°C$
High-silica glass (vitreous silica)		96	—	—	—	—	—	—	3	Made by removing the alkalines from borosilicate glass after melting and shaping. Very low thermal expansion; very high T_g, can be used continuously at 800 °C

5.18.1 *Thermal tempering*

In this process the glass is heated up to its annealing temperature and the surface is cooled rapidly (chilled) by an air blast, as shown in Fig. 5.7. This rapid cooling results in the surface of the glass becoming rigid whilst the interior is still plastic. Under these conditions the surfaces will develop tensile stresses. As the bulk of the glass cools down to room temperature contraction occurs, which forces the already solidified surfaces to contract as well. Internal tensile stresses are developed, and these are not only sufficiently great to cancel out the initial tensile stresses in the surfaces of the glass but are sufficiently great to introduce compressive stresses into the surfaces of the glass. Compressive surface stresses improve the strength of the glass, as they do in most materials. It should be noted that the higher the coefficient of thermal expansion of the glass and the lower its thermal conductivity, the greater will be the residual stresses and the stronger the glass becomes. It may seem strange that introducing stresses into the glass can make it stronger. However, since glass usually fails at its surface due to *induced tensile forces*, the fact that the surface is *prestressed in compression* results in a considerable increase in strength. The glass will not break until the induced tensile stresses have overcome the compressive prestressing. Tempered glass has a strength and impact resistance three to five times greater than annealed glass whilst retaining the same appearance, clarity, hardness and coefficient of thermal expansion as the original glass. Once tempered, the glass cannot be cut, machined or ground as this would upset the stress system leading to disintegration of the glass. All forming and cutting must occur before tempering. 'Toughened' glass windscreens are made from tempered glass and, because of the energy stored in the residual stresses, they shatter into small but relatively harmless granules when broken. Thermal tempering is a relatively quick process (minutes) and can be applied to most types of glass.

Fig. 5.7 *Residual stress in tempered glass plate and stages involved in inducing compressive surface residual stress and improved strength (adapted from Kalpakjian, Manufacturing Engineering and Technology, Addison Wesley)*

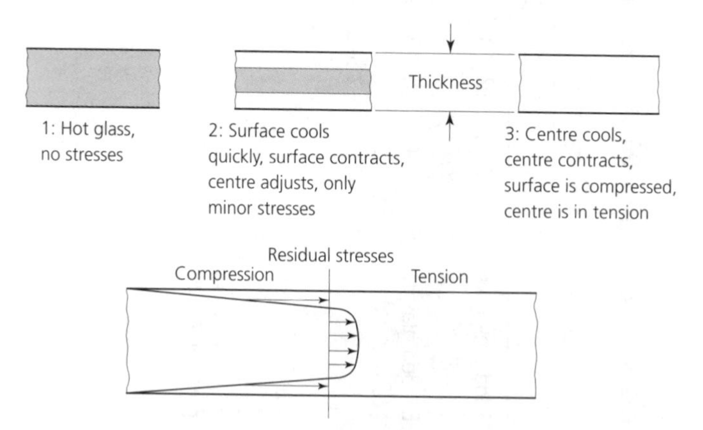

5.18.2 *Chemical tempering*

In this process, the glass is heated in a bath of molten potassium nitrate (KNO_3), potassium sulphate (K_2SO_4) or sodium nitrate ($NaNO_3$), depending upon the type of glass. An ion exchange takes place during which larger atoms from the molten salt bath replace smaller atoms at the surface of the glass. The 'wedging' action of the larger atoms causes the smaller surface atoms to crowd together, introducing compressive stresses in the surface of the glass. This is shown diagrammatically in Fig. 5.8. The time required for chemical tempering is about one hour, which is considerably longer than for thermal tempering. Chemical tempering can be carried out at various temperatures. At low temperatures, product distortion is minimal and complex shapes can be treated. At elevated temperatures there may be some distortion, but the product can itself be used at *higher temperatures without loss of strength*. It is in the ability of the glass to retain its strength at elevated temperatures, that the more costly chemical-tempering process scores over the lower cost thermal-tempering process.

Fig. 5.8 *Chemical tempering of glass: (a) normal 'packing' of glass surface atoms; (b) the infusion of larger atoms at the surface of the glass causes 'crowding' of the glass atoms, thus introducing compressive stresses in the surface of the glass*

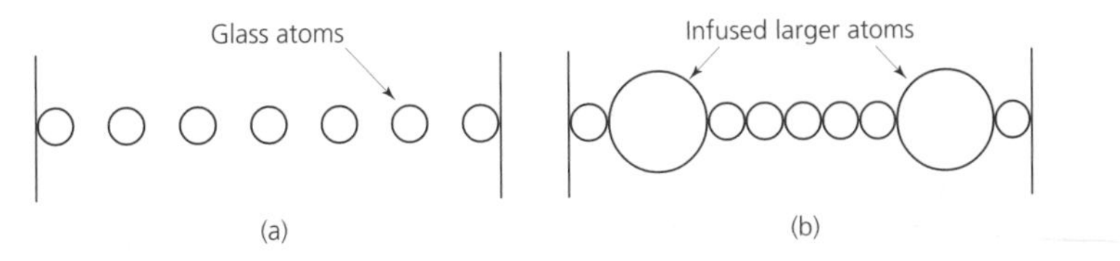

5.18.3 *Annealing*

In the annealing process glass is heated to a critical temperature depending upon the type of glass and cooled slowly, particularly through the glass transition range of 100–200 °C, after which it can be cooled relatively rapidly to room temperature. The cooling time may range from a few minutes for very small parts to several months for very large parts – for example, the glass blank from which the mirror for a 600 mm reflector telescope could be ground. Unlike the tempering process that deliberately introduced stresses into the glass, annealing aims to remove *all stresses* from the glass. Although it is not as strong as tempered glass, annealed glass has the advantage that it can be finished by cutting, drilling, grinding and polishing, without shattering due to the sudden release of internal stress energy.

5.19 Lamination

In the lamination process a sheet of tough transparent plastic is sandwiched between two flat sheets of glass to which it is bonded. When laminated glass is broken, its pieces are held together by the plastic sheet. Care must be taken in the selection of the plastic film. It must retain its transparency under the service conditions for which the laminated glass is being used. In the early days of lamination, the plastic film tended to 'yellow' and lose its transparency with age. Nowadays, through improvements in its manufacture laminated glass is increasingly being preferred to toughened glass for the windscreens of motor vehicles.

5.20 Mechanical properties of glass

The mechanical properties of glass have little in common with crystalline materials such as the metals. Normal tests for tensile and impact strengths show glass to be brittle at room temperature. Glass is elastic to the point of failure and shows no previous yield or plastic deformation, whereas the most brittle metals show some plastic deformation. The failure of glass is always due to a tensile stress component in its loading, even when the applied load is essentially compressive.

It has already been explained that whilst plastic flow occurs in a metal due to 'slip', glass lacks slip planes and only viscous flow can occur. The ease with which viscous flow in glass can occur increases as the temperature rises. However, viscous flow in glass at room temperature is negligible, although large areas of glass have been known to increase in thickness at their lower edges over a long period of time, running into many years.

5.21 Vitreous silica

Vitreous silica occurs naturally as quartz and, unlike the glasses previously described, it has a tetrahedal crystalline structure similar to the ceramic materials described earlier in this chapter. Vitreous, or crystalline, silica has a very high purity and is manufactured by fusing pure quartz crystals or glass sand in electric arc furnaces or by means of an oxy-hydrogen flame. No fluxes are used and there are no other ingredients present to produce a low-temperature eutectic. Thus the fusion temperature is about 1750 °C. It is extremely difficult to produce a homogeneous glass free from blowholes, despite the high viscosity of the glass. Vitreous silica is available as a translucent solid when the minimum silica content is 99.6 per cent, and is available as a transparent solid when the minimum silica content is 99.9 per cent.

Transparent silica (fused quartz) has a high transparency to ultraviolet and infrared radiations. It is much stronger, more impervious to gases, and more resistant to devitrification than translucent silica. The high fusion point of fused quartz together with its low and regular thermal expansion makes it highly resistant to thermal shock and capable of continuous operation at high temperatures, the upper limit being about 1100 °C. Typical applications for transparent vitreous silica are optical instruments requiring a high transparency to a wide range of radiation frequencies, and high-temperature applications such as the envelopes of tungsten–halogen electric lamps. The cheaper, translucent, vitreous silica is used for chemical laboratory equipment, electrical insulating materials, and cooking hobs with flush, easy-clean surfaces.

Polycrystalline glass (Pyroceram) is produced by adding nucleating agents to a conventional or special-purpose glass batch whilst the glass is molten. After processing to the desired form, the polycrystalline glass is then heat treated. Submicroscopic crystallites are formed and, although polycrystalline glass is not ductile, it has much greater strength and hardness than commercial glasses.

5.22 Composite materials

Ceramic materials generally show brittle characteristics due to crack propagation when subjected to tensile loading. The use of composite materials, as shown in Fig. 5.9, overcomes this problem by preventing any crack from running. In its simplest form a composite material consists of two independent and dissimilar materials in which one material forms the matrix to bond together with the other, reinforcing, material. The matrix and the reinforcement are chosen so that their desirable mechanical properties complement each other, whilst their deficiencies are neutralised. In particular circumstances there may be more than one type of reinforcement present at the same time.

Fig. 5.9 *Principles of reinforced composite materials: (a) crack propagation in a non-reinforced ceramic material; (b) behaviour of a laminated composite when in tension; (c) fibre reinforcement*

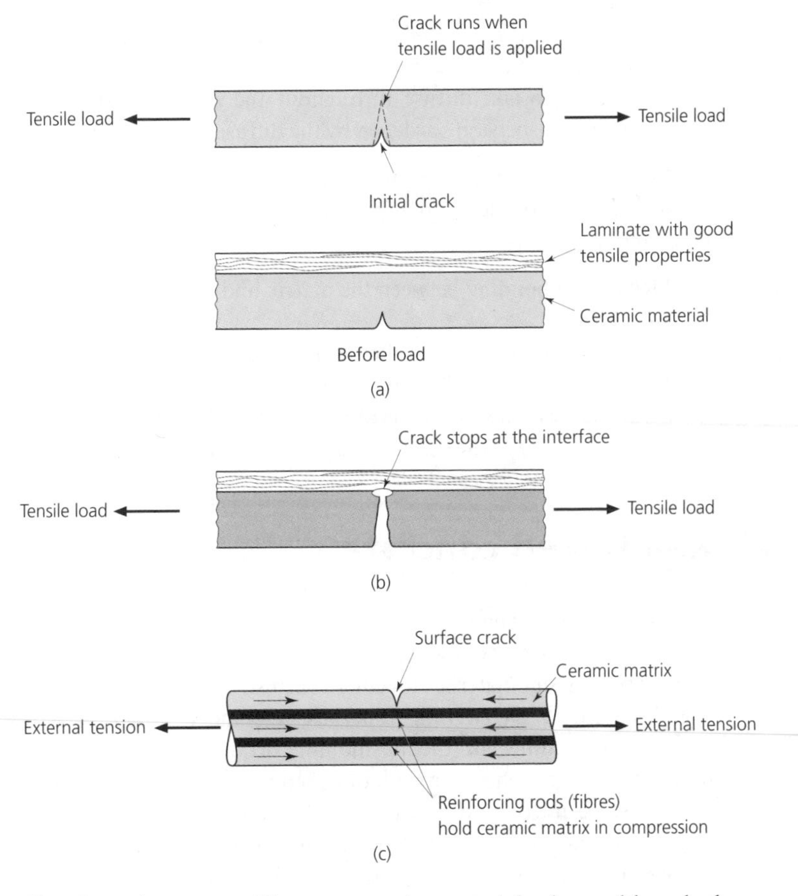

Metallic alloys do not qualify as composite materials since, although there may be hard, strong particles present in a softer matrix in the solid state, these are derived from a single homogeneous liquid. In a true composite material the constituent components are always separate and distinguishable.

The reinforcement in a composite material may be in many different forms. For example, fibres in glass-reinforced polyesters (GRP); steel rods and mesh in reinforced concrete; nodules in the form of natural stones in concrete; particles in cermets and cemented carbides. In fact composite materials can be classified according to the type of reinforcement.

- *Fibre reinforcement* This is present in such materials as natural wood, glass-reinforced polyester and epoxy resins, and reinforced concrete.
- *Particle hardening* This is present in materials called *cermets* where the toughness of metals is combined with the hardness and strength of ceramics. The particles increase the strength of the composite directly since they suffer elastic strains when the material is stressed and contribute to the loadbearing capacity of the composite. They also increase the strength and hardness of the composite indirectly by interfering with dislocations along the slip planes. (Rather like sliding one sheet of sand paper over another.)
- *Dispersion hardening* Unlike particle hardening where the matrix is caused to flow between the reinforcing particles, the reinforcing particles in dispersion hardening are much smaller and diffuse throughout the matrix in the solid state – for example, aluminium dispersion hardened by the diffusion of aluminium oxide particles.

No matter which mechanism of reinforcement is used in a composite material, cohesion between the matrix and the reinforcement is essential and must occur in one or more of the following ways:

- Mechanical bonding between the matrix and the reinforcement by mechanical 'keying' and friction.
- Physical bonding between the matrix and the reinforcement by van der Waals forces acting between the surface molecules of the various constituents.
- Chemical bonding by chemical reactions at the interfaces of the various constituents. However, some of the compounds formed in this manner can be weak.

5.23 Reinforced concrete

This is a composite material that combines the *particle reinforcement* of the aggregate with the *fibre reinforcement* of the reinforcing rods or mesh. A hydraulic cement matrix binds the various constituent materials together. The use of parallel fibres and rods for reinforcement was discussed in *Engineering Materials*, Volume 1. Not only is the reinforcement area fraction important, but the positioning of the reinforcement is also important. For instance, since the tensile strength of concrete is virtually non-existent, the steel-reinforcing bars must be introduced as closely as possible to the point of maximum tensile stress. Figure 5.10 shows a simple reinforced concrete beam. It can be seen that the reinforcement is concentrated near the surface on the tension side. Although concrete is very strong in compression, it is also usual in practice to introduce some reinforcement on the compression side of the beam to guard against reversed stresses. These can occur whilst lifting the beam into position, as shown in Fig. 5.11. If a simple beam, reinforced only on the tension side, is lifted by a crane the stresses are reversed by the beam's own weight and the beam fractures as shown (the reinforcing bars bend). Therefore, in a practical reinforced concrete beam, additional reinforcement is added as shown in Fig. 5.11(b).

Fig. 5.10 *Simple reinforced concrete beam*

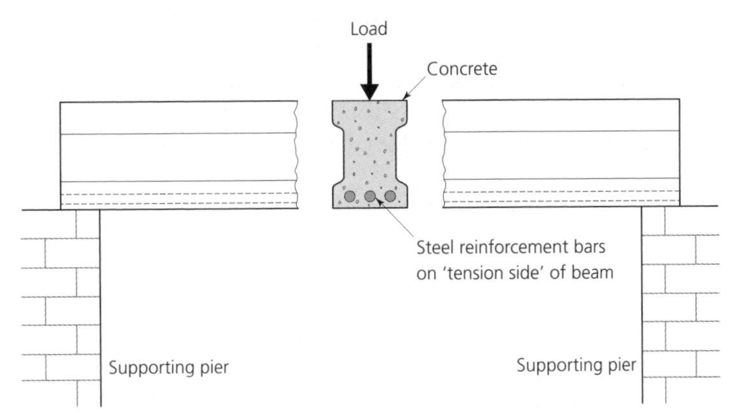

Fig. 5.11 *Additional reinforcement to resist handling loads*

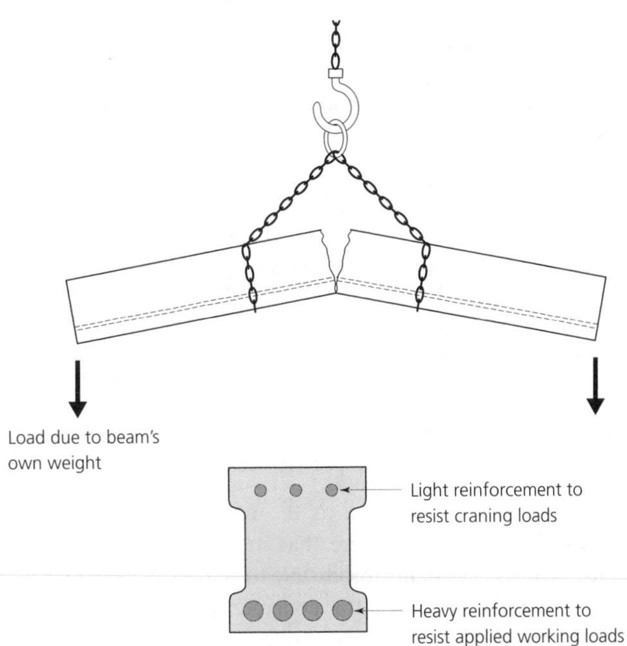

Simple reinforcement, such as has been considered so far, is never used for highly stressed structural members. When a load is applied to a beam it commences to deflect until the stresses in the beam balance the applied load. In a homogeneous beam, such as a steel girder, this deflection is of little importance providing it is kept within reasonable limits. However, in a concrete beam, bending of the reinforcement (which is flexible and elastic) can be disastrous since it would permit the concrete matrix to crack, as shown in Fig. 5.12.

To overcome this problem the steel reinforcement is stressed in tension by means of hydraulic jacks whilst the concrete is cast into the mould. When the concrete has set, the hydraulic jacks are released and the prestressed reinforcement tries to return to its original length and places the concrete in compression. Members made in this manner are referred to as *prestressed concrete*. The design load and prestressing are such that the concrete is always in compression and no cracking occurs. In the event of accidental overloading, any cracks in the concrete are closed immediately the excess load is removed providing the reinforcement is not stressed beyond its elastic limit. Figure 5.13 shows a section through a prestressed concrete beam in the course of manufacture.

Fig. 5.12 *Cracking of simple reinforced beam in service*

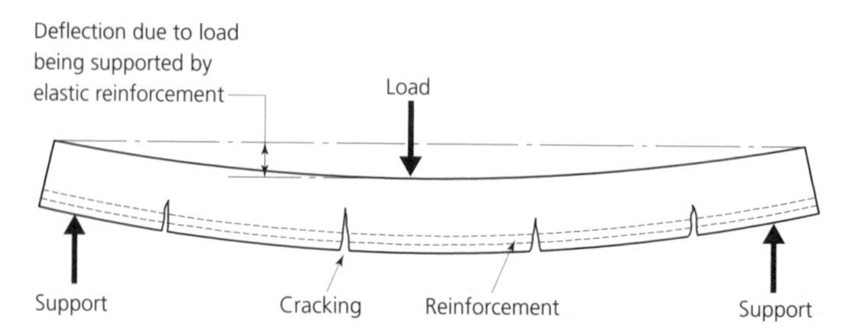

Fig. 5.13 *Prestressing concrete beams: (1) the reinforcement bars are stressed in tension by the hydraulic jacks; (2) the concrete is poured into the mould; (3) when the concrete is set, the jacks are released; the reinforcement, behaving in an elastic manner, tries to shrink back to its original length and compresses the beam*

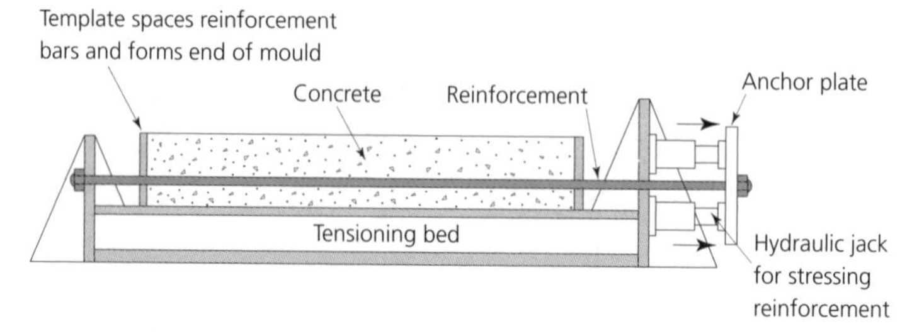

The success of simple and prestressed reinforcement depends upon two main factors:

- Bonding of the concrete to the reinforcement to ensure transmission of the load without slip and to ensure that the beam and reinforcement are subjected to equal strain. This is virtually impossible to achieve in practice.
- Compatibility between the concrete and the reinforcement so that the steel is not attacked chemically and corroded. A number of concrete structures have become unsafe or even have collapsed from this cause.

An alternative to prestressing is *post-tensioned concrete*. Here, the reinforcement is applied and tensioned *after* construction. This allows the degree of tensioning of the reinforcement to be closely controlled at the time of installation and for the tension to be corrected from time to time thereafter. Since the reinforcement passes through precast ducts, it is not in intimate contact with the concrete and corrosion due to chemical incompatibility is unlikely. Post-tensioning is used for large structures such as bridges. High-tensile steel hawsers are passed through ducts that are cast into the concrete. The hawsers are stretched hydraulically until the required stress is achieved. They are then locked in position by taper plugs. Figure 5.14 shows various systems of post-tensioning.

Fig. 5.14 *Post-tensioning concrete beams*

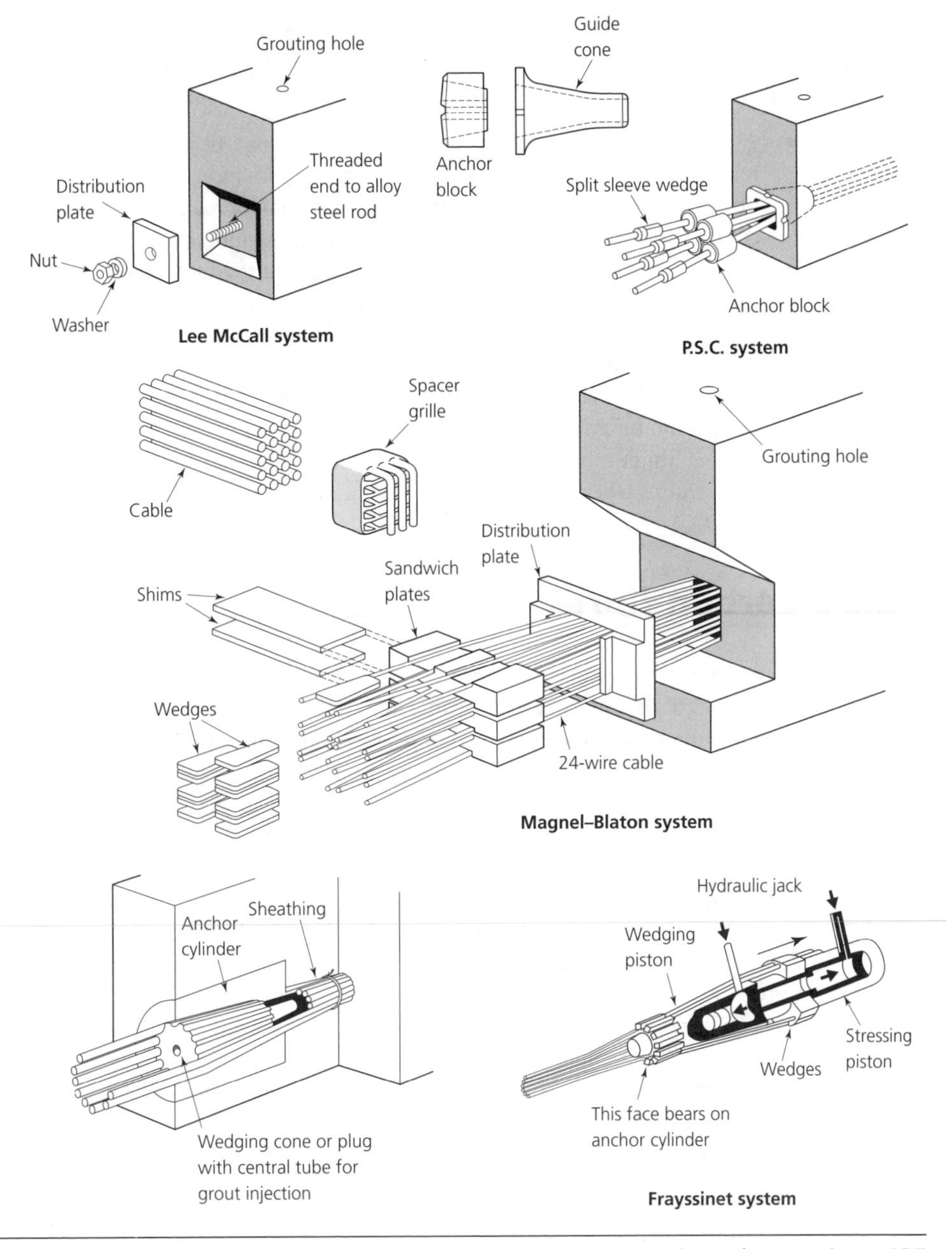

5.24 Glass fibres

The use of glass fibres to reinforce a polyester matrix has already been introduced in *Engineering Materials*, Volume 1. Glass, in the form of fibre, is considerably stronger than when in the 'bulk' condition previously discussed and can be used in engineering as a structural material. The high strength of glass fibres is due partly to the fact that they are free from scratches and other surface defects, and partly to the surface tension effects resulting from the high surface/volume ratio of the fine fibres.

The glasses used for fibre manufacture do not have the same compositions as the bulk glasses previously described. The grades of glass used for reinforcing fibres are as follows:

- *E-glass* (electrical grade) has good insulation properties and is used for making glass-reinforced printed circuit boards for the electronics industry. It is not attacked by water or alkalis and has high strength. As well as for electrical components, it is the composition most widely used for general-purpose mouldings.
- *C-glass* (chemical grade) is used for chemical plant equipment. It is low in aluminium oxide and calcium oxide and has good resistance to acid attack. It is also used for fibreglass surfacing mats that must resist environmental attack and protect the substrate and its reinforcement.
- *S-glass* (high-strength grade) fibre is produced in continuous filaments for weaving into mats and fabrics for pressure vessels and boat hulls.
- *M-glass* (high-modulus grade) is an expensive material reserved for very high-strength applications where exploitation of its special properties offsets its high cost.

5.25 Carbon fibres

These have a higher elastic modulus and a lower density than glass fibres and, therefore, can be used to reinforce composite materials having a higher strength/weight ratio. Carbon fibres are produced by the *pyrolysis* of polyacrylonitrile filaments in an inert atmosphere. *Pyrolysis* is the decomposition of substances at high temperatures. The polymer chains decompose so that only a skeletal structure of carbon atoms remains after the other atoms in the original polymer have been driven off. The fibres are kept in tension during pyrolysis in order to maintain their special properties, which would be lost if the fibres became deformed. For fibres produced by this process, the mechanical properties of the product are influenced by the final heating. This may be greater than 2000 °C, and the product may be classified as high-strength fibres or high-modulus fibres.

The fibres so produced are polycrystalline and consist of large numbers of very small crystallites. Carbon fibre is used as a reinforcing material in polymeric materials to produce lightweight composites of high strength. Its uses range widely over such diverse applications as fan blades for gas turbines, sports equipment, and racing car body panels. Table 5.4 compares the properties of some glass fibre and carbon fibre reinforcing materials.

Table 5.4 *Properties of common reinforcement fibres*

Material	*Relative density*	*Tensile strength (GPa)*	*Tensile modulus (GPa)*	*Specific strength* (GPa)*	*Specific modulus† (GPa)*
E-glass	2.55	3.5	74	1.4	29
S-glass	2.50	4.5	88	1.8	35
Carbon (high strength)	1.74	3.0	230	1.8	130
Carbon (high modulus)	2.00	2.1	420	1.1	210
Steelwire (for comparison)	7.74	4.2	200	0.54	26

*specific strength $= \dfrac{\text{tensile strength}}{\text{relative density}}$

†specific modulus $= \dfrac{\text{tensile strength}}{\text{relative density}}$

Composites vary in electrical conductivity, depending upon their constituent materials. Carbon fibre-reinforced materials can be good conductors of electricity. This was shown when an angler received a fatal electric shock when his carbon fibre fishing rod got too close to high-voltage overhead cables and caused a flashover.

Both glass and carbon fibres can be made into a composite material with all the fibres aligned in one direction. This will produce a product with optimum strength and stiffness, but only when the loading arrangement is perpendicular to the fibres. Properties will be poor when loading is parallel to the fibres. This effect of having different properties in different directions is called *anisotropy*. Anisotropy can be overcome by using multiple layers of the fibres with varying orientations, similar to plywood as described in *Engineering Materials*, Volume 1, Chapter 9. The fibres can be woven into fabrics with various orientations, thus reducing anisotropy. Obviously strength and stiffness in any one direction will not be as great as with unidirectional fibre reinforcement composites. Figure 5.15 shows a sample of woven carbon fibre fabric.

Fig. 5.15 *Sample of woven carbon fibre fabric (reproduced courtesy of the University of Luton)*

5.26 'Whiskers'

'Whiskers' are single hair-like crystals with a high *aspect ratio* and very high strengths. The aspect ratio of a fibre is the ratio of the fibre length (l) to the fibre diameter (d):

$$\text{aspect ratio} = \frac{l}{d}$$

The aspect ratios of whiskers might typically be between 20 and 1000.

The crystal diameter may range from 0.5 to 2 micron in diameter by up to 20 mm long. In general, composites with high aspect ratio reinforcement have improved properties over those with low aspect ratio reinforcement. The tensile strength of a carbon whisker can be as high as 21 GPa, compared to a carbon fibre that has a tensile strength of 3 GPa. This is due to the relative freedom of dislocations in such crystals; in fact they usually have only a single dislocation running along the longitudinal axis. The properties of some whiskers are listed in Table 5.5. Such whiskers are difficult and costly to manufacture and are only used for special applications. It should be noted that there are health and safety implications when handling these materials, owing to carcinogenic risks. Boron and carbon whiskers in a polymeric matrix are in most common use, but alumina whiskers are also being used to reinforce the metal nickel, forming a composite that retains its strength at high temperatures.

Table 5.5 *Properties of common reinforcement 'whiskers'*

Material	*Relative density*	*Tensile strength (GPa)*	*Tensile modulus (GPa)*	*Specific strength* (GPa)*	*Specific modulus* (GPa)*
Alumina	3.96	21.0	430	5.3	110
Boron carbide	2.52	14.0	490	5.6	190
Carbon (graphite)	1.66	20.0	710	12.0	430

*See Table 5.4.

5.27 Particle-hardened materials

Before dealing with the specialised materials known as cermets, common particle-hardened materials found in general engineering are 'tough-pitch' copper and 'precipitation age-hardened' alumium–copper alloys of the *duralumin* type. Tough-pitch copper is less refined than high-conductivity copper and contains particles of copper oxide left over from the extraction process. These oxide particles greatly increase the strength of the copper but lower its electrical conductivity. In the case of precipitation age-hardened duralumin type alloys, particles of aluminium–copper intermetallic compounds are formed which lead to increased strength but substantially lower ductility.

More sophisticated particle-hardened materials are commonly known as *cermets* since they are metals hardened and strengthened by particles of ceramic materials uniformly distributed throughout the metallic matrix (similarly to the aggregate in concrete). The principles of particle hardening were considered earlier in the chapter. Cermets are widely used for cutting tools and will be considered further in Section 10.16. For example, cemented carbides consist of particles of tungsten carbide or mixtures of tungsten and titanium carbides in a matrix of metallic cobalt. The preformed compact of the powdered ingredients is then sintered at a temperature above the recrystallisation temperature of the cobalt. Such materials combine the toughness and ductility of the matrix with the hardness, wear resistance and strength at high temperatures of the ceramic particles. Alternatively, cermet materials can be manufactured by infiltrating the spaces between the solid ceramic particles with molten metal. Whichever method is used, it is essential that a strong bond should exist at the interface of the metal and the ceramic particles. Some typical cermets are listed in Table 5.6.

Table 5.6 *Common cermets: compositions and applications*

Type	*Ceramic particles*	*Metal matrix*	*Applications*
Borides	Titanium boride	Cobalt–nickel	Mostly cutting tool tips
	Molybdenum boride	Chromium–nickel	
	Chromium boride	Nickel	
Carbides	Tungsten carbide	Cobalt	Mostly cutting tool tips
	Titanium carbide	Cobalt or tungsten	Tool tips and abrasives
	Molybdenum carbide	Cobalt	
	Silicon carbide	Cobalt	
Oxides	Aluminium oxide	Cobalt–chromium	Disposable tool tips; refractory sintered components, e.g. spark plug bodies, rocket and jet engine parts
	Magnesium oxide	Cobalt–nickel	
	Chromium oxide	Chromium	

5.28 Dispersion-hardened materials

The use of alumina 'whiskers' to harden and strengthen metallic nickel has already been discussed. Alumina in the form of spheroidal particles can be used to increase the tensile strength rather than the hardness of composites. If aluminium is ground to a fine powder in the presence of oxygen under pressure, aluminium oxide (alumina) is formed on the surface of the particles. The grinding process causes some of the surface alumina to disintegrate and become distributed throughout the mass of the aluminium powder. The mixture is then sintered and consists of about 6 per cent alumina particles dispersed through a matrix of aluminium. This material is known as Sintered Aluminium Powder (SAP) and is stronger than pure aluminium. Although not as strong as duralumin alloy at room temperature,

SAP holds its strength at much higher temperatures, as shown in Fig. 5.16. Alumina particles may be used to dispersion strengthen metals other than aluminium, for example, silver and nickel.

Fig. 5.16 *Comparative effects of temperature on duralumin and sintered aluminium powder (SAP)*

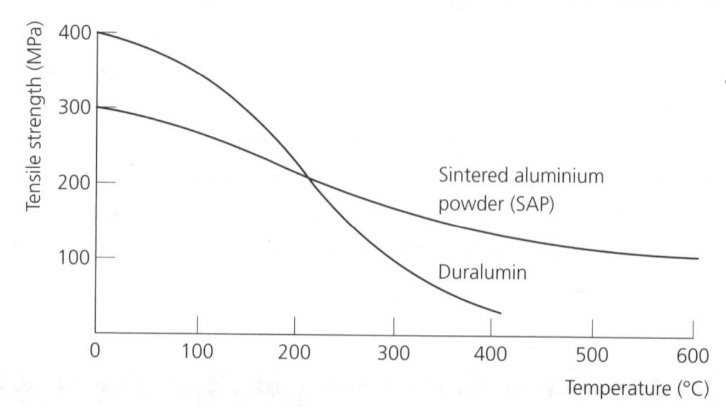

SELF-ASSESSMENT TASK 5.2

1. Discuss the reasons why the properties of composite materials may not be uniform in all directions. State the name of this effect.
2. Explain why particulate-reinforced composites have more uniform properties than fibre-reinforced composites.
3. Describe the ways by which a reinforcing material can be bonded to its matrix material.
4. Calculate the aspect ratios of whiskers that are 5 mm long by 0.02 mm diameter, and 1 mm long by 0.02 mm diameter. State which will provide the most beneficial reinforcement in a composite material.

EXERCISES

5.1 List the basic properties of ceramic materials and explain how these properties make them important engineering materials.

5.2 In each instance, state the name, properties and a typical application for a ceramic material that is representative of each of main groups: crystalline ceramics, amorphous ceramics and bonded ceramics.

5.3 Explain the essential differences in properties and behaviour under load between metallic crystals and ionically bonded ceramic crystals.

5.4 With the aid of sketches, explain the essential differences between ceramic structures, chain structures, sheet structures and framework structures, and describe the properties of ceramic materials possessing such structures.

5.5 For the following groups of ceramic materials, name a typical application and a typical material suitable for that application, giving reasons for your choice: clay refractories, common refractories and high-grade refractories.

5.6 In each instance, describe **two** typical engineering applications for metallic oxides, borides, nitrides and carbides, giving reasons for your choice based upon the properties, cost and availability of these materials.

5.7 Describe **three** processes by which ceramic materials may be shaped prior to firing or sintering.

5.8 (a) Select **two** types of glass and describe their composition, properties and typical applications.
(b) Describe the heat-treatment processes available for making glass less susceptible to fracture.
(c) Describe a chemical treatment process available for making glass less susceptible to fracture.

5.9 Compare the properties of E-glass, S-glass and M-glass as used in producing reinforcement fibres and give a typical application appropriate to each.

5.10 Compare and contrast the advantages and limitations of the following systems of reinforcing concrete:

(a) simple reinforcement
(b) prestressed reinforcement
(c) post-tensioned reinforcement

5.11 Explain what is meant by the particle hardening of a composite material and the dispersion hardening of a composite material. In each case give an example of such a material, together with a typical application.

5.12 Describe the general properties and describe suitable applications for:

(a) carbon fibre
(b) 'whiskers'

6 Polymeric (plastic) materials

The topic areas covered in this chapter are:

- Polymerisation: addition, condensation and rearrangement.
- Copolymers and terpolymers.
- Blending.
- Biodegradable plastics.
- Flammability of polymers.
- Ageing and weathering.

6.1 Polymerisation

Polymeric (plastic) materials were introduced in *Engineering Materials*, Volume 1. In that book we considered the basic 'building blocks' and how they can be combined together into polymers to produce the 'plastic' materials that we use in so many different ways. We learned that we could conveniently classify polymeric materials as:

- *Thermoplastics* that soften every time they are heated.
- *Thermosetting plastics* (thermosets) that are polymerised ('cured') during the moulding process and can never again be softened by the application of heat.

Finally we looked at typical examples of commercially available thermoplastics and thermosetting plastics together with their properties and some typical applications. Let's now look at some of these issues in rather more detail.

The word *polymer* is made up of two words, *poly* meaning many and *mer* from the Greek word *meros* meaning part, thus a polymer has many 'parts'. In reality such materials consisted of long-chain molecules (polymers) made up from many individual molecules (monomers). A simple example is the *polyethylene* molecule that consists of a long chain of repeating ethylene (CH_2) groups, viz.

$$—CH_2—CH_2—CH_2—CH_2—CH_2—CH_2—CH_2—$$

The chain lengths vary, but for commercial polymer chains there could be anything between 1000 and 10,000 of the CH_2 groups. Here we have taken many units or *mers* of low molecular weight and built them up into a long-chain molecule (*polymer*) with a high molecular weight. These materials of high molecular weight are called *high polymers* or *macromolecules*.

Note that the *monomer* is ethylene (C_2H_4), the *polymer* is polythene $(CH_2)_n$ and the repeating group or *mer* is $—CH_2—$ which cannot exist by itself as a molecule.

There are, essentially, three ways in which monomers may be converted into polymers synthetically:

- Addition polymerisation.
- Condensation polymerisation.
- Rearrangement polymerisation.

6.1.1 *Addition polymerisation*

We have just seen that if we *add* together ethylene groups into a long chain we get the 'plastic' material called polyethylene. We have achieved *addition polymerisation*. Painless wasn't it. In addition polymerisation, a monomer which, in this context, is a low molecular weight molecule and possesses a double bond, is induced to break the double bond so that the resulting free valences are able to join up to other similar molecules. For example our ethylene molecule (C_2H_4) can be considered as consisting of two CH_2 groups linked together by a double bond as shown below:

$$\underset{\text{Ethylene}}{n \cdot \mathrm{CH_2{=}CH_2}} \xrightarrow[\text{(heat + pressure + catalyst)}]{\text{Polymerisation}} \underset{\text{Polyethylene}}{\sim\!\sim\!\left(\mathrm{CH_2—CH_2}\right)_n\!\sim\!\sim}$$

If we can induce the ethylene molecules to break their double bonds they can use the free valences created to link together to form the long-chain polymer molecule called *polyethylene*. The formation of some further examples of addition polymerisation is shown in Fig. 6.1.

Fig. 6.1 *Some further polymers formed by addition polymerisation*

Monomer **Polymer repeating unit**

$$\begin{matrix} H & & H \\ | & & | \\ C & = & C \\ | & & | \\ H & & CH_3 \end{matrix} \qquad \sim\!\sim\!\left(\begin{matrix} & H & & H & \\ & | & & | & \\ — & C & — & C & — \\ & | & & | & \\ & H & & CH_3 & \end{matrix}\right)_n\!\sim\!\sim \ \text{Polypropylene}$$

$$\begin{matrix} H & & H \\ | & & | \\ C & = & C \\ | & & | \\ H & & Cl \end{matrix} \qquad \sim\!\sim\!\left(\begin{matrix} & H & & H & \\ & | & & | & \\ — & C & — & C & — \\ & | & & | & \\ & H & & Cl & \end{matrix}\right)_n\!\sim\!\sim \ \text{Polyvinyl chloride (PVC)}$$

$$\begin{matrix} H & & H \\ | & & | \\ C & = & C \\ | & & | \\ H & & C_6H_5 \end{matrix} \qquad \sim\!\sim\!\left(\begin{matrix} & H & & H & \\ & | & & | & \\ — & C & — & C & — \\ & | & & | & \\ & H & & C_6H_5 & \end{matrix}\right)_n\!\sim\!\sim \ \text{Polystyrene}$$

$$\begin{matrix} Fl & & Fl \\ | & & | \\ C & = & C \\ | & & | \\ Fl & & Fl \end{matrix} \qquad \sim\!\sim\!\left(\begin{matrix} & Fl & & Fl & \\ & | & & | & \\ — & C & — & C & — \\ & | & & | & \\ & Fl & & Fl & \end{matrix}\right)_n\!\sim\!\sim \ \text{Polytetrafluoroethylene (PTFE) (Teflon)}$$

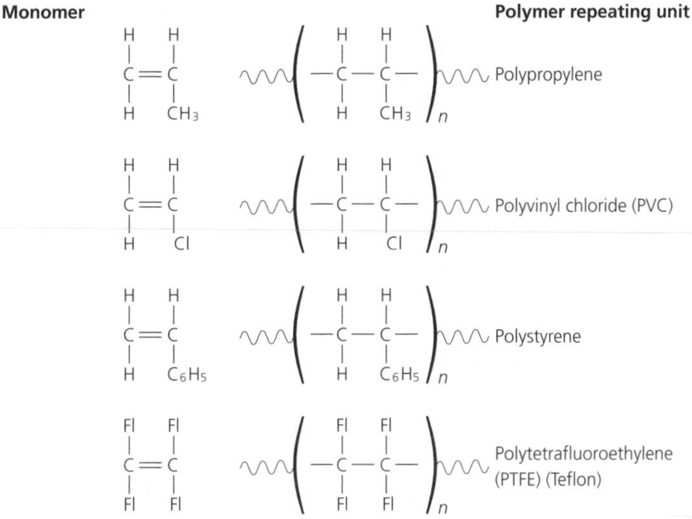

The molecular weight of a polymer is the sum of the molecular weights of the individual '*mers*'. The higher the molecular weight of any given polymer, the longer is its chain length. Since polymerisation is a random event, not all the polymer chains produced are of equal length. However, as for any random events, the chain lengths produced do fall into a normal distribution curve. Therefore, we can determine and express the average molecular weight of a polymer on a statistical basis by averaging. The spread of the molecular weight distribution is referred to, not surprisingly, as the *molecular weight distribution* (MWD). Molecular weight and MWD have a significant influence on the properties of the polymer. For instance, the tensile and impact strength, resistance to cracking in the solid state and viscosity in the liquid state all increase with increasing molecular weight, as shown in Fig. 6.2.

Fig. 6.2 *Effect of molecular weight and degree of polymerisation on the strength and viscosity of polymers (adapted from Kalpakjian Manufacturing Engineering and Technology, Addison Wesley)*

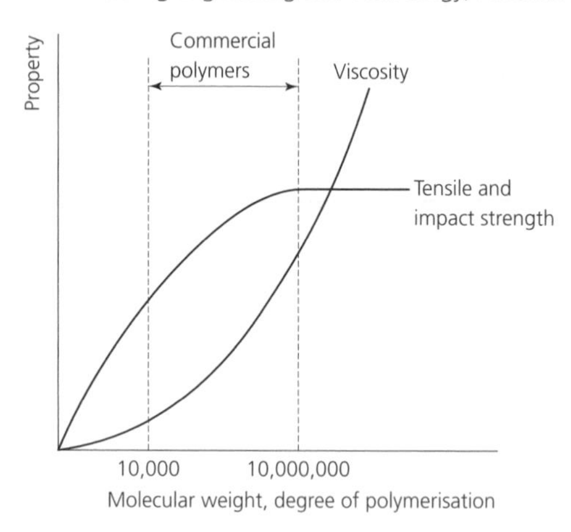

Sometimes it is more convenient to express the size of a polymer chain in terms of its *degree of polymerisation* (DP). This can be defined as the ratio of the molecular weight of the polymer to the molecular weight of the repeating units, the '*mers*'. For instance, the mer weight of PVC is 62.5 whereas the molecular weight of the polymer could be 50,000. Thus the degree of polymerisation $(DP) = 50,000/62.5 = 800$. The DP is important when considering the economics of plastic processing. The higher the DP, the more viscous will be the polymer, and the more difficult and costly it will be to process. Let's see why this is.

During polymerisation, the monomers are linked together by covalent bonds to form the polymer chain. Because of their strength, covalent bonds are also referred to as *primary bonds*. The chains so formed are held together by much weaker *secondary bonds* such as van der Waals bonds, hydrogen bonds, and ionic bonds. The longer the chains, the greater will be the secondary bonding between them and the greater will be the energy required to overcome the bond strength during processing. For example, ethylene monomers and polymers having a DP of 1, 6, 35, 140 and 1350 will be in the form, respectively, of gas, liquid, grease, wax and solid plastic at room temperature.

The chain-like polymers we have been considering so far are called *linear polymers*. With such materials it is possible for the long-chain molecules to slide past each other under shear forces above such a temperature that the molecules have enough energy to overcome the intermolecular attractions. That is, above a certain temperature, such polymers are capable of plastic flow. Above this temperature the polymer is plastic; below this temperature the polymer is solid. Polymers that behave in this manner are called *thermoplastics*.

Although thermoplastics are essentially linear polymers, some have more than one type of structure and may contain some *branched* chains. Branching causes changes in the properties of a plastic. In branched polymers, side-branched chains are attached to the main chain during synthesis of the polymer. Branching interferes with the relative movement of the molecular chains and this affects the resistance to deformation and also the stress–crack resistance. Although less resilient, such plastics will have increased strength. Another effect of branching is to lower the density of the plastic since branching interferes with the *packing efficiency* of the polymer chains.

Taking branching a stage further, some polymers are cross-linked and even networked to form three-dimensional structures. Polymers with highly cross-linked and network structures are very rigid and cannot be softened by heating. They are called *thermosetting plastics* or *thermosets* for short. Figure 6.3 shows, schematically, linear chains, branched chains, cross-linked chains and networks. Since thermosets are not synthesised by addition polymerisation but by *condensation polymerisation*, let's move on.

Fig. 6.3 *Schematic illustration of polymer chains: (a) linear structure – thermoplastics such as acrylics, nylons, polyethylene, and polyvinyl chloride have linear structures; (b) branched structure, such as in polyethylene; (c) cross-linked structure – many rubbers or elastomers have this structure, vulcanisation of rubber produces this structure; (d) network structure – this is basically highly cross-linked, examples are thermosetting plastics, such as epoxies and phenolics (adapted from Kalpakjian, Manufacturing Engineering and Technology, Addison Wesley)*

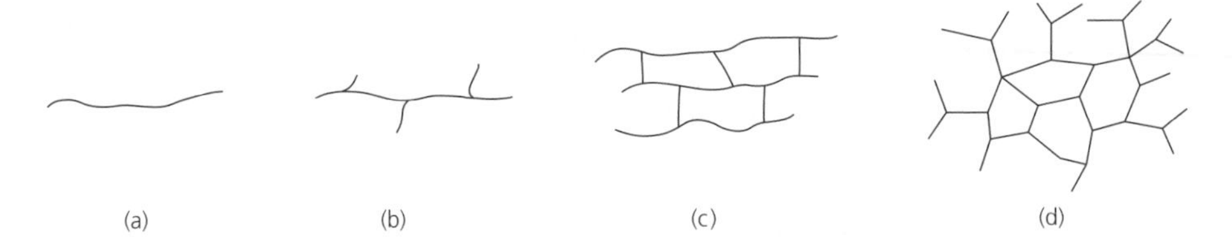

6.1.2 *Condensation polymerisation*

When long-chain molecules in a polymer are cross-linked in a three-dimensional arrangement, the structure becomes one giant molecule with strong covalent bonds. Such polymers are called *thermosetting polymers* or *thermosets*, because during polymerisation the network is competed and the shape of the moulding is permanently set.

One of the earliest thermosetting plastics was 'Bakelite' (phenol–formaldehyde). It is synthesised by condensation polymerisation – that is, during condensation polymerisation the chemical compounds react between themselves and release some small molecules such as water. The release of these small molecules is called 'condensation'. Figure 6.4 shows the condensation polymerisation of phenol-formaldehyde. We have to be a little careful in generalising, for some *thermoplastics* such as *nylon* are produced by *condensation polymerisation* and some thermosets do not undergo condensation polymerisation during the final cure. However, the *majority* of thermoplastics are addition polymerised, and the majority of thermosets are condensation polymerised.

Fig. 6.4 *Condensation polymerisation of phenol–formaldehyde*

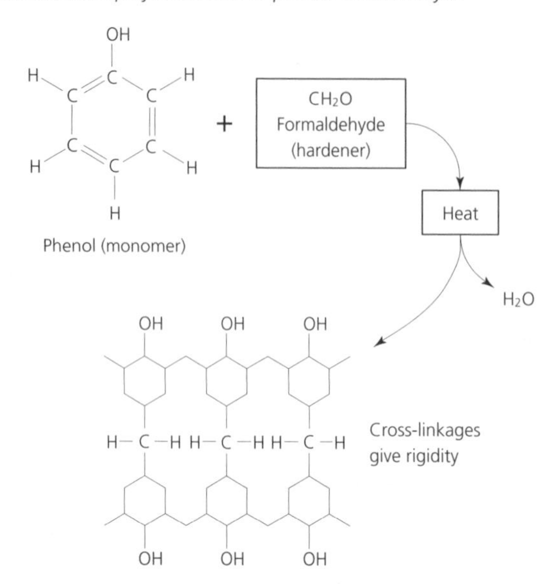

A completely polymerised thermoset is no longer fusible and is incapable of being shaped except by machining. Therefore, such materials must be supplied for processing in some intermediate stage of polymerisation. They are classified according to the degree of polymerisation they have already undergone.

- *A stage* – fusible and soluble; that is, thermoplastic.
- *B stage* – fusible and partly soluble, suitable for moulding.
- *C stage* – fully polymerised (cured) and incapable of ever again being softened by heating; however, if heated to a sufficiently high temperature they will char, degrade and burn up.

Thermosets in the A stage can be used for the manufacture of some adhesives or are converted to the B stage. Thermosets in the B-stage are used for moulding powders and granules after being compounded with additives to modify their properties. The finished moulding is in the C stage of polymerisation.

Phenol–formaldehyde (Bakelite) resins can be produced by a one-stage or two-stage process. In the one-stage process, as shown in Fig. 6.5, phenol is reacted with an excess of formaldehyde, so that the phenol/formaldehyde (P/F) ratio is less than 1. The mixture of phenol and formaldehyde is heated in the presence of an alkaline catalyst such as sodium hydroxide or ammonia. The reaction is interrupted before condensation is complete and the resin is only at the A stage or at the B stage.

The A-stage resin, called *resol*, has low molecular weight, short linear chains with some ether cross-linkages. It is completely soluble in the alkaline solution present as a catalyst. At the B-stage, called *resitol*, it has long linear polymer chains and a slight amount of cross-linking between the chains. It is insoluble in the alkaline solution but is readily soluble in organic solvents. When cooled the resin is hard and brittle, but when heated it becomes soft again. At this stage it still has the characteristics of a thermoplastic. The resin A or B or a mixture of the two can be used for adhesives, casting, moulding and for the manufacture of laminates. Further heating in the presence of curing agents causes extensive cross-linking and the formation of a hard rigid solid. It is now at the °C stage and has become a *thermoset*.

Fig. 6.5 *Single-stage production of phenol–formaldehyde resin*

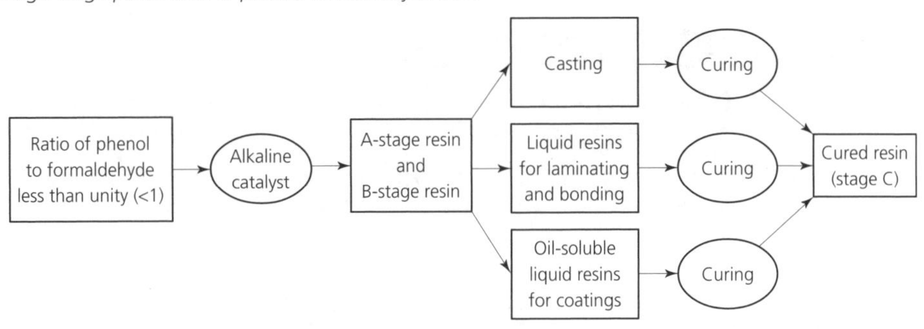

The two-stage process is shown in Fig. 6.6. Here only a portion of the formaldehyde is introduced at the start so that the P/F ratio is greater than 1. The material is heated in the presence of an acid catalyst, and the reaction is allowed to proceed until no further chemical changes occur. The resin, called *Novalac*, is fusible and soluble in organic powders; corresponding to the B-stage resin, it is hard and brittle and can be ground into a powder. Various additives such as fillers, pigments and lubricants are mixed with resin powder to improve its mechanical and moulding properties. An inhibitor is also often added to ensure that further polymerisation does not take place under normal ambient conditions prior to moulding. Hexamethylenetetramine $[(CH_2)_6N_4]$ is added as a source of formaldehyde and ammonia. Remember that only part of the formaldehyde required for full polymerisation was added at the start. During moulding, the process temperature required results in additional formaldehyde being released to convert the soluble, fusible Novalac resin into a hard, infusible, insoluble solid – the thermoset phenol–formaldehyde. This is the *curing* process and polymerisation is now complete. Since some condensation polymerisation takes place during moulding at temperatures above the boiling point of water, venting of the moulds is essential to release any steam generated. Also, allowance must be made for shrinkage of the moulding.

Fig. 6.6 *Two-stage production of phenol–formaldehyde resin*

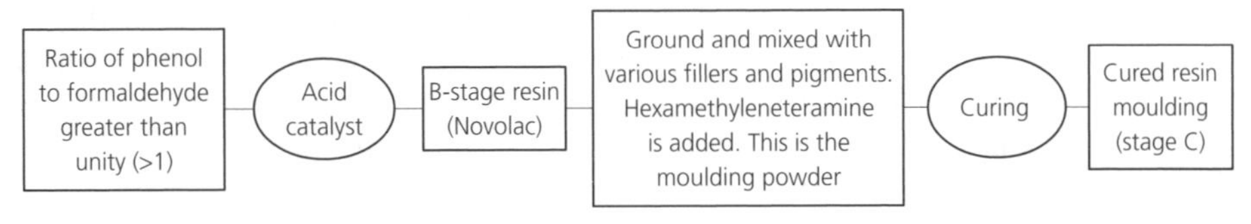

6.1.3 *Rearrangement polymerisation*

This is a little more complicated. There is no universally accepted description for this mechanism of polymerisation but the terms *rearrangement polymerisation* and *polyaddition* are often used. In many respects this mechanism for polymerisation lies part way between addition polymerisation and condensation polymerisation. Like the former mechanism there is no molecular expulsion, but the kinetics of the polymerisation reactions are akin to the latter. A commercial example is the manufacture of polyurethanes and polyurethane foams by the interaction of di-alcohols and glycols (diols) with di-isocyanates.

SELF-ASSESSMENT TASK 6.1

1. Explain the essential differences between polymerisation by addition, condensation and rearrangement. Give an example of each polymerisation mechanism.
2. State the essential differences between thermoplastics and thermosets.

6.2 Copolymers and terpolymers

So far we have only considered repeating units of the same type in the polymer chain. A molecule of this type is called a *homopolymer*. Just as metals can be combined together to form alloys in order to produce a material with special properties tailored to a particular application, so it is possible to combine two or three monomers together. For example, it is possible to improve the impact resistance strength and formability of plastics in this manner.

When two polymers are combined together as in styrene–butadiene the resulting material is referred to as a *copolymer*. When three polymers are combined together the resulting material is referred to as a *terpolymer*. An example of a terpolymer is acrylonitrile–butadiene-styrene (ABS) used for crash helmets and automobile body parts. This material is often referred to as *high-impact-resistant* polystyrene. The introduction of the elastomer, butadiene, gives the terpolymer resilience and the ability to resist impact loads.

6.3 Blending

The glass transition temperature (T_g) was introduced in *Engineering Materials*, Volume 1. The glass transition temperature for any given thermoplastic can be determined by the abrupt change in its tensile modulus, as shown in Fig. 6.7. Below the glass transition temperature polymeric materials show a relatively high tensile modulus, with little extension and a high level of rigidity. Above the glass transition temperature, the tensile modulus is lower, the level of rigidity is lower and the extension is very considerably increased.

Fig. 6.7 *Glass transition temperature (T_g)*

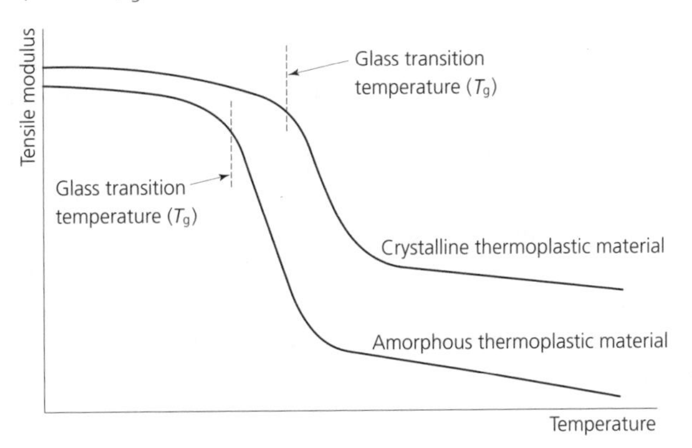

To improve the brittle behaviour of amorphous polymers below their glass transition temperature we can mix them with another polymer, usually a small amount of an elastomer. This is called 'blending'. Unlike a copolymer where the various homopolymers combine together chemically to form a complex molecule, a blend simply consists of tiny particles of the elastomer dispersed physically throughout the mass of the amorphous polymer. This enhances its toughness and impact strength by improving its resilience and resistance to crack propagation. Blends such as these are referred to as *rubber-modified polymers*. Another development is the introduction of *polyblends*. These combine, and make use of, the favourable properties of several different polymers. *Miscible blends* (mixing without the separation of the phases, similar to the alloying of metals) have also been developed enabling polymer blends to become more ductile. Polymer blends account for about 20 per cent of all polymer production.

6.4 Biodegradable plastics

One-third of all plastic production is used for disposable products such as bottles, bags and packaging. The disposal of this amount of plastic waste at a time when there is a growing awareness of environmental issues is a major problem. Although initial efforts were unsuccessful in producing true biodegradable plastics, three different bioplastics are now available: starch-based bioplastics, lactic-based bioplastics and bioplastics made from the fermentation of sugar. These three types of bioplastic degrade over periods of time, ranging from a few months to a few years. They are all designed to degrade completely when exposed to micro-organisms in soil or in water. No toxic by-products are created as a result of the degradation process.

The starch-based system is the most widely used commercially. Here, the starch granules are processed into a powder, which is heated until it becomes a sticky liquid. The liquid is then cooled, formed into pellets and processed in a conventional processing plant. The lactic-based system uses fermenting corn or other feedstocks to produce lactic acid. This is then polymerised to form a polyester resin. In the third system, organic acids are added to a sugar feedstock and the resulting reaction produces a highly crystalline and very stiff polymer which, after further processing, behaves in a manner similar to more conventional polymers produced from petroleum feedstocks.

Another approach is to incorporate a photo-activator into the polymer. This absorbs UV light and, instead of converting it into heat, it generates highly reactive chemical intermediates that destroy the polymer. One such photo-activator is iron dithiocarbamate. Once this reduces the molecular weight of the polymer to about 9000, the polymer becomes biodegradable.

Biodegradable plastics are comparatively new and their long-term performance both during their useful life-cycle and in land-fill sites during degradation have yet to be fully assessed. One major concern is that research into biodegradability will divert attention from the issue of recycling plastics. Unlike biodegradability, recycling conserves materials and energy.

6.5 Flammability of polymers

If the temperature is sufficiently high and an adequate supply of oxygen is present, most polymers will burn. The flammability (ability to burn) of polymers varies considerably, depending upon their composition. For example, polymethyl methacrylate will continue to burn once ignited, whilst polycarbonate self-extinguishes. The flammability of polymers can be reduced either by making them from less flammable raw materials, or by incorporating *flame retardants* into their formulation. Compounds of chlorine, bromine and phosphorus are all flame retardants. Figure 6.8 shows how replacing one of the hydrogen atoms in a highly flammable ethylene monomer with a chlorine atom we get fire-resistant polyvinyl chloride (PVC).

Fig. 6.8 *Chlorinated plastics: H = hydrogen, C = carbon, Cl = chlorine*

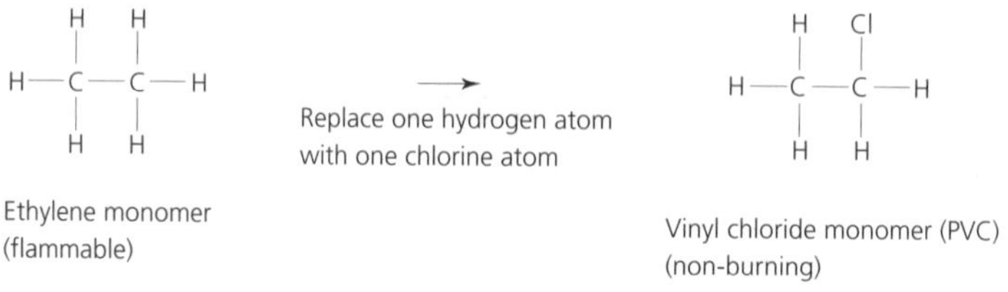

```
   H   H                                          H   Cl
   |   |                                          |   |
H—C—C—H              ——→                   H—C—C—H
   |   |      Replace one hydrogen atom           |   |
   H   H      with one chlorine atom              H   H
```

Ethylene monomer (flammable)

Vinyl chloride monomer (PVC) (non-burning)

A number of disasters have been reported over recent years both on a domestic and on a public scale resulting from fire involving plastic materials used in upholstery, drapes, clothing and décor. These have resulted from burning of the plastic, melting and dripping of the burning plastic, coupled with dense smoke and toxic fumes. As a result, research into the propagation of fires associated with polymers has shown that the burning of plastics is a complex multi-stage process.

- *Primary thermal processes* during which an external source of heat energy is applied to the polymer causing a rise in temperature. Since polymers are poor conductors of heat (good thermal insulators) the temperature build-up can be quite rapid.
- *Primary chemical processes* in which external heat source may supply free radicals that can accelerate combustion. The polymer being heated might also be activated by autocatalytic and auto-ignition mechanisms as its temperature rises.

- *Decomposition* of the polymer becomes rapid once a certain temperature has been reached and a variety of products, such as combustible and non-combustible gases and liquids, charred solids and smoke, may also be produced. These can either accelerate or retard further decomposition depending upon the composition of the polymer and also upon the environmental conditions at the scene of the fire.
- Ignition will occur when the ratio of any combustible gases given off and the oxygen in the air present reach the correct proportion and are above the ignition temperature.
- Combustion follows ignition and the ease with which combustion is maintained is dependent upon the cohesive energy of the polymer bonds present.
- Such combustion may be accompanied by flame propagation and also by non-flammable degradation and physical changes such as shrinkage, melting and charring. Smoke and toxic gases may be given off in large volumes. Records show that more fatalities are caused by suffocation and poisoning by toxic gases than by the victims being burned to death. One worrying development is that many of the additives incorporated into polymers as flame retardants actually increase the amount of smoke generated as the rate of flame propagation decreases. Rigorous testing, legislation, and research programmes are all being actively pursued to try to overcome the hazards associated with the widespread and increasing use of polymers in domestic and public buildings.

6.6 Ageing and weathering

The ageing and weathering of plastics are important considerations when selecting them for products requiring an extended service life. For example, the author inherited a galvanised steel watering can that is still serviceable after more than 50 years, whereas a plastic watering can only 5 years old has recently become brittle and broken up. No doubt it was made from a relatively cheap plastic devoid of any anti-ageing additives. The long-term behaviour of plastics is dependent upon many factors, for example:

- Chemical environments such as atmospheric oxygen, acidic fumes and water.
- Heat.
- Ultraviolet light.
- High-energy radiation.

Remember, a chain is only as strong as its weakest link and this applies to the long-chain molecules of most polymeric materials. Weak links, particularly terminal weak links, can be the point where an 'unzipping' reaction can commence. The removal of the end monomer can make the rest of the chain unstable and the polymer depolymerises (degrades). There are four ways in which this phenomenon can be moderated.

- Use highly purified monomers to prevent the formation of weak links in the first place.
- Polyacetals are particularly prone to depolymerisation and their chain ends must be capped by a stable grouping to prevent the occurence of 'unzipping' reactions.
- Forming copolymers with a small amount of a second polymer that will obstruct the 'unzipping' reaction. For example, industrially, methyl methacrylate, which is susceptible to 'unzipping', can be copolymerised with small amounts of ethyl alcohol,

and formaldehyde can be copolymerised with ethylene dioxide or 1,3-dioxalane to provide stability.
- By the use of certain additives that divert or moderate any degradation reactions. A wide range of antioxidants and UV filters is available for blending with the polymer.

Commercial plastics contain a number of additives and ingredients that may be affected by the weathering and ageing agencies. They may also react with each other as well as with the polymer and may accelerate or retard degradation. Since the various additives and polymers react so differently to chemicals and radiant energy, weathering behaviour cannot be generalised. It is very type specific. Another major problem facing the plastics industry is the need to know the weathering and ageing characteristics of a plastic over long periods of time – approximately 20 years in the case of plastic products used in the building industry. Unfortunately, accelerated testing has so far only met with limited success, due to the fact that the reactions of heat and light, or oxygen and light, may be quite serious over a period of time in combination but may have a negligible effect on the polymer when studied individually. It is very difficult to devise a test that can subject the polymer not only to the full range deterioration agencies to be met with in service, but also to all the varying combinations of those agencies at any given time. Again, how can the service conditions be accelerated? Simply increasing the intensity of radiation or increasing the temperature is not a solution since such measures will change the nature of the reactions that occur in practice.

SELF-ASSESSMENT TASK 6.2

1. Briefly explain the essential differences between copolymers, terpolymers and blends. Give an example of each.
2. Discuss the advantages and disadvantages of biodegradable plastic waste disposal compared with reclamation recycling.
3. Discuss the hazards of flammable plastic materials and how flammability may be reduced.

EXERCISES

6.1 Discuss the meaning of stages A, B and C as applied to thermosetting plastic resins. Explain why a thermosetting moulding powder will be based on a B-stage resin.

6.2 Name a typical copolymer and a typical terpolymer and describe how their properties are superior to the homopolymers on which they are based.

6.3 Name a typical polymer blend and how blending improves its properties.

6.4 Explain what is meant by the term 'biodegradable plastic' and state how its special properties are achieved.

6.5 Research an account of a major fire disaster involving plastic decorative material and furnishings and comment upon the effects of such materials in exacerbating the disaster.

6.6 Research and summarise the legislation and testing methods concerning the flammability of plastic materials.

6.7 Describe the effects of ageing and weathering on plastic materials and how these effects can be reduced.

6.8 Research the attempts that are currently being made into accelerated tests for ageing and weathering and describe the difficulties in developing a satisfactory test protocol.

6.9 Explain, with examples, what is meant by a linear chain, branching chain, cross-linked chain and a network as applied to polymers and describe how these phenomena affect the properties of a polymer.

6.10 Explain what is meant by an 'unzipping' reaction as applied to a polymer chain, how such a reaction is initiated, how it affects the stability of the polymer, and the steps that can be taken to reduce the possibility of such a reaction occurring.

7 Synthetic adhesives

The author is indebted to the Loctite Corporation for their support during the writing of this chapter; the publishers wish to thank Loctite for permission to adapt and reproduce their copyright material. Sections 7.2, 7.6, 7.7, 7.8, 7.9, 7.10, 7.12, 7.13, and 7.16 are derived, wholly or in part, from the *Loctite World Wide Design Handbook*.

The main topic areas covered in this chapter are:

- The adhesive bond.
- Thermoplastic adhesives.
- Thermosetting adhesives.
- The curing of adhesives.
- Joint design and surface preparation.
- Factors affecting adhesion.
- Selection of adhesives.

7.1 Introduction

Polymeric materials for moulding and extrusion were introduced in *Engineering Materials*, Volume 1. Many of these materials are capable of being used directly as adhesives, or as a base for adhesives. *Adhesives* can be divided into two main categories:

- Natural adhesives.
- Synthetic adhesives.

The natural adhesives are vegetable and animal derivatives that may be used directly with little modification. Compared to the majority of synthetic adhesives, they are relatively weak but have the advantage of not being toxic. Unfortunately, they are adversely affected by damp and soften when raised much above room temperature. Natural adhesives may be subdivided further into *gums* and *glues*.

- *Gums* are made from vegetable matter, resins and rubbers being extracted from the sap of trees and starch derivatives being extracted from the by-products of flour milling.
- *Glues* are derived from the horns, hooves and bones of animals and the bones of fish. Derivatives of milk and blood are also used. Such glues soften at the boiling point of water and are largely used for joining wood in the furniture and toy-making industries.

Natural glues and gums are still widely used for low-strength applications, but they are being supplanted increasingly by high-strength synthetic adhesives and the range of applications of adhesive bonding is ever increasing. Table 7.1 lists some of the main groups of adhesives, whilst Table 7.2 lists some of the more important advantages and limitations of adhesive bonding as compared with mechanical and thermal jointing processes. This chapter will be concerned with the characteristics and applications of synthetic adhesives.

Table 7.1 ***Main categories of adhesives***

Origin	*Basic type*	*Adhesive material*
Natural	Animal	Albumen, animal glue (inc. fish), casein, shellac, beeswax
	Vegetable	Natural resins (gum arabic, tragacanth, colophony, Canada balsam, etc.); oils and waxes (carnauba wax, linseed oils); proteins (soyabean); carbohydrates (starch, dextrines)
	Mineral	Inorganic materials (silicates, magnesia, phosphates, litharge, sulphur, etc.); mineral waxes (paraffin), mineral resins (copal, amber); bitumen (inc. asphalt)
	Elastomers	Natural rubber (and derivatives, chlorinated rubber, cyclised rubber, rubber hydrochloride)
Synthetic	Elastomers	Synthetic rubbers and derivatives (butyl, polyisobutylene, polybutadiene blends (inc. styrene and acrylonitrile), polyisoprenes, polychloroprene, polyurethane, silicone, polysulphide, polyolefins (ethylene vinyl chloride, ethylene polypropylene) Reclaimed rubbers
	Thermoplastic	Cellulose derivatives (acetate, acetate-butyrate, caprate, nitrate, methyl cellulose, hydroxy ethyl cellulose, ethyl cellulose, carboxy methyl cellulose) Vinyl polymers and copolymers (polyvinyl-acetate, alcohol, acetal, chloride, polyvinylidene chloride, polyvinyl alkyl ethers) Polyesters (saturated) (polystyrene, polyamides (nylons and modifications) Polyacrylates (methacrylate and acrylate polymers, cyano-acrylates, acrylamide) Polyethers (polyhydroxy ether, polyphenolic ethers) Polysulphones
	Thermosetting	Amino plastics (urea and melamine formaldehydes and modifications) Epoxides and modifications (epoxy polyamide, epoxy bitumen, epoxy polysulphide, epoxy nylon) Phenolic resins and modifications (phenol and resorcinol formaldehydes, phenolic-nitrile, phenolic-neoprene, phenolic-epoxy) Polyesters (unsaturated) Polyaromatics (polyimide, polybenzimidazole, polybenzothiazole, polyphenylene) Furanes (phenol furfural)

Table 7.2 *Advantages and limitations of bonded joints*

Advantages

1. The ability to join dissimilar materials, and materials of widely different thicknesses
2. The ability to join components of difficult shape that would restrict the application of welding or riveting equipment
3. Smooth finish to the joint which will be free from voids and protrusions such as weld beads, rivet and bolt heads, etc.
4. Uniform distribution of stress over entire area of joint. This reduces the chances of the joint failing in fatigue
5. Elastic properties of many adhesives allow for flexibility in the joint and give it vibration-damping characteristics
6. The ability to electrically insulate the adherends and prevent corrosion due to galvanic action between dissimilar metals
7. The joint will be sealed against moisture and gases
8. Heat-sensitive materials can be joined

Limitations

1. The bonding process is more complex than mechanical and thermal processes, i.e. the need for surface preparation, temperature and humidity control of the working atmosphere, ventilation and health problems caused by the adhesives and their solvents. The length of time that the assembly must be jigged up whilst setting (curing) takes place
2. Inspection of the joint is difficult
3. Joint design is more critical than for many mechanical and thermal processes
4. Incompatibility with the adherends. The adhesive itself may corrode the materials it is joining
5. Degradation of the joint when subject to high and low temperatures, chemical atmospheres, etc.
6. Creep under sustained loads

7.2 The adhesive bond

A bonded joint relies on an adhesive bridge between the substrates of the materials being joined, as shown in Fig. 7.1. A typical bonded joint is shown in Fig. 7.2(a) together with the terminology used for the various features of the joint. The strength of the bond depends upon two factors:

- Adhesion.
- Cohesion.

7.2.1 *Adhesion*

Adhesion is the ability of the bonding material (*adhesive*) to stick (*adhere*) to the materials being joined (*adherends*). There are two ways in which the bond can occur and these are shown in Figs. 7.2 (b) and 7.2(c). Physical forces of attraction and absorption – described as 'van der Waals forces' – have the greatest significance in adhesive bonding. The range of these intermolecular forces is considerably lower if the adhesive material does not come into intimate contact with the bonding sites due to such factors as surface roughness of mechanically treated surfaces. Therefore the adhesive must have good 'wetting' properties (see Fig. 7.3) so that it will penetrate into the surface roughness and wet the complete surface.

Fig. 7.1 *The adhesive bridge (reproduced courtesy of Loctite Corporation)*

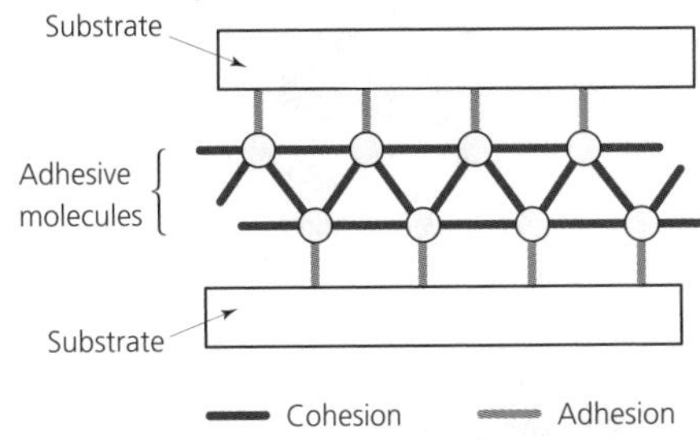

Fig. 7.2 *Bonded joints: (a) elements of the bonded joint; (b) a simple cemented joint in which the adhesive penetrates the pores of the adherends to form the bond (this occurs with rough or porous surfaces); (c) the adhesive and the adherends react together chemically so that an intermolecular bond is formed*

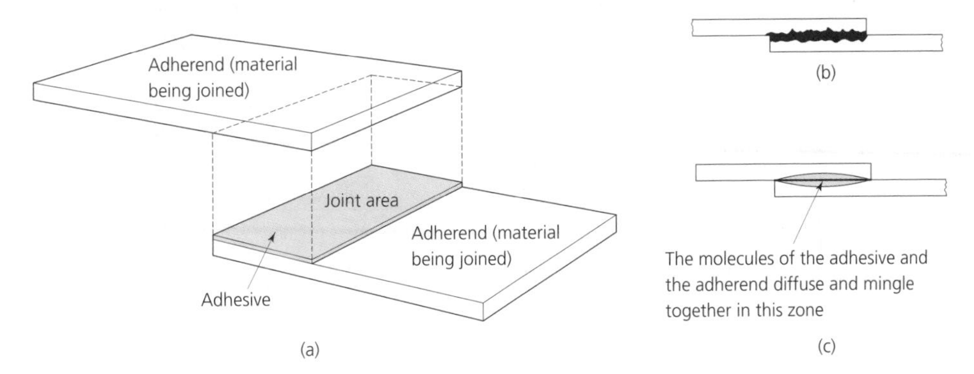

The strength of the adhesive force thus depends upon the thoroughness of the surface wetting (to achieve the fullest intermolecular exchange) and also upon the adhesive capacity of the surface. For any given surface tension for the adhesive, wetting depends on the surface energy of the substrate and the viscosity of the adhesive. Wetting is reduced if surface contaminants are present. The success of any bonded joint is dependent on the surface preparation of the components being joined. The relationship between the surface energy of the adhesives and the material is a direct function of the *contact angle*, as shown in Fig. 7.3. An adhesive wets a solid surface adequately only if its surface energy 'k' is the same or lower than that of the substrate. Table 7.3 shows that metals can be bonded easily whereas plastics are often more difficult. Surface energy can be improved by suitable pretreatment (see Section 7.12).

Fig. 7.3 *Adhesives only adequately wet the surfaces of bonded joints if their surface energy is equal to or lower than the surface energy of the substrate (reproduced courtesy of Loctite Corporation)*

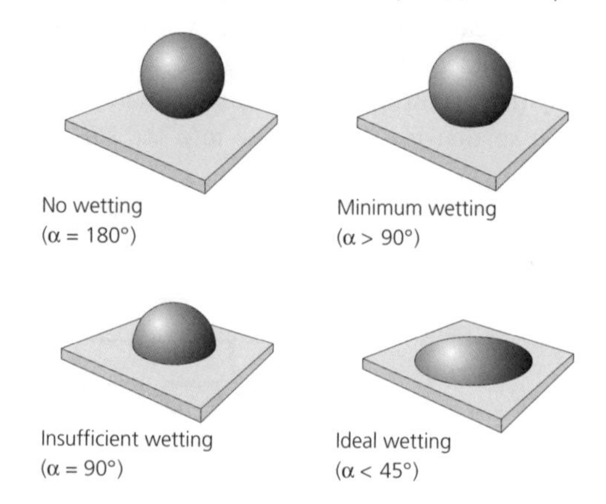

Table 7.3 *Critical surface energies of various materials*

(At room temperature)

Material	*Surface energy*
PTFE (Teflon)*	18 mN/m
PVC	40 mN/m
Polyamide 6/6	46 mN/m
Iron	2030 mN/m
Tungsten	6800 mN/m
By comparison:	
Loctite adhesives	30–47 mN/m

*Teflon is a registered trademark of E.I. DuPont de Nemours Co., Inc.
Source: Loctite Corporation.

7.2.2 Cohesion

Cohesion is the force prevailing between the molecules within an adhesive that keeps the materials together. These forces include:

- Intermolecular forces of attraction (van der Waals forces).
- Interlocking of the polymer molecules amongst themselves.

In accordance with the rule that a chain is only as strong its weakest link, the forces of adhesion and cohesion in a bonded joint should be equal. Figure 7.4 shows three ways in which a bonded joint can fail. These failures can be prevented by careful design of the joint, correct selection of the adhesive, careful preparation of the joint surfaces, and control of the working environment (cleanliness, temperature and humidity).

Fig. 7.4 *Failure of bonded joints*

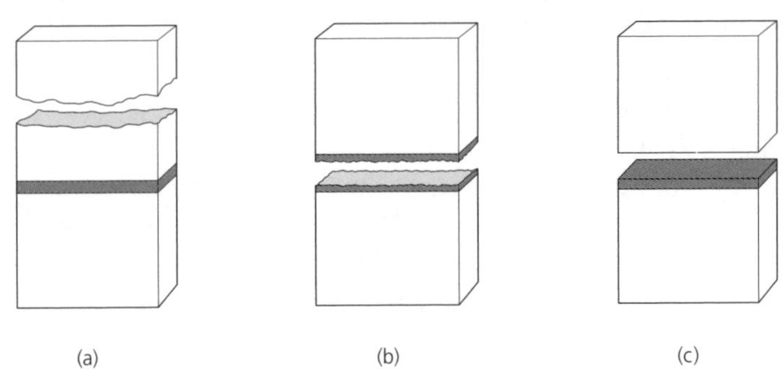

No matter how effective the adhesive and how carefully it is applied, the joint will be a failure if it is not correctly designed and executed. It is bad practice to apply adhesive to a joint originally proportioned for bolting, riveting, soldering or welding. The joint must be proportioned to exploit the properties of adhesives.

Most adhesives are relatively strong in tension and shear, but weak in cleavage and peel (these terms are explained in Fig. 7.5). As previously stated, the adhesive must 'wet' the joint surfaces thoroughly, otherwise voids will occur and the actual bonded area will be less than the designed area. Such a reduction in area will seriously weaken the strength of the joint. Figure 7.6 shows the effects of 'wetting' on the formation of the joint.

Fig. 7.5 *The stressing of bonded joints: (a) tension; (b) cleavage; (c) shear; (d) peel*

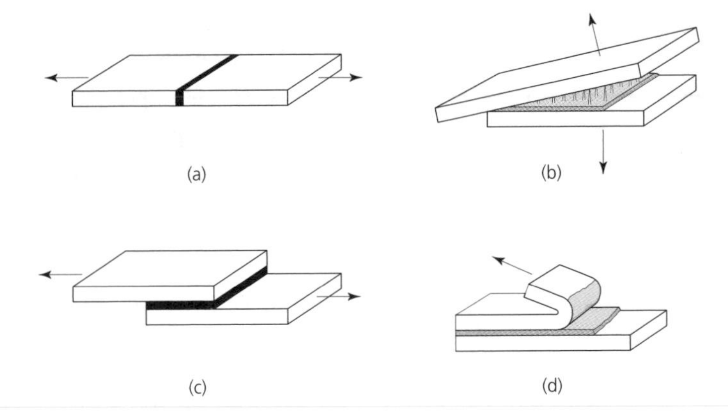

Fig. 7.6 *Wetting capacity of an adhesive: (a) an adhesive with a poor wetting action does not spread evenly over the joint area, reducing the effective area and weakening the joint; (b) an adhesive with a good wetting action will flow evenly over the entire joint, ensuring a sound joint of maximum strength*

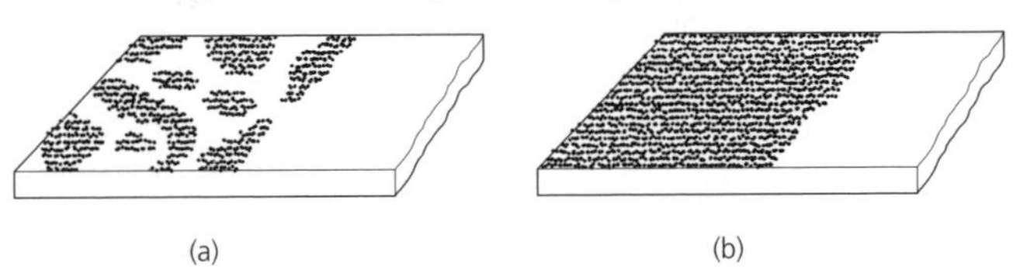

SELF-ASSESSMENT TASK 7.1

1. Discuss the essential differences betwen gums, glues and synthetic adhesives, and suggest typical examples where each would be used, giving reasons for your choice.
2. (a) Explain what is meant by the terms 'adherend' and 'adhesive'.
   (b) Explain what is meant by the terms 'adhesion' and 'cohesion'.
   (c) Explain what is meant by the terms: 'mechanical bond' and 'specific bond'.
3. With the aid of diagrams, show how bonded joints may fail in service.

7.3 Thermoplastic adhesives

As with all thermoplastic materials, thermoplastic adhesives soften when they are heated and harden again when they are cooled. They may be classified into four categories.

7.3.1 *Heat-activated adhesives*

Heat-activated adhesives are softened by heating until they are fluid enough to spread freely over the whole joint surface. Upon cooling to room temperature the adhesive sets and adheres to the materials being joined so that a bond is achieved.

7.3.2 *Solvent-activated adhesives*

Solvent-activated adhesives are softened by the use of suitable solvents, and the bond is achieved by the solvent evaporating. Because evaporation is essential to the setting of the adhesive, a sound bond is almost impossible to achieve at the centre of a large joint area, as shown in Fig. 7.7. This is particularly the case when joining non-absorbent materials.

Fig. 7.7 *Solvent-activated adhesive fault: joints made between non-porous adherends (such as metal or plastic) with solvent-activated adhesives may fail due to lack of evaporation of the solvent; the solvent around the edge of the joint sets off, forming a seal and preventing further evaporation of the solvent, this reduces the effective area of the joint and reduces its strength*

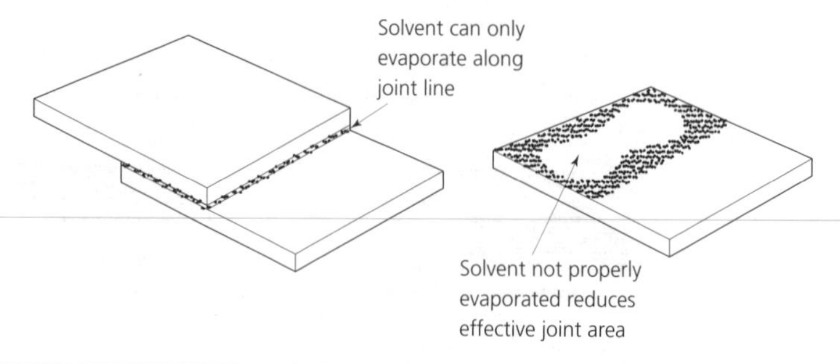

7.3.3 *Impact adhesives*

These are solvent-activated adhesives that are spread separately on the two joint faces and then left to dry by evaporation. When dry, the treated faces are brought together whereupon they form a bond by intermolecular attraction. Figure 7.8 shows the steps in making an impact joint. This overcomes the problems of evaporation associated with joints having large contact areas.

Fig. 7.8 *The use of an impact adhesive: (a) adhesive is spread thinly and even on both joint surfaces and left to dry by evaporation (avoiding the problem in Fig. 7.6); (b) when dry, the surfaces are brought into contact – they form an immediate intermolecular bond*

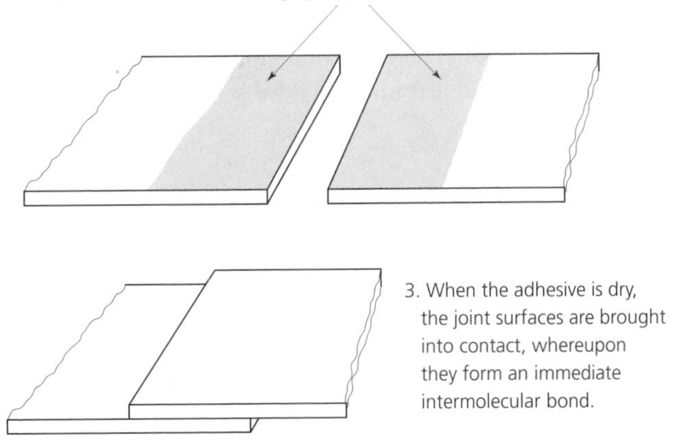

7.3.4 *Solvent cements (solvent welding)*

These are solvents which, when applied to thermoplastics, soften the joint surfaces and form a bond when the joint surfaces are brought together under pressure. Gap-filling properties of the cement are improved by 'bodying' the solvent with some of the plastic material being joined. This reduces shrinkage as the solvent evaporates and prevents the formation of stresses in the joint.

Thermoplastic adhesives are based upon synthetic materials such as polyamides, vinyl, acrylics and cellulose derivatives. They can also be derived from such natural materials as resin, shellac, mineral waxes and recycled rubber. Thermoplastic adhesives are not as strong as thermosetting plastics but, being more flexible, they are more suitable for joining non-rigid materials.

7.4 Thermosetting adhesives

As with all thermosetting plastic materials, thermosetting adhesives require heat to make them set. The setting (curing) process causes chemical changes to take place within the adhesive. Once set (cured) they cannot be softened again by the reapplication of heat. This makes them less temperature sensitive than thermoplastic adhesives.

The heat necessary to cure the adhesive can be applied externally by means of an oven (autoclave) or by radiant heat (for example, when phenolic resins are used), or internally by adding a chemical hardener (for example, when epoxy resins are used). The hardener is a chemical that reacts with the adhesive to generate heat internally (an *exothermic* reaction). Since the setting process is a chemical reaction and not dependent upon solvent evaporation, the area of the joint can be made as large as is necessary to achieve the required joint strength.

Thermosetting adhesives are much stronger than thermoplastic adhesives and can be used for making structural joints between high-strength materials such as metals. The body shells of motor cars and stressed members in aircraft are increasingly dependent upon adhesives for their joints in place of spot welding and riveting. The stresses are more uniformly distributed, and the joints are sealed against corrosion. Further, the relatively low process temperatures involved in adhesive bonding do not affect the crystallographic structure of the metal. Thermosetting adhesives tend to lack flexibility when cured and, therefore, are not suitable for joining flexible (non-rigid) materials.

7.5 Further curing mechanisms

The specialised adhesives developed by the Loctite Corporation are reactive polymers. They change from liquids to solids through various chemical polymerisation reactions. This company has developed numerous adhesives with special curing properties for unique situations. These can be classified on the basis of their curing properties:

- Anaerobic reaction.
- Exposure to ultraviolet (UV) light (also secondary curing options).
- Activation systems (modified acylics).
- Moisture curing (silicones, urethanes).

7.6 Adhesives cured by anaerobic reaction

Anaerobic adhesives are single-component materials that cure at room temperature when deprived of contact with oxygen. The curing component in the liquid remains inactive as long as it is in contact with atmospheric oxygen. If the adhesive is deprived of oxygen (by bringing the adhesive-coated components together), then curing occurs rapidly – especially with simultaneous metal contact. We can imagine the cure as follows. When atmospheric oxygen is excluded, free radicals are formed under the effect of metal ions (Cu, Fe). These free radicals initiate the polymerisation process, as shown in Fig. 7.9.

In Fig. 7.9(a) the liquid adhesive is kept stable through a constant supply of oxygen. When the adhesive is enclosed in the joint gap and separated from the oxygen supply, as shown in Fig. 7.9(b), the peroxides are changed into free radicals through a reaction with the metal ions. The free radicals then initiate the formation of radical chains as shown in Fig. 7.9(c). Finally, the cured state (Fig 7.9 (d)) shows a solid structure with 'cross-linked' polymer chains.

Fig. 7.9 *The curing process of adhesives cured by anaerobic reaction (reproduced courtesy of Loctite Corporation)*

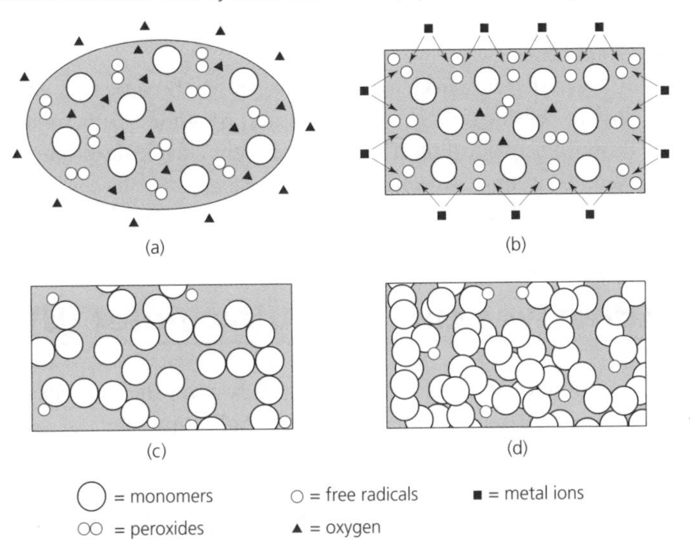

Table 7.4 *Effect of substrate material on anaerobic cure rate*

Active materials (speedy cure)	*Passive materials (slower cure)**
Steel	High-alloy steel
Brass	Aluminium
Bronze	Nickel
Copper	Zinc
Iron	Tin
	Silver
	Gold
	Oxide films
	Chromate films
	Anodic coatings
	Plastics
	Ceramics
	Stainless steel

*Use surface primer for fast fixture.
Source: Loctite Corporation.

The capillary effect of the liquid adhesive carries it into even the smallest gaps to fill the joint. The cured adhesive is then 'keyed' to the surface roughness of the parts. This holds cylindrical parts together. The curing process is also stimulated by contact between the adhesive and the metal surfaces that act as catalysts (see Table 7.4). Passive materials have only a slight catalytic effect, if any, requiring activators to be used for rapid, complete, curing. In this case the liquid activator is applied to one or both bonding faces before the adhesive is applied. Anaerobics may also be heat cured (e.g. 30 minutes at 120 °C). Adhesives cured by anaerobic reaction may generally be described as follows:

- Very high shear strength.
- Good temperature resistance (from $-55\,^{\circ}C$ to $+230\,^{\circ}C$ (max)).
- Rapid curing.
- Easy to dispense from automatic dispensers because they are single component.
- Micro-finishing of parts not necessary; roughness between 8 and 40 μm acceptable.
- Simultaneous sealing effect with excellent chemical resistance.
- Good resistance to vibration.
- Good resistance to dynamic loads.

7.7 Adhesives cured by ultraviolet (UV) light

The cure times of these adhesives depends upon the intensity and wavelength of the UV light. Polymerisation initiated by UV light thus always requires coordination of the product and the source of UV radiation. The photo-initiators are split by the UV radiation to form free radicals. The free radicals formed in turn start the polymerisation process, as shown in Fig. 7.10. Typically, the UV-curing process demands are divided into three types:

- Depth curing by UV radiation.
- Surface curing by UV radiation.
- Secondary curing systems.

Fig. 7.10 *The curing process in UV-curable adhesives (reproduced courtesy of Loctite Corporation)*

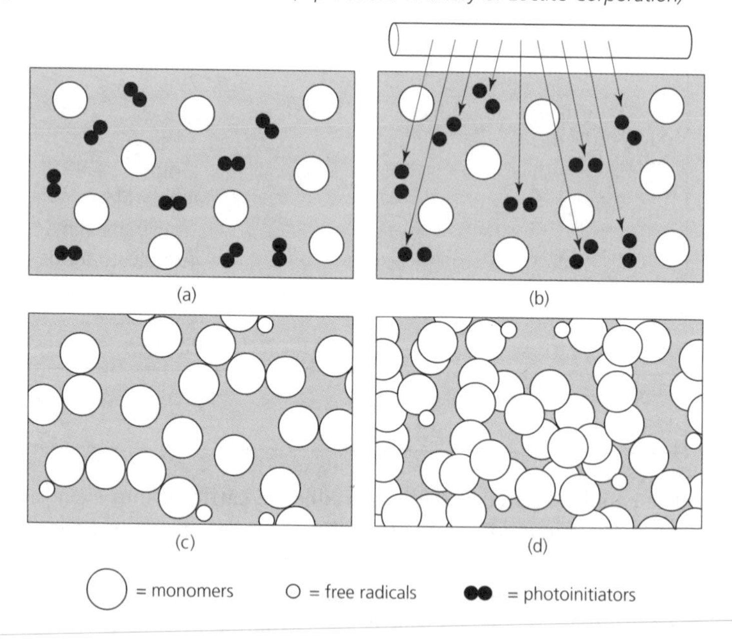

The curing process in UV-curable adhesives is as follows. When the adhesive is in the liquid state the monomers and the photo-initiators coexist without reacting with each other, as shown in Fig. 7.10(a). When exposed to UV light, as shown in Fig. 7.10(b), the photo-initiators turn into free radicals. These free radicals initiate the formation of monomer chains, as shown in Fig. 7.10(c). When curing is complete and the adhesive is solid, cross-linked polymer chains will have been formed, as shown in Fig. 7.10(d).

7.7.1 *Depth curing*

Ultraviolet systems that emit high-intensity light on wavelengths in the band 300–400 nm (longer UV wavelengths) are better to yield higher cure depths, as shown in Fig. 7.11.

Fig. 7.11 *Typical curing behaviour of UV adhesives during depth curing; intensity of radiation source: 100 mW/cm² (reproduced courtesy of Loctite Corporation)*

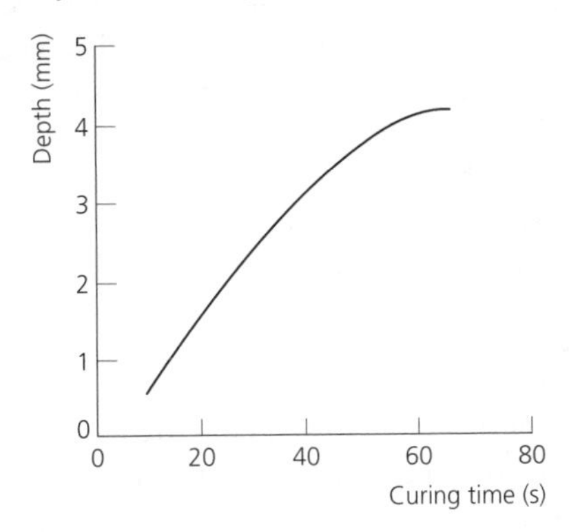

7.7.2 *Surface curing*

Surface curing is especially important when potting or bonding with UV materials. If inadequate UV lamp systems are used, a tacky surface may remain. To prevent this, the surface of UV radiation should produce high-intensity emissions in the wave band below 300 nm. This is largely effective in overcoming the unwanted reaction of the adhesive surface with the atmospheric oxygen that inhibits the curing of the adhesive at the surface, as shown in Fig. 7.12.

Fig. 7.12 *Typical curing behaviour of UV adhesives during surface curing (reproduced courtesy of Loctite Corporation)*

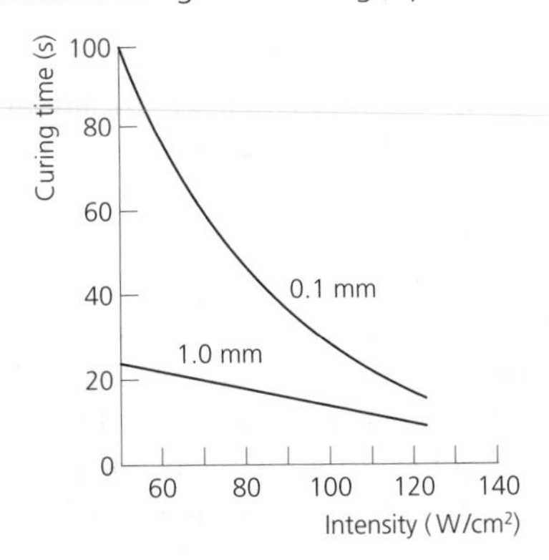

7.7.3 *Secondary curing*

Secondary curing systems are used when the UV radiation does not reach all locations wetted by the adhesive. Complete polymerisation can be obtained in shadow areas with several secondary systems:

(a) Anaerobic curing.
(b) Heat.
(c) Activators.
(d) Ambient moisture.
(e) Atmospheric oxygen.

The secondary systems under (a) to (c) generally occur in combination; (d) and (e) usually occur alone as additional curing systems. The degree of curing required for complete polymerisation of the adhesive through exposure to UV light is dependent on the type of adhesive. It ranges from 0 per cent for adhesives with a distinct anaerobic component up to 100 per cent for adhesives that can only be cured by exposure to UV light. Some adhesives react with activators as a secondary curing system.

Adhesives cured by UV light can be described as having:

- High strength.
- High gap-filling capacity.
- Very short curing times to handling strength.
- Good to very good environmental resistance.
- Good dispensing capacity with automatic application systems as single-component adhesives.

7.8 Adhesives cured by anionic reaction (cyanoacrylates)

Single-component cyanoacrylate adhesives polymerise on contact with slightly alkaline surfaces. In general, ambient humidity in the air and on the bonding surface is sufficient to initiate curing to handling strength within a few seconds. The best results are achieved when the relative humidity value is 40–60 per cent at the workplace at room temperature. Lower humidity leads to slower curing; higher humidity accelerates it, but may impair the final strength of the bond. The curing process is shown in Fig 7.13.

In cyanoacrylate adhesives the acid stabiliser molecules indicated in Fig. 7.13(a) prevent the adhesive molecules from reacting, thus keeping the adhesive liquid. Surface moisture neutralises the acid stabiliser, as shown in Fig. 7.13(b). This initiates the polymerisation process, as shown in Fig. 7.13(c). Finally, many polymerisation chains that are interwoven with each other are formed, as shown in Fig. 7.13(d).

Dry air does not impair the strength of the bond. However, the longer curing times associated with dry air slows down production. With the help of an air-treatment system, favourable humidity levels can be kept constant in the bonding workplace. Acidic surfaces (pH value < 7) may delay or prevent curing, whereas alkaline surfaces (pH values < 7) accelerate curing.

Fig. 7.13 *The curing process in cyanoacrylate adhesives (reproduced courtesy of Loctite Corporation)*

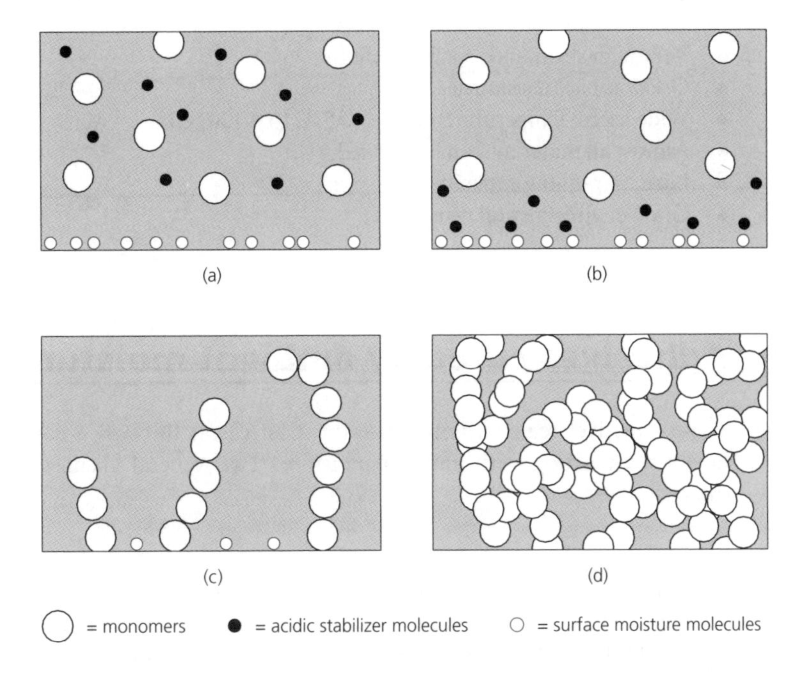

After adhesive application, parts must be joined quickly since polymerisation begins in only a few seconds. The open time is dependent upon the relative humidity, the humidity of the bonding surfaces and the ambient temperature. Due to their very fast curing times cyanoacrylate adhesives are particularly suitable for bonding small parts. Cyanoacrylate adhesives should be applied to only one surface and the best bond is achieved if only enough adhesive is applied to fill the joint gap. Activators such as isopropanol, acetone and heptane may be used to speed the curing process and cure excess adhesive. Features of these adhesives are:

- Very high shear and tensile strength.
- Very fast curing speed (in seconds).
- Minimum adhesive consumption.
- Almost all materials may be bonded.
- Simple dispensing by the good ageing resistance of single-component adhesives.
- Simultaneous sealing effect.

7.9 Adhesives cured with activator systems (modified acrylics)

These adhesives cure at room temperature when used with activators such as acetone and heptane/isopropanol. Adhesives and activators are applied separately to the bonding surfaces. These components of the adhesive system are not premixed, so it is not necessary to be concerned about 'pot life'. The characteristic properties of modified acrylics are:

- Very high shear and tensile strengths.
- Good impact resistance.
- Wide useful temperature range (−55 °C to +120 °C).
- Almost all materials can be bonded.
- Large gap-filling capacity.
- Good environmental resistance.

7.10 Adhesives cured by ambient moisture

These adhesives/sealants polymerise (in most cases) through a condensation effect that involves a reaction with ambient moisture. Two general chemistry types fall into this category.

7.10.1 *Silicones*

These materials vulcanise at room temperature by reacting with ambient moisture (RTV). The solid rubber silicone is characterised by the following properties:

- Excellent thermal resistance.
- Flexible, tough, low modulus, high elongation.
- Effective sealants for a variety of fluid types.

7.10.2 *Urethanes*

Polyurethanes are formed through a mechanism in which water reacts (in most cases) with a formulative additive containing isocyanate groups. These products are characterised by the following properties:

- Excellent toughness and flexibility.
- Excellent gap filling (up to 5 mm).

It has only been possible to give a brief review of the main types of these specialised adhesives and the reader is referred to the technical literature published by the Loctite Corporation for detailed information on adhesive types and applications.

SELF-ASSESSMENT TASK 7.2

1. Briefly explain what is meant by an 'anaerobic reaction' and describe how this can be used to cure (solidify) certain classes of adhesives.
2. Explain the differences between 'depth curing' and 'surface curing' when using UV-sensitive adhesives and give an example of an appropriate application for each of these curing mechanisms.
3. Briefly explain the mechanism of anionic curing mechanisms as applied to cyanoacrylates.
4. Briefly explain where the use of phenolic resin adhesives would be appropriate and state how this group of adhesives are cured.

7.11 Joint design

It has already been shown that adhesive-bonded joints are strong in tension and shear but weak in cleavage and peel. Therefore, all joints designed for use with adhesives should subject the bond only to tension or shear. Mechanical interlocking is also desirable for highly stressed joints, particularly if it places the bond in compression.

Figure 7.14 shows a simple lap joint. It can be seen that the applied tensile forces are not axially aligned and that the joint is subjected to a distorting couple. If the distortion becomes sufficiently severe, it results in the adhesive bond no longer being in pure shear but being subjected to some cleavage and peel as well. Figure 7.15 shows some alternative lap joints in which the tensile forces are aligned so that the adhesive bond is only subjected to shear forces. As has just been described, the forces acting upon bonded joints are not always as simple as they may seem. Figure 7.16 shows correct and incorrect designs for some further joints.

Fig. 7.14 *Distortion and failure of a simple lap joint*

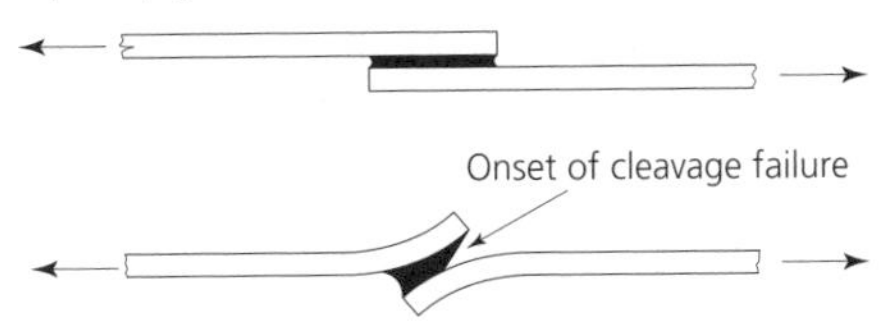

Fig. 7.15 *Alternatives to the simple lap joint*

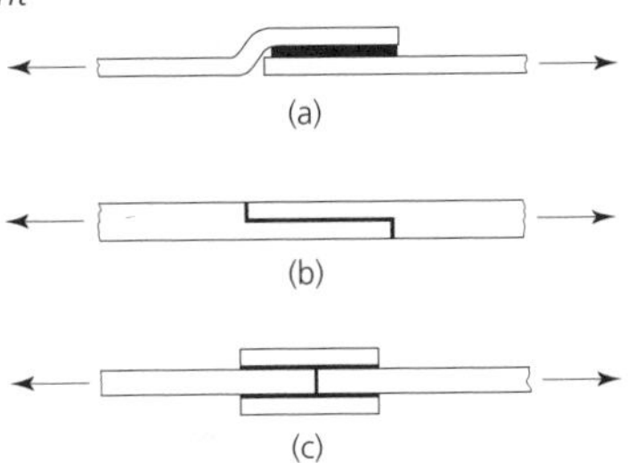

Fig. 7.16 *Joint design for adhesive bonding*

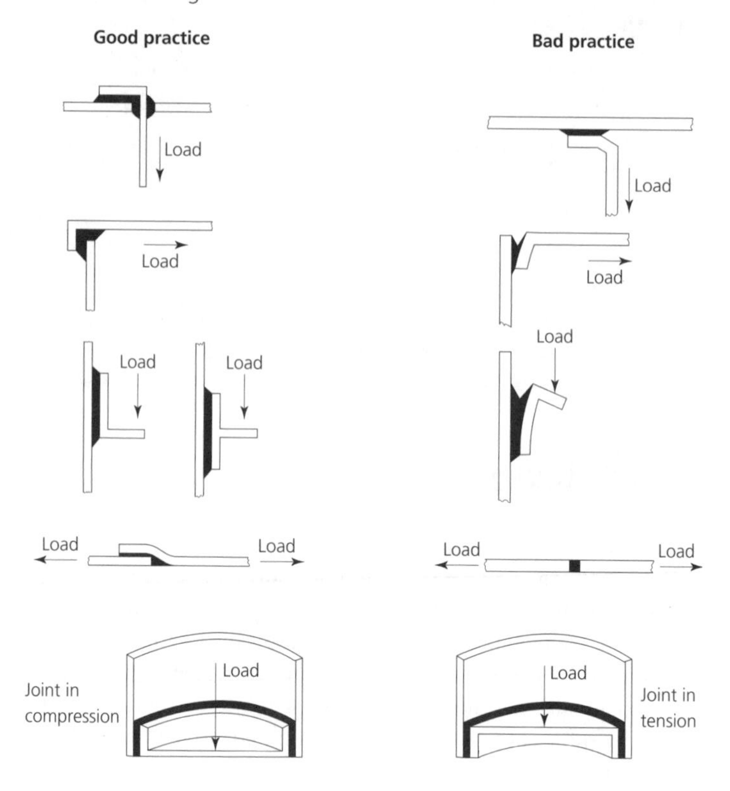

7.12 Surface preparation

Correct surface pretreatment is necessary for optimum bonding. Bond strength is determined to a great extent by the adhesion between the joint surfaces and the adhesive. It is important to understand that adhesive joints are stronger the more thoroughly the surfaces are cleaned, as shown in Fig. 7.17. Adhesion can be improved by removing unwanted surface films by degreasing or mechanical abrasion, and, if needed, by coating with primers.

Fig. 7.17 *Contamination of the surface of the substrate reduces adhesion (reproduced courtesy of Loctite Corporation)*

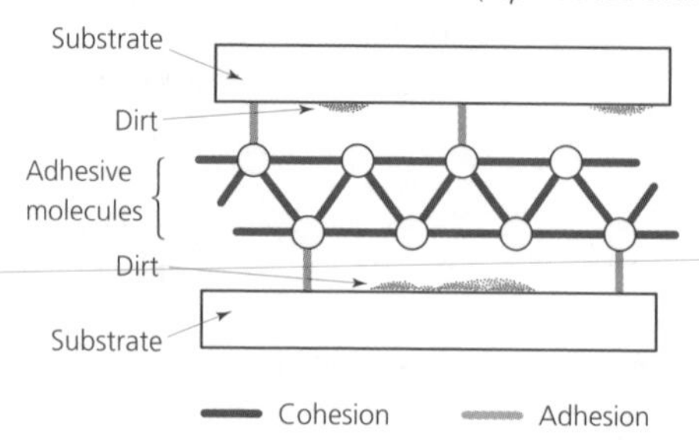

7.13.1 *Effect of the plastic substrate surface layer*

Plastic substrates often confront us with the problem that the volume properties (properties of the basic material itself) do not correspond to the surface properties. The reasons for this may be the composition of the plastic substrate as well as the manufacturing process. Weak surface layers will result in low 'bond' strength regardless of the adhesive selection.

7.13.2 *Plastics with low molecular weight constituents*

Many plastics contain low molecular weight constituents. These include stabilisers, non-reactive components, solvent residues, plasticisers and filler materials. All these components can influence bonding if they are present on the surface. Many tend to move towards the surface (migration) and concentrate there. Thus, a separate layer forms on the surface of the base material, considerably reducing the potential strength of the bond or even preventing a bond.

7.13.3 *Internal and external mould release agents*

Internal and external mould release agents are used to guarantee release of moulded or pressed plastic parts from the forming tools. Mould release agents are described as *internal* if they are already mixed into the granules and take effect during the processing of the plastic. They often produce surfaces that are difficult or impossible to bond. These mould release agents may be distributed throughout the whole plastic body, so that even grinding the surfaces may not be effective.

External mould release agents, on the other hand, are sprayed into the open mould. They are based on paraffins, soaps and oils (e.g. silicone oil). Because of the manner of treatment, these mould release agents may be found not only on the surface but also in layers near the surface. The most suitable pretreatment for such surfaces is mechanical roughening (e.g. grinding).

7.13.4 *Surface properties resulting from processing*

In the course of injection moulding or compression moulding (pressing) of plastic parts, surface structures and thus *surface properties* not identical with the *volume properties* may be formed. In practical terms these are called injection and pressing skins, which are very smooth, compressed surfaces, usually with stress concentrations. The further such an injection or pressing skin has developed, the poorer are its adhesive properties. Their effect is comparable to that of a protective coating covering the base material. The simplest and most effective pretreatment is the destruction of this protective surface coating by mechanical removal, e.g. by grinding and abrasion. The unwanted characteristics for bonding and the surface treatment methods to overcome them are summarised in Table 7.7.

7.12.1 *Degreasing surfaces to be bonded*

The complete removal of oil, grease, dust and other residual dirt from the surface is required for the best possible adhesive joint. Solvents that evaporate without residues are suitable for this purpose. The most important solvents and their cleaning capacity are listed in Table 7.5.

Table 7.5 Common solvents used for cleaning bonded surfaces

Solvent	*Cleaning capacity*	*Inflammable or combustible*
Hydrocarbons (e.g. isoparaffins)	Good	Yes
Ketones (acetone)	Good	Yes
Alcohols (isopropanol)	Moderate	Yes

Source: Loctite Corporation.

Alkaline or acid-based aqueous cleaning systems almost always contain corrosion inhibitors. If these remain on the cleaned bond surfaces, they may reduce the strength of the adhesive. If such cleaning systems are to be used, then preliminary tests should always be carried out. In every case the substrates must be thoroughly rinsed or wiped off. The cleaning and degreasing of the bond surfaces is the same as for the finishing processes described in Section 4.20. If special degreasing baths are used for large-scale production runs, it is advisable to preclean very dirty surfaces so that the cleaning bath is not contaminated.

For many applications, pretreatment of the surfaces with solvent is sufficient. It removes oils, greases, loose particles of dirt and other contamination and thus prepares the surface for bonding. When cleaning with solvents, it is possible to assist the chemical degreasing process for separating dirt from the surface by mechanical action (rubbing with a cleaning rag or brush) thus obtaining a better cleaning result. With grey iron castings, and nodular cast iron, additional surface cleaning is necessary to remove graphite from the bond surfaces.

7.12.2 *Mechanical pretreatment*

Soiled metal surfaces are often covered with an oxide coating that cannot be removed by degreasing. Where a heavy oxide film (scale) has to be removed from hot worked material, acid pickling with a suitable inhibitor is employed (see also Section 4.18). In less extreme cases, mechanical surface treatment is used such as grit blasting, grinding or wire brushing.

Grit blasting is a good way to clean large surfaces. The surface roughness achieved by this method provides very good bonding results, provided the grit used is not too coarse. Grinding achieves equally good surface roughness (e.g. 300–600 for aluminium, 100 for steel). After grit blasting, as well as grinding or brushing, all parts should be degreased to remove residual grit. Very dirty parts should also be degreased before mechanical treatment so that the grit or abrasive used does not just smear the surface contaminates. In practice, mechanical pretreatments are very simple to use and generally provide adequate bonding strength as well.

Paint should be stripped or ground from painted parts before the actual pretreatment, so that bond strength will not be reduced by the relatively low adhesion of the paint to the substrate. If plastic or rubber parts are to be bonded, the surface film or vulcanisation film should first be removed mechanically. For plastics, fused cast iron or aluminium oxide abrasives have proved effective. Rubber surfaces require cleaning to remove mould release agents, either with solvents or by grinding.

7.12.3 *Surface ionisation pretreatment*

Surface ionisation pretreatment changes the polarity of the surfaces and their energy, just as wet chemical pretreatment does. Depending on the material, geometry of the workpiece, production sequence and number of parts, the processes listed in Table 7.6 may be used.

Table 7.6 *Pretreatment methods for difficult-to-bond plastics*

Pretreatment method	*Materials*
Flame treatment	Predominantly PE, PP
Corona process	Plastics with low surface energy
Low-pressure plasma	Plastics with low surface energy

Source: Loctite Corporation.

7.12.4 *Primers*

Bonding plastics like polypropylene (PP), polyethylene (PE), polytetrafluoroethylene (PTFE), silicones and many thermoplastic elastomers require surface ionisation pretreatment of the bond surfaces. Primers have been developed that activate such plastics for successful bonding.

An important advantage of the use of a primer compared to surface ionisation pretreatment methods is its simple handling: as thin a layer as possible is simply sprayed or brushed on the substrates. After a brief drying time (10–60 seconds), the adhesive is applied as usual and the parts joined. Primer should not be used with plastics that can be easily bonded without pretreatment (e.g. PVC, ABS, NBR). If, for example, PVC is to be bonded to PP, only the PP surface should be treated with primer. Of course, the surfaces must be precleaned as with every other adhesive joint.

7.12.4 *Wettability test*

Cleaning processes can be evaluated with the *water break test*. Several drops of pure water are applied to the cleaned surfaces. On an inadequately cleaned surface, the sperical form of the drop is largely retained, and the surface must be cleaned again. If the water runs on the treated surface, then wetting has been satisfactory and the bond face is sufficiently clean, as shown in Fig. 7.18. This method is not suitable for anodic coatings on aluminium and magnesium.

Fig. 7.18 *Surface preparation can be tested with the 'water break' test: (a) inadequately prepared bonding face; (b) suitably prepared bonding face (reproduced courtesy of Loctite Corporation)*

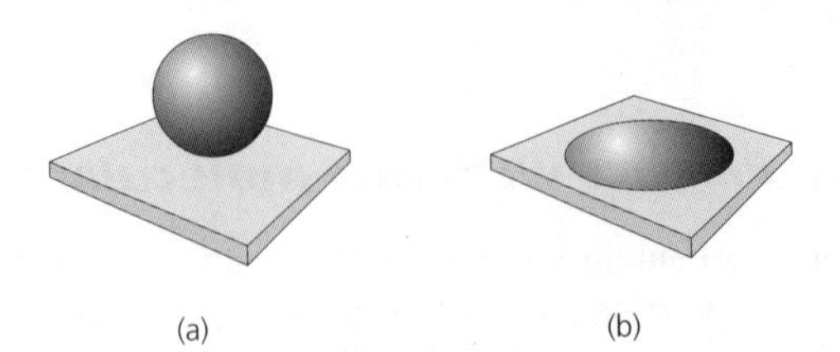

The advantage of the water break test is the easy availability of the 'test fluid', the water. This advantage, however, is limited since the variation in water hardness affects the surface tension. In some cases even distilled water does not produce reliable results with the water break test. In such cases, comparable fluids with defined surface tensions, are recommended. Note that the test only covers wettability and not adhesive bonding capacity.

7.13 Special requirements for bonding plastic substrates

In *Engineering Materials*, Volume 1, and in Chapter 6 of this text, we have already classified polymer ('plastic') materials as:

- Thermosets.
- Thermoplastics.
- Elastomers.

However, this simple classification is inadequate for defining assembly properties. The varying chemical structure of individual plastic substrates and the resulting physical properties are deciding factors for adhesive engineering.

As with all materials that are to be bonded, two preconditions must be met:

1. The adhesives must be able to wet the plastic, i.e. the surface tension of the plastic must be greater than, or the same as, that of the adhesive.
2. The surfaces of the plastic must have properties conducive to adhesion, i.e. the chemical and physical interaction between the adhesive and the substrate at the interface.

If one of these conditions cannot be met, then the plastics in question are often unsuitable for bonding. If neither condition is met, then the plastic cannot be bonded without pretreatment.

Table 7.7 Unwanted characteristics for bonding

Unwanted characteristics for bonding	*Surface treatment methods*
Low-molecular constituents on the surface	• Clean with suitable solvents or cleaning agents • Remove mechanically (grinding)
Internal mould release agents on the surface	• Clean with aqueous, alkaline cleaning agents
External mould release agents on the surface	• Remove mechanically (grinding) • Wash with suitable cleaning agents
Injection or pressing skin	• Remove mechanically

Source: Loctite Corporation.

7.13.5 *Stress cracking in thermoplastics*

Amorphous, thermoplastic unfilled substrates tend to form cracks when brought into contact with certain liquids (solvents). This is often called *stress cracking*. The plastics most susceptible are polycarbonates (PC), polymethyl methacrylates (PMMA), acrylonitrile–butadiene–styrene copolymers (ABS) and polystyrene (PS). As the term *stress cracking* suggests, cracks are formed through the interaction of two conditions:

- Certain stresses must exist in the substrate. In most plastic parts these are present as so-called 'frozen-in' stresses caused by processing, or they are produced by the effect of external forces.
- A low molecular weight medium must act on the part (e.g. acetone, alcohol). As a rule, adhesives may also cause stress cracking whilst in a liquid state.

7.13.6 *Ways of preventing cracking*

The cracking of plastics during bonding can largely be eliminated with the following processes or by changing the choice of plastic material:

- Post-bake the plastic parts, thus reducing the internal stress.
- When joining, avoid squeezing and deforming as this can generate stress from the outside.
- Use fast-curing adhesives since they minimise exposure to the solvent effect of liquid adhesive and, thus, stress cracking.
- When using cyanoacrylates, do not apply excess adhesive to ensure that there are no open adhesive residues on the edges of the joint.
- When using adhesives cured by UV light, make sure that the joint is cured by means of UV light *immediately* after the adhesive has been applied. Shadow areas in which UV adhesive remains liquid are to be avoided.
- Anaerobics are not suited for unfilled amorphous thermoplastic materials.

7.13 7 *Bonding plastic substrates (summary)*

When several unwanted surface effects occur at the same time, mechanical surface treatment has proved to be the most affective and most comprehensive solution. This processing method changes the surface structure in a way that is conducive to bonding, and resultantly the effective bonded surface becomes greater because of its roughness.

Physical and chemical surface pretreatment methods are used in cases where plastics are difficult or impossible to bond by any other method in order to allow the adhesive to make better contact with the surface.

SELF-ASSESSMENT TASK 7.3

1. Discuss the importance of the pretreatment of the bond surfaces on the quality of the joint produced.
2. Describe the 'water break' test for wettability and list the advantages and limitations of this test.
3. Discuss the more important methods of pretreatment of the bond surfaces.
4. Explain the difficulties encountered in the adhesive bonding of some plastic materials and state how these difficulties may be overcome.

7.14 Factors affecting adhesion

We have just spent some time considering the special problems associated with the bonding of plastic materials. Let's now return to a more general consideration of the factors affecting adhesion, assuming that the joint surfaces have been correctly prepared prior to application of the adhesive.

7.14.1 *Contact time*

Adhesives do not achieve their full bond strength instantaneously. Therefore it is necessary to keep the joint faces in contact and under pressure for a prescribed time as recommended by the manufacturers of the adhesive for a given application.

7.14.2 *Pressure*

As stated above, the joint must be kept under pressure for a prescribed time whilst the bond achieves its required strength. The pressure must be sufficient to ensure uniform contact over the whole joint area and sufficient to squeeze out pockets of air and excess adhesive. However, the pressure must not be excessive otherwise the adhesive film will be too thin, or even removed altogether and the joint strength will be adversely affected. Further, excess pressure can introduce surface stresses into the components being joined.

7.14.3 *Temperature*

This depends upon the type of adhesive used. In the case of thermosetting adhesives the temperature of the joint must be sufficient to ensure complete curing of the adhesive. The joint is usually heated in an autoclave or, if the assembly is too large, radiant heaters are used. Even 'cold setting' adhesives, which use the exothermic reaction of a hardener to ensure curing, must be kept at a sufficiently high ambient temperature to ensure that the exothermic reaction occurs in a satisfactory manner. In the case of thermoplastic adhesives the ambient temperature must be sufficiently high to ensure complete evaporation of the solvent. Usually a warm, dry, working environment around 20 °C is suitable for 'cold-setting' both thermosetting adhesives and thermoplastic adhesives.

7.14.4 *Thickness of adhesive film*

With correctly prepared surfaces, the adhesion at the interface is usually greater than the strength of the adhesive itself and, under most conditions, failure occurs within the adhesive film. Failure of the adhesive film is usually caused by the propagation of cracks that are accelerated by the presence of discontinuities and flaws. Therefore thin layers of adhesive, providing the joint faces are adequately covered, provide the strongest joints.

7.14.5 *Molecular weight*

The bonding together of two materials by means of an adhesive is dependent upon two factors: the mechanical adhesion in which the surface roughness or absorption properties of the adherends provides a 'key' for the adhesive to grip; and the specific adhesion that depends principally upon the van der Waals forces between the molecules of the adhesive and the surface molecules of the adherends.

Mechanical adhesion is more likely to occur when bonding materials such as wood, but for hard, smooth materials such as metals, plastics and ceramics, specific adhesion is the more important. Since the van der Waals forces are greater for large molecules than for small molecules, it follows that adhesives should have large molecules. Thus most adhesives are organic compounds composed of very large and complex molecules having a high density of polar groups.

7.15 Selection of adhesives

The selection of an adhesive for a particular application is a complex exercise dependent upon a large number of factors. There are a number of specialist directories and handbooks that can be consulted in order to draw up a 'short list' of possible adhesives. The final choice invariably involves consultation with the manufacturers of the adhesives under consideration. This section merely reviews the more general factors involved in the selection of an adhesive and reviews a small number of adhesive/adherend combinations.

7.15.1 *Joint requirements*

The type and method of assembly of components using bonded joints are amongst the most important factors influencing the selection of an adhesive. Primarily the adhesive is selected to provide a bond of adequate strength under service conditions and for the duration of the service life of the assembly. The joint characteristics must be reproducible and satisfy any quality control requirements. In addition a number of secondary factors influencing the choice of adhesive have to be considered, including:

- Its suitability for the application process demanded by the economics and method of assembly, size of components and batch quantity.
- Its ability to act as a sealant against liquids and gases.
- Its ability to act as a thermal or electrical insulator.
- Its ability to resist vibration and fatigue.
- Its ability to prevent corrosion in joints involving metal components.

7.15.2 *Materials to be bonded*

The mechanical and physical properties of the materials being bonded, and the economic and practical restraints upon joint surface pretreatment are important factors in the selection of an adhesive. Reference to Fig. 7.1 on p. 159 shows that failure can occur in the adherend, or at the adherend/adhesive interface, or cohesively within the adhesive. Usually it is preferable for cohesive failure to occur. There is nothing to be gained from using a high-strength adhesive when bonding low-strength adherends. Further, a rigid adhesive is unsuitable when bonding flexible or semi-rigid materials. Adhesion must also be considered. For example, epoxy adhesives have a much higher cohesive strength than a solvent cement such as polystyrene cement. However, the polystyrene cement will provide the stronger bond when joining polystyrene mouldings since the epoxy resin will have difficulty in adhering to the joint surfaces.

7.15.3 *Compatibility*

Adhesives and adherends must be mutually compatible.

- They must have similar thermal expansion properties to prevent stresses developing in the joint with changes of temperature.
- The adhesive must not cause corrosion of the adherend. Some acidic adhesives will attack metallic adherends.
- Any solvents or volatiles in the adhesive must not affect the adherends.
- The adherends must not be adversely affected by plasticisers in flexible adherends. They must resist the migration of such plasticisers.

7.15.4 *Bond Stress*

Usually, the adhesive selected should have similar strength characteristics to the adherends being bonded together. An exception would be where bonding is only temporary pending the application of some other joining process. In this case bond strength need only be sufficient to withstand handling during the final joining process and should not interfere with that process.

As well as the magnitude of the stresses created in the bond by the application of external forces, the direction of loading and the conditions under which the load is applied must also be considered. The joint may be subjected to tension, compression, shear, peel or cleavage, or a combination of these loads depending upon the design of the joint. The bond loading may be constant, intermittent or vibratory. Most adhesives show optimum strength characteristics when in tension or compression closely followed by shear. A small joint area requires an adhesive that is strong in tension or shear. However, such an adhesive may be weak in peel or cleavage. A lower strength adhesive may have better cleavage and peel characteristics. In this case its lower tension and shear strength characteristics can be accommodated by increasing the joint area.

Often the high strength, thermosetting adhesives form brittle bonds that are adversely affected by vibration and impact loading which causes the bond to crack or shatter. Under such conditions a slightly weaker but more resilient adhesive may perform more satisfactorily. Other adhesives may show satisfactory strength characteristics under test conditions but will tend to 'creep' under sustained loads in service.

Joint thickness has already been considered earlier. With high moduli adhesives, optimum tensile and shear strengths are obtained when the adhesive film thickness lies between 0.06 and 0.12 mm. Thinner films lead to adhesive starvation and catastrophic failure of the joint. Thicker films may lead to early cohesive failure. Adhesives based upon the elastomers generally achieve their optimum performance characteristics with a film thickness in excess of 0.12 mm as this enhances their peel strengths by allowing more 'give' in the joint. Such elastomer adhesives and joint proportions are usually superior under vibratory loads for the same reason.

7.15.5 *Processing*

There is often a considerable difference between the requirements for an adhesive that is suitable for manual application in the case of prototype and small quantity production compared to an adhesive that is to be applied and spread by machine under quantity production conditions. Other factors include the environmental conditions in the workshop, working life of the adhesive once it has been mixed, drying time, curing temperature and the method of achieving that temperature, the effect of the curing temperature on the adherend material, and such safety considerations as flammability, toxicity and odour. Again, pretreatment processes for the joint surfaces that are suitable under experimental prototype and small quantity production conditions may be quite impracticable under quantity production conditions.

7.15.6 *Cost*

The cost of the adhesive is often only a small part of the total cost of achieving a satisfactorily bonded joint. Therefore the process costs should include such items as the cost of pretreatment, the cost of processing equipment, the cost of ensuring a correct working environment, the cost of safety requirements, the labour cost, and the setting time. In the case of a lap joint the cost of adherend material overlap required to give an adequate joint area must also be considered when comparing strong, and potentially more expensive adhesives, with weaker and cheaper adhesives that require a larger joint area.

7.15.7 *Service conditions*

Adhesives that prove satisfactory under short-term test conditions may fail in service. Such failure may be due to changes in temperature and humidity, degradation of the adhesive due to ultraviolet radiation, load characteristics such as unexpected vibrations and shock loads, and environmental pollutants. The service life of the adhesive must match the service life of the assembly as a whole. Therefore, the rate and pattern of degradation of the adhesive must match the rate and pattern of degradation of the adherend material.

The above factors, which should be considered when selecting an adhesive for a particular application, are in no way exhaustive but are intended merely to give an indication of the problems to be considered in arriving at an initial 'short list' suitable for further consideration.

Figure 7.19 lists some typical solvent cement adhesives and the adherend materials for which they are appropriate; and Fig. 7.20 lists some thermoplastic and elastomer adhesives and the adherend materials for which they are appropriate, together with some thermosetting adhesives and the adherend materials for which they are appropriate.

Fig. 7.19 *Bonding of plastics with solvent cements (adapted from Shields J., Adhesive Handbook, Butterworth)*

The surfaces to be joined are softened, by the application of a suitable solvent and then pressed together to effect a bond. Gap-filling properties of the cement are improved by bodying the solvent with some of the thermoplastic: this reduces at the shrinkage at the joint which would otherwise result in stress formation.	**Solvent cements**																													
Plastics	Acetic acid (glacial)	Acetone	Acetone:ethyl acetate:cellulose acetate butyrate (40:40:20)	Acetone:ethyl lactate (90:10)	Acetone:methoxyethyl acetate (80:20)	Acetone:methyl acetate (70:30)	Butyl acetate:acetone:methyl acetate (50:30:20)	Butyl acetate:methyl methacrylate monomer (40:60)	Ethyl acetate	Ethyl acetate:ethyl alcohol (80:20)	Ethylene dichloride	Ethylene dichloride:methylene chloride (50:50)	Glycerine:water (15:85)	Methyl acetate	Methylene chloride	Methylene chloride: methyl methacrylate monomer (60:40)	Methylene chloride:methyl methacrylate monomer (50:50)	Methylethyl ketone	Methyl isobutyl ketone	Methyl methacrylate monomer	Tetrachloroethylene	Tetrachloroethane	Tetrahydrofuran:cyclohexanone (80:20)	Toluene	Toluene:ethyl alcohol (90:10)	Toluene:methylethyl ketone (50:50)	(1,1,2) Trichloroethane	Trichloroethylene	Xylene	Xylene:methyl isobutyl ketone (25:75)
Acrylonitrile butadiene styrene																		✓	✓							✓				✓
Cellulose acetate fim		✓		✓	✓	✓	✓		✓					✓																
Cellulose acetate butyrate		✓	✓	✓	✓	✓	✓		✓					✓																
Cellulose propionate						✓																								
Cellulose nitrate		✓							✓					✓																
Ethyl cellulose										✓															✓					
Polyamide (nylon)	✓																													
Polymethyl methacrylate											✓					✓	✓			✓										
Polycarbonate											✓	✓				✓						✓					✓			
Polystyrene									✓						✓			✓			✓			✓				✓		
Polyvinyl chloride and copolymers (acetate)																		✓	✓				✓						✓	
Styrene acrylonitrile								✓	✓									✓												
Styrene butadiene		✓																✓	✓											
Polyvinyl alcohol													✓																	
Polyphenylene oxide																														✓

Fig. 7.20 *Complementary adhesives and adherends (adapted from Shields J., Adhesive Bonding, National Design Council)*

In general, any two adherends may be bonded together if the chart shows that they are compatible with the same adhesive

Adherends	**Natural**	Animal glues	Starch	Dextrine	Casein	**Elastomers**	Acrylonitrile butadiene	Polychloroprene	Polyurethane	Silicone rubber	Polybutadiene	Natural rubber	Butyl	**Thermoplastics**	Cellulose nitrate	Polyvinyl alcohol	Polyvinyl acetate	Polyacrylate	Silicone resin	Cyanoacrylate	**Thermosets**	Phenolic formaldehyde	Urea formaldehyde	Resorcinol formaldehyde	Melamine formaldehyde	Polyesters (unsaturated)	Epoxy resins	Polyimides	Phenolic-vinyl formal	Phenolic-polyvinylacetal	Phenolic nitrile	Phenolic epoxy	**Inorganic**	Sodium silicate
Metals							✓	✓				✓					✓			✓							✓	✓	✓		✓	✓		
Glass, ceramics		✓						✓							✓		✓			✓							✓			✓		✓		✓
Wood					✓							✓			✓		✓					✓	✓	✓	✓		✓							
Paper		✓	✓	✓	✓										✓	✓	✓																	✓
Leather		✓					✓	✓				✓			✓																			
Textiles, felt		✓					✓					✓			✓		✓																	
Elastomers																																		
Polychloroprene (Neoprene)								✓																										
Nitrile							✓													✓														
Natural							✓					✓								✓									✓					
Silicone										✓																								
Butyl							✓						✓																					
Polyurethane							✓	✓	✓																									
Thermoplastics																																		
Polyvinyl chloride (flexible)							✓	✓	✓																									
Polyvinyl chloride (rigid)							✓	✓	✓																		✓							
Cellulose acetate									✓						✓					✓														
Cellulose nitrate									✓						✓					✓														
Ethyl cellulose															✓					✓							✓				✓			
Polyethylene (film)								✓			✓							✓																
Polyethylene (rigid)																											✓				✓			
Polypropylene (film)								✓			✓							✓																
Polypropylene (rigid)																											✓				✓			
Polycarbonate									✓																		✓							
Fluorocarbons												✓							✓			✓					✓							
Polystyrene									✓											✓							✓							
Polyamides (nylon)								✓														✓		✓			✓				✓			
Polyformaldehyde (acetals)									✓											✓						✓	✓							
Methyl pentene								✓												✓														
Thermosets																																		
Epoxy																				✓		✓		✓			✓							
Phenolic									✓											✓			✓				✓				✓			
Polyester																										✓	✓							
Melamine								✓	✓																		✓							
Polyethylene teraphthalate							✓	✓											✓							✓								
Diallyl phthalate							✓																			✓	✓							
Polyamide																											✓	✓						

7.16 Evaluation of adhesive joint failure

Some important criteria of an adhesive joint failure may be determined by visual evaluation of the bonded parts. Thus it may be possible to determine whether adhesion or cohesion failure has led to the failure of the joint, or even whether the bonded parts have been destroyed.

- *Adhesion failure* The adhesive can be separated completely from the face of one substrate.
- *Cohesion failure* The adhesive breaks and remains of the adhesive are to be found on both substrates.

Table 7.8 lists some causes and methods of overcoming adhesion and cohesion failure.

Table 7.8 ***Adhesive failures: preventative techniques***

Type of failure	*Methods to increase strength*
Adhesion failure	Obviously the weak point of the bond is the boundary layer between the bonded part and the adhesive. Either the material is unsuited for bonding, or the bonding face was dirty. In both cases the strength can be increased by a suitable pretreatment of the surface
Cohesion failure	The adhesive is overstressed through external action (e.g. stress spikes, temperature, ageing, etc.). Remedy: design changes in the bonding geometry, and/or the type of adhesive is unsuitable for this application

Source: Loctite Corporation.

7.16.1 *Evaluation and methods of improvement*

The appearance of the joint face only tells us where the weak point of the bonded joint is, not the cause of failure. To correct the problem, it is essential to find the causes of failure. Some causes and solutions for adhesive joint failures are listed in Table 7.9.

Table 7.9 ***Adhesive failures: causes and remedies***

Possible causes	*Solutions*
Faulty substrates	Check tolerances, gaps and materials, and monitor more carefully
Dirty surfaces	Check pretreatment for suitability and modify accordingly (e.g. cleaning agents, cleaning procedures, subsequent intermediate storage, etc.)
Faulty or incorrect execution of the bond	Check all process parameters, execution of the bond optimise type and duration of fixing, check whether all curing conditions have been met in fixed state
Insufficient curing of the adhesive	Check curing preconditions (e.g. gap, air tightness, temperature, humidity, etc.). Observe curing times in accordance with data sheet. Check whether shelf life of adhesive was exceeded
Mechanical overstress or unfavourable stress (peeling)	Enlarge bonding face and/or modify joint geometry or force application. Check suitability of adhesive for type of stress (tensile, tensile shear, etc.)
Thermal overstress	Select adhesive with greater temperature resistance
Corrosion or infiltration and destruction of the adhesive coating through liquid and gaseous media	Protect joint gap at the contact faces to the medium by means of a suitable coating or design the bonded parts in such a way that there is no contact with the medium

Source: Loctite Corporation.

It has only been possible to give a brief overview of the widespread and complex process of adhesive bonding and some of the problems associated with them. Since adhesive bonding processes are becoming more and more widely used in engineering and associated industries, a thorough knowledge of this branch of materials science is of increasing importance. All the major manufacturers of adhesives offer comprehensive engineering support for their products and they should be consulted over all new applications.

EXERCISES

7.1 Describe how you would evaluate an adhesive joint failure, and discuss the causes and solutions for such failures.

7.2 Research manufacturers' literature and describe:

(a) the methods of disassembly of adhesive joints
(b) the ageing of adhesive joints

7.3 Research manufacturers' literature and discuss the advantages and limitations of retaining metal cylindrical assemblies by the use adhesives compared with using thermal (shrink) or mechanical (press) fits.

7.4 With the aid of diagrams, show how a welded assembly of your choice should be redesigned to make use of adhesive bonding.

7.5 Describe how the substrate joint surface should be prepared for adhesive bonding.

7.6 Name an example of each of the following types of thermoplastic adhesives and in each instance suggest a typical application giving reasons for your choice:

(a) heat-activated adhesives
(b) solvent-activated adhesives
(c) impact adhesives

7.7 Explain the essential differences in processing and properties between thermosetting adhesives and thermoplastic adhesives. Suggest typical applications where thermosetting adhesives should be used, naming the type of adhesive chosen. Give reasons for your choice.

7.8 Discuss, in detail, the factors that affect adhesion.

7.9 For **two** examples of your choice, explain the factors influencing your selection of a suitable adhesive and state the adhesive type chosen.

7.10 Discuss the health and safety hazards, and the precautions that must be taken when using synthetic adhesives on an industrial scale.

Part C
Materials in service

8 Deformation and failure of materials

The topic areas covered in this chapter are:

- The structure of materials.
- The deformation of materials.
- Theory of fracture.
- Fatigue failure.
- Creep failure.
- Material specification.

8.1 Structure of Materials

An understanding of the structure of materials is essential to the understanding of the behaviour of such materials in service. The crystalline structure of metals and crystallinity in some polymeric materials were discussed in *Engineering Materials*, Volume 1, together with the solidification of molten metal and casting defects. Let's now develop these matters further.

In Chapter 11 you will find that crystal orientation will be referred to using the notation (100) or (111), for example. These numbers are derived from the *Miller indices*. The Miller indices are used to describe the planes and directions of atoms in crystals. Miller indices are proportional to the reciprocals of the intercepts that the planes make with the three principal crystal axes, namely the x, y and z axes.

Figure 8.1 shows how these indices are derived. The plane PQR is shown shaded and it can be seen that the intercepts that it makes with the three axes are $x = 2$ unit lengths, $y = 4$ unit lengths and $z = 4$ unit lengths. The reciprocals of these intercept values are $\frac{1}{2}$, $\frac{1}{4}$ and $\frac{1}{4}$ respectively. These reciprocals are then converted to the smallest integers that are in the same ratio, that is: 2, 1, 1. Thus the Miller indices for the plane PQR in Fig. 8.1 is (211). A plane on the 'opposite side' of the origin would have negative intercepts on the axes and its indices would be denoted as $(\bar{2}\bar{1}\bar{1})$.

Directions through a crystal can be specified in a similar manner by the use of indices that give the integral coordinates of a point on a line drawn between the point and the origin. Direction indices are enclosed in square brackets [ ] to distinguish them from Miller indices which are enclosed in parentheses ( ). Figure 8.2 shows the Miller indices for some typical planes in a simple cubic system, whilst Fig. 8.3 shows some important directions in a simple cubic cell.

Fig. 8.1 *Derivation of Miller indices for plane PQR*

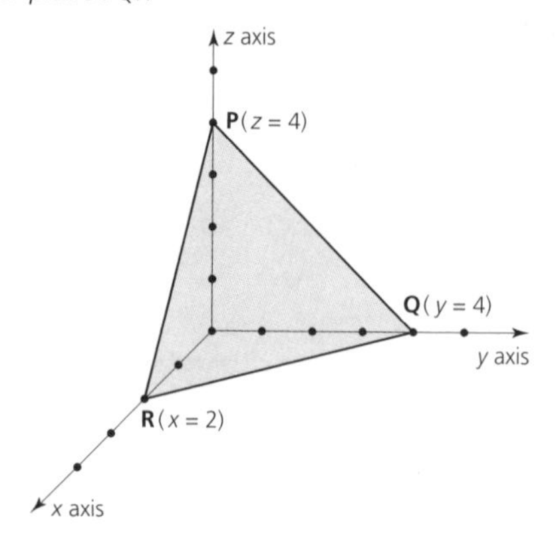

Fig. 8.2 *Miller indices for some typical planes in a simple cubic system (adapted from Higgins, The Properties of Engineering Materials, Arnold)*

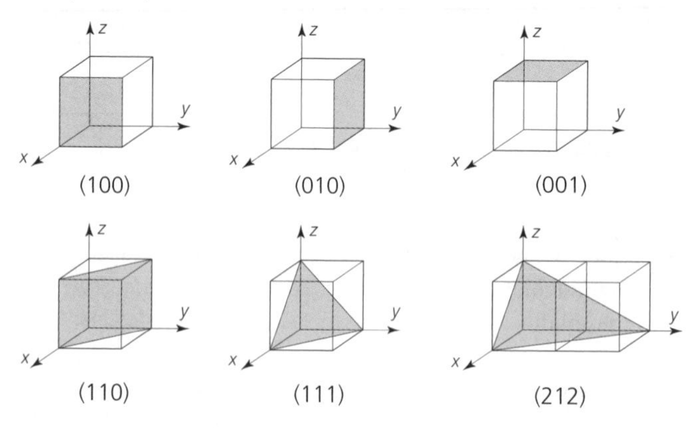

Fig. 8.3 *The important directions for a simple cubic cell with sides of unit length*

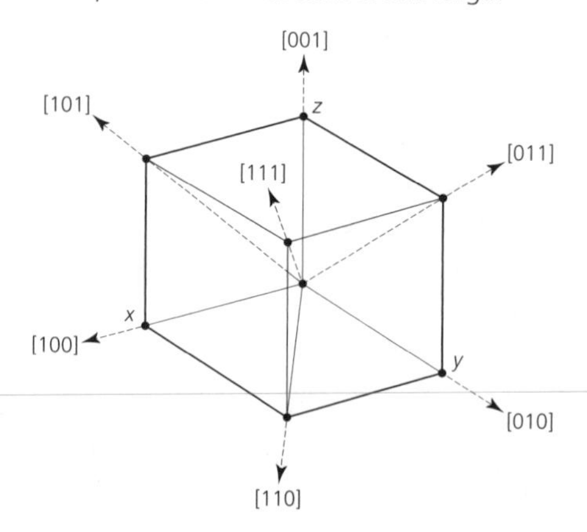

The properties of metals is also dependent upon the metallic bond that results from the fact that metallic atoms have only few electrons in their valency subshells. In the crystal lattice of any metal the metallic ions conform to some regular geometric pattern, as has been shown previously in Fig. 5.3, whilst the associated electrons form a cloud dispersed between and surrounding the ions. These electrons can be said to be *delocalised* as no particular electron or group of electrons belongs to any particular atom. These shared electrons (negatively charged) bind the metallic ions (positively charged) tightly into the lattice (attraction of unlike charges) so that most metals have high melting and boiling points compared to most non-metals. The mobility of the electron cloud is responsible for the high electrical and thermal conductivity of metals compared to non-metals.

Non-metals also form crystals – for example, ice crystals when water freezes and crystals of sodium chloride (common table salt). These crystals are much weaker than metallic crystals and the reason for this was shown in Fig. 5.3.

Some materials are allotropic – that is, they can exist in more than one form, with each form having its own special properties. The non-metallic element carbon, for example, can exist in the form of diamond or in the form of graphite. The unit cell for a diamond crystal is a tetrahedron. As the crystal grows, not all the 'fringe' atoms are linked, so the diamond can be considered to have a *cubic* structure. Figure 8.4 shows the stages in the crystal growth for a diamond.

Fig. 8.4 *The crystal structure of diamond (adapted from Higgins, The Properties of Engineering Materials, Arnold)*

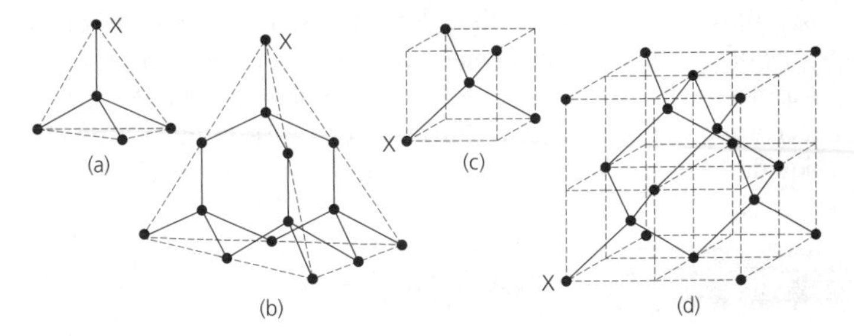

On the other hand, carbon can assume a *layer structure*. The carbon atoms within the layers are covalently bonded together and are thus relatively strong, whilst the layers themselves are only held together by relatively weak van der Waals forces. For this reason the layers or plates can easily slide over each other, which makes graphite an excellent lubricant.

Reference to Fig. 8.5 shows that only three of the four atoms available in the valency sub-shell of graphite are used in forming the covalent bond. The fourth electron in each case can be considered as being 'shared', and these surplus (shared) electrons form an electron cloud in a similar way to the electron cloud in a metallic bond. For this reason graphite is an electrical conductor, whilst diamond is an electrical insulator. Since the number of electrons available to form the cloud is limited compared to the number available in a metallic material, graphite has only limited conductivity (high resistance) compared to metals.

Fig. 8.5 *The layer structure of graphite (adapted from Higgins, The Properties of Engineering Materials, Arnold)*

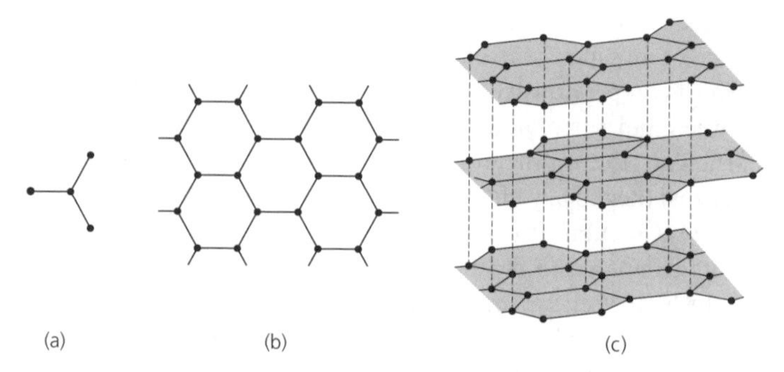

Crystallinity on the molecular scale occurs in many polymeric materials (plastic) and this phenomenon was discussed in *Engineering Materials*, Volume 1. Since molecules as well as atoms are self-contained units, it is possible for molecules to arrange themselves into crystal lattices. The atoms within the molecules are held together by covalent or ionic bonds but the molecules are held together by the weaker van der Waals forces. Such materials usually have a mixture of amorphous and crystalline regions (micelle regions). Although usually found in polymeric materials, some metals (for example, tellurium) also form long-chain molecules with areas of crystallinity on the molecular scale.

Amorphous (without shape) substances are not true solids since all true solids must, by definition, be crystalline. Some amorphous substances such as glass and pitch appear to be solid and will shatter when subjected to sudden impact. However, when lightly loaded they will 'flow' over a period of time. Pitch will settle to the shape of its container and a vertical pane of glass in a window will eventually become thicker at the lower end than at the top. Further, such amorphous materials do not show a clearly defined melting point but become increasingly softer as their temperature is raised, and eventually resemble high-viscosity liquids. Many polymeric materials behave in this manner.

SELF-ASSESSMENT TASK 8.1

1. Work out the Miller indices for the plane that cuts the three principal axes as follows: $x = 2$ units, $y = 1$ unit, $z = 2$ units. Sketch this plane in a simple cubic system.
2. Calculate values of x, y and z (i.e. the positions of intercept on the three principal axes) for the plane having Miller indices (200).
3. Metals are normally excellent electrical and thermal conductors. Explain why this is so.
4. Explain what is meant by the following terms:
   (a) allotrope
   (b) van der Waals forces
   (c) amorphous

(Note: These terms have been explained previously in Volume 1).

8.2 Deformation of materials (metallic)

The crystalline structure of metals has already been discussed (Section 8.1) and it is the strict geometrical symmetry of the crystal lattice which allows plastic deformation to occur in solid metallic materials. When plastic deformation occurs, planes of atoms slip past each other, as shown in Fig. 8.6(a). These planes of movement are called *slip planes*, and usually lie between, and parallel to, the planes of greatest atomic density, as shown in Fig. 8.6(b).

When a ductile material is subjected to an applied force, movement of the lattice structure can occur along the slip planes, as shown in Fig. 8.7. Figure 8.7(a) shows the slip planes before the application of an external force: Fig. 8.7(b) shows the movement of the same slip planes after the application of an external force; and Fig. 8.7(c) shows the microstructure of a sample of copper that has been polished, etched and squeezed between the jaws of a vice. (Notice the slip bands across many of the grains.)

Fig. 8.6 *Slip planes: (a) between planes of high atomic density; (b) the orientation of slip planes*

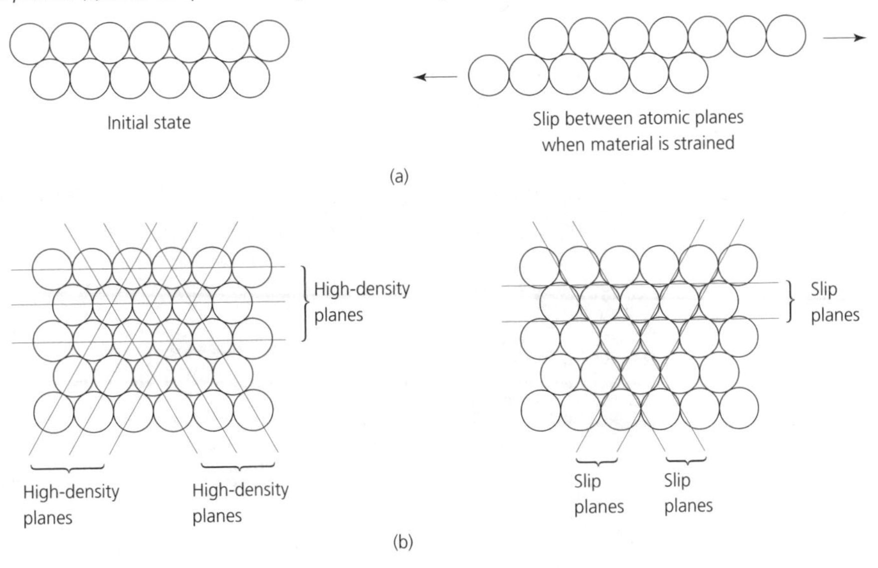

Obviously movement does not occur along all the slip planes available in a crystal, but only in those planes that are at a suitable angle to the applied force. Further, slip can only occur where the crystal is not constrained, thus slip cannot cross the existing grain boundaries. Since slip can only occur within a grain, the bigger the grain the greater the amount of slip that can take place. This is borne out in practice, since fine-grained materials are generally less ductile and malleable than when the same material has been processed to enlarge its grain structure. The difference between elastic and plastic deformation in a ductile material is shown in Fig. 8.8. When the applied force creates a stress in the material that is below that of its elastic limit, distortion of the crystal lattice occurs, but there is no slip. Such

Fig. 8.7 *Formation of slip bands during plastic deformation of a metal: (a) slip planes before the application of a force; (b) movement of slip planes after the application of a force; (c) micrograph of squeezed copper showing slip bands (reproduced courtesy of the University of Luton)*

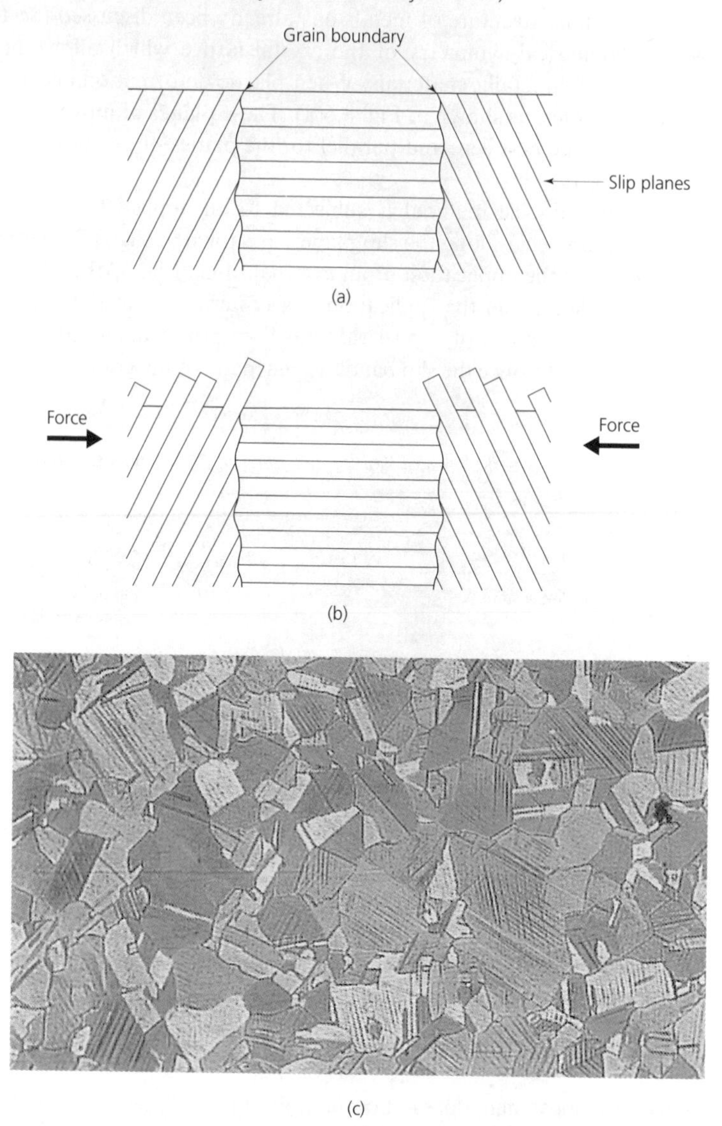

distortion is called *strain*. When elastic deformation occurs, the strain disappears when the applied force is removed and the material returns to its original shape. If the applied load is increased so that the metal is stressed beyond its elastic limit, both elastic and plastic strain will be present and slip will occur. When the applied force is removed, the elastic element of the deformation disappears (some 'spring-back' occurs) but the plastic deformation due to slip remains, as shown in Fig. 8.8(c).

Fig. 8.8 *Deformation of a metallic crystal: (a) elastic deformation; (b) plastic deformation*

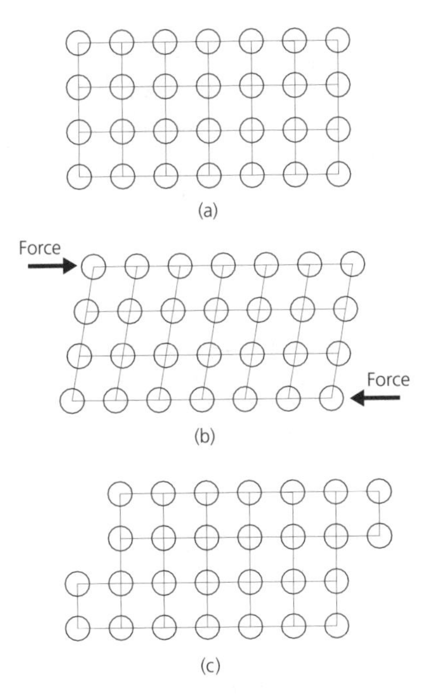

8.2.1 *Dislocation*

Figure 8.8(c) implies that deformation is due to 'block-slip' and that all the atoms in a slip plane move simultaneously. If this were actually to happen, then the force needed to cause plastic deformation would be very much greater than it is in reality. It is now known that slip occurs through a system of dislocations rather like moving a carpet along a floor a little at a time by bunching it and moving the ruck along, as shown in Fig. 8.9. Plastic deformation due to dislocation is possible because crystals are not as perfect as has been implied so far. Whilst the majority of atoms will follow the general lattice pattern, there will also be a variety of faults and deficiencies, referred to as 'point defects', leading to irregularities and distortions in the surrounding crystal lattice. Such point defects may be interstitial or substitutional solute atoms, vacancies, coherent precipitates, and impurity atoms.

Fig. 8.9 *'Carpet analogy' of dislocation: (a) the equivalent of block slip; (b) the equivalent of deformation by dislocation*

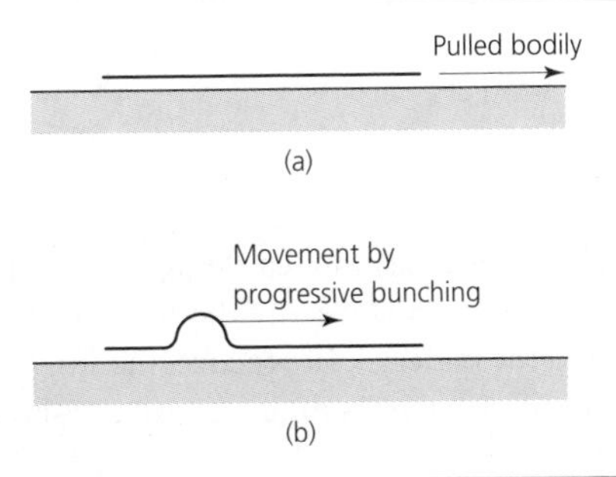

Edge dislocations and screw dislocations are referred to as *line defects*. When considering plastic deformations, *edge dislocations* are very important. Figure 8.10(a) shows a typical edge dislocation, whilst Fig. 8.10(b) to 8.10(d) shows how the dislocation moves progressively through the crystal, under the action of an applied force, until a *slip step* is formed. The presence of an edge dislocation is indicated, diagrammatically by the sign ⊥. The applied force must be sufficient to stress the material beyond its elastic limit and dislocation movement will cease when the force is removed. The dislocation may also be halted when it meets some other fault or reaches the crystal boundary.

Fig. 8.10 *Deformation by dislocation: (a) crystal showing an edge dislocation; (b) and (c) application of a force causes dislocation movement; (d) dislocation (slip) complete*

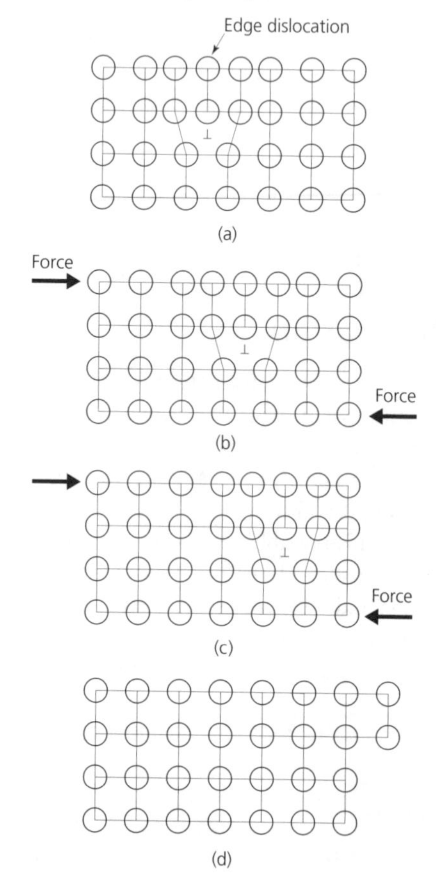

A *screw dislocation* may also cause slip as shown, in principle, in Fig. 8.11. Slip resulting from a screw dislocation is usually the result of the application of offset or 'shear' forces being applied to the material. In practice, the situation is often more complex with edge and screw dislocations occurring simultaneously, as shown in Fig. 8.12.

Fig. 8.11 *The movement of a screw dislocation (adapted from Higgins, The Properties of Engineering Materials, Arnold)*

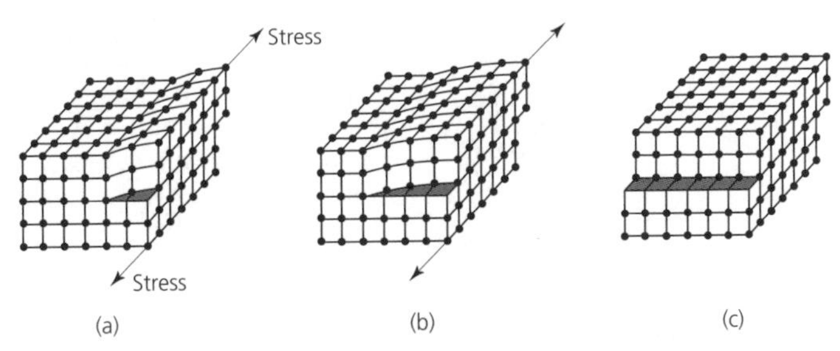

Fig. 8.12 *Slip by the movement of a dislocation loop over a slip plane (adapted from Higgins, The Properties of Engineering Materials, Arnold)*

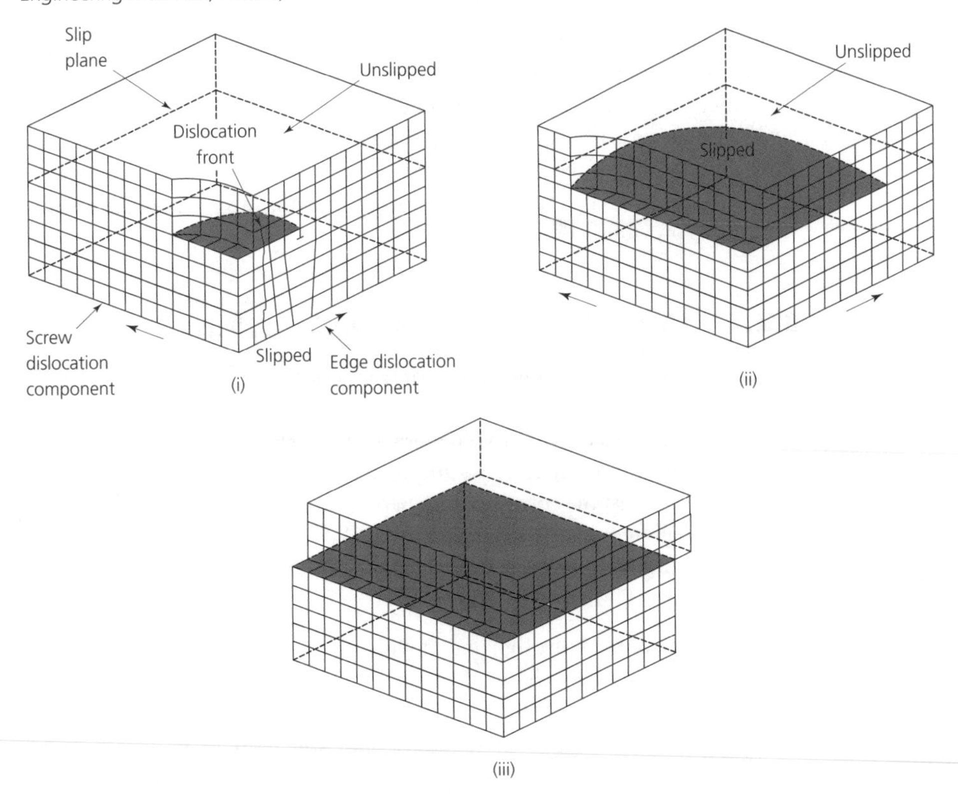

Another mechanism by which deformation can take place is *twinning*. The principle of deformation occurring by twinning is shown in Fig. 8.13. Unlike slip, where all the atoms in a block move the same distance, in twinning each successive plane of atoms moves a different distance. When twinning is complete, the deformation of the crystal lattice will result in one half of the twin becoming the mirror image of the other half, as shown in Fig. 8.13(b). Like slip, twinning proceeds by a series of dislocations. The force required to produce twinning dislocations is generally greater than that required to produce slip.

Fig. 8.13 *Principle of twinning: (a) crystal lattice prior to deformation; (b) crystal lattice after deformation*

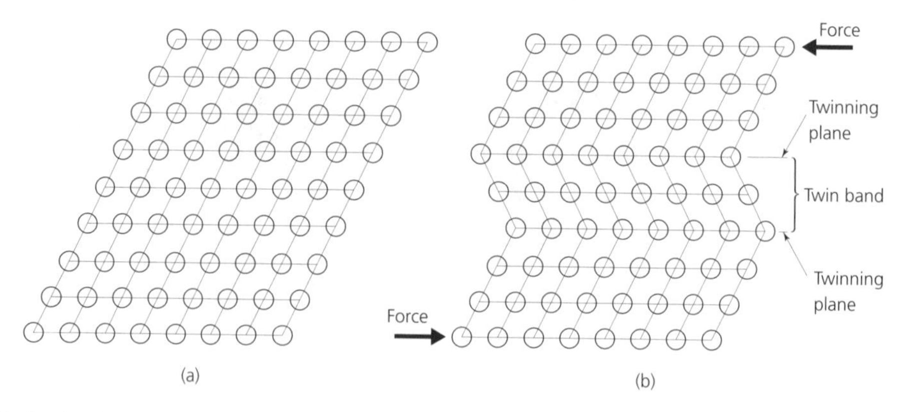

Twinning is not as common as block slip and occurs mainly when metals are shock loaded at low temperatures. For example, cold-heading rivets made from α-iron (which has body-centred-cubic crystals) causes them to develop thin lamellar twins referred to as *Neumann bands*. Twinning can also be caused by heat treatment and is common in copper, brass alloys, bronze alloys and austenitic steel alloys when these metals are annealed. Such twins are the result of recrystallisation and grain growth, rather than the result of mechanical stressing as previously described

8.2.2 *Interaction between dislocations*

The extra half-plane of atoms or ions introduced by a simple edge dislocation results in an increase in strain energy in the vicinity of the dislocation. When two or more dislocations occur in the same vicinity they will interact with each other. The additional half-plane of an edge dislocation introduces crowding of the atoms, resulting in a region of compressive strain energy, whilst the corresponding separation of the atoms beyond the edge dislocation results in a region of tensile strain energy, as shown in Fig. 8.14.

There are three possible results from the interaction of dislocations that are in close proximity to each other.

1. If they are of the same sign and moving on the same or an adjacent slip plane, as shown in Fig. 8.14(a), they will tend to repel each other to reduce the crowding of the atoms and lower the level of strain energy in the region. Therefore the stress required to move them towards each other (produce slip) will need to be increased, resulting in an increase in the total strain energy. Since tensile strength is a measure of the stress required to produce slip in a material, the interaction between multiple edge dislocations in close proximity will result in a corresponding increase in tensile strength for the material. If the applied stress is of sufficient magnitude, dislocations of the same sign will move through the crystal lattice until they are obstructed by the grain boundary. It has already been stated that slip cannot proceed beyond a grain boundary.
2. If the dislocations are of opposite signs but lying on the same slip plane or adjacent slip planes, as shown in Fig. 8.14(b), they will tend to be attracted to each other. Thus,

Fig. 8.14 *Multiple dislocations: (a) dislocations of the same sign repel each other to reduce crowding; (b) dislocations of opposite sign attract and annihilate each other; (c) dislocations of opposite sign and not on the same slip plane results in the formation of voids (vacancies)*

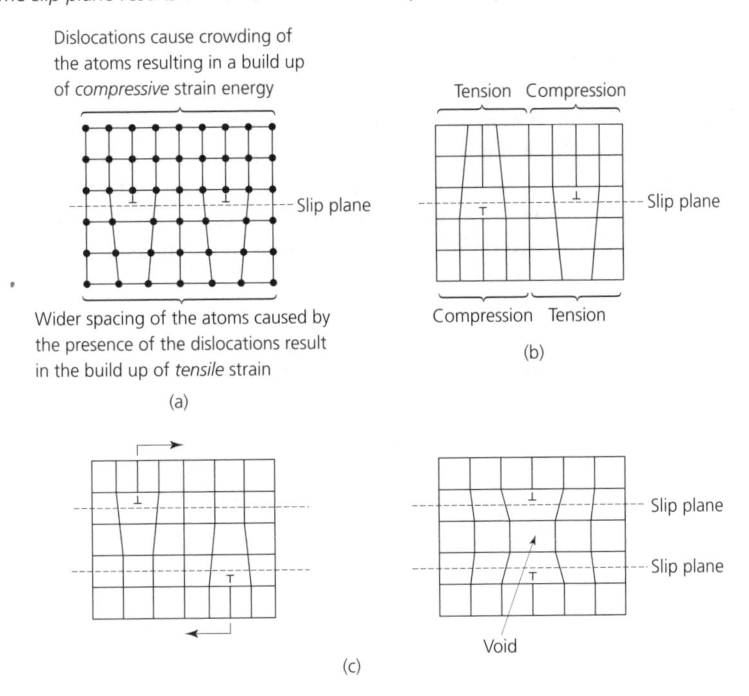

when they become coincident they will cancel each other out (*annihilate* each other) with a corresponding reduction in strain energy. Under such conditions the applied stress required to produce a given degree of slip will be of reduced magnitude compared to that required in (1) above.

3. If the same condition apply as in Fig. 8.14(b), but the slip plane are separated as shown in Fig. 8.14(c), then the dislocations will still annihilate each other but a void type discontinuity will be left in the crystal lattice. Again, there will be a reduction in strain energy in the region.

8.2.3 *Generation of dislocations*

So far only the movement of existing dislocations which have resulted from the solidification process have been considered. However, when metals are formed by plastic deformation not only does slip take place along a number of planes, but these slip planes are visible under quite low magnification. Therefore, to be visible under low magnification not only must the slip planes be quite widely spaced, but the displacement must be of the order of 400 atoms. Since a dislocation running out at the end of a slip plane is only one atomic spacing in depth, a large number of dislocations along the same plane are required to produce a visible 'step' of some 400 atom spacings in magnitude. Therefore, the initial edge dislocation must be capable of reproducing itself many times if it is to produce a magnitude of slip which is visible under low magnification. A possible mechanism by which the multiplication of dislocations can occur is the *Frank–Read Source*. In Fig. 8.15 the initial dislocation is shown restricted at its ends, possibly by other dislocations, grain boundaries or lattice imperfections, so that when a force is applied the ends of the dislocation cannot move. This causes the dislocation to bow outwards. Initially the radius of the bow will be large but as the applied force is increased the radius is reduced. Once the bow has become a semi-circle the applied force has reached a maximum value, after which the force required to maintain growth of the loop can be reduced progressively. Eventually proximity of the strain regions P and Q will result in their annihilation and the dislocation loop will be complete. Since the original dislocation line still remains, the process can repeat itself continuously, as long as an adequate external force is applied, thus allowing the extensive slip necessary for forming by plastic flow to take place.

Fig. 8.15 *The Frank–Read source: (a) dislocation with anchored ends; (b) dislocation starts to bow; (c) dislocation starts to sweep round anchored ends; (d) dislocation loop continues to expand; (e) dislocation loop breaks away; (f) dislocation loop complete but original dislocation line AB remains to reproduce itself again*

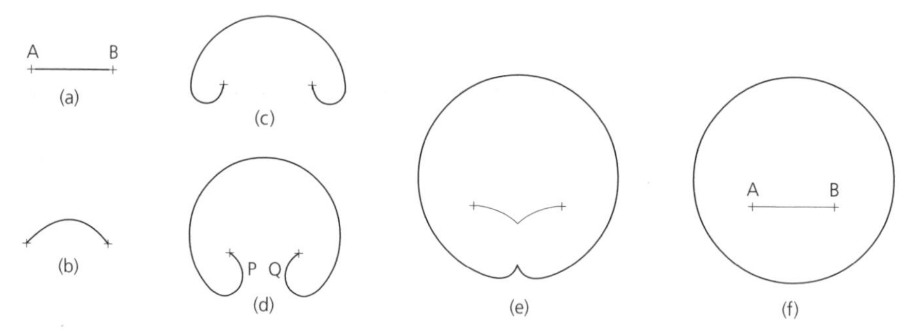

Another mechanism resulting in the self-generation of dislocations is caused by *misorientation* of the crystal lattice at the grain boundaries during solidification of the molten metal. This has already been considered in *Engineering Materials*, Volume 1. When the angle of disorder between adjacent crystals is small, the two crystals may join together along their common boundary resulting in some elastic strain and the generation of edge dislocations.

8.2.4 *Work hardening*

When metals are formed by plastic flow, movement of dislocations along the slip planes (or by twinning) occurs and the number of slip planes available become progressively reduced. As movement of the dislocations becomes blocked, additional force has to be applied to generate new dislocations and to move these until they too become blocked. Eventually, no further slip or twinning can take place and the metal has reached a state of maximum tensile strength and work hardness but minimum ductility. Any further increase in applied force can only result in fracture.

In this condition, the metal is in a state of considerable mechanical stress resulting from internally balanced (locked-up) elastic stain energy. To avoid fracture in service or to allow further processing by plastic flow, the metal needs to be stress relieved by heat treatment.

8.2.5 *Dispersion hardening*

Any mechanism that interferes with the movement of dislocations along the slip planes of a metal results in raising the yield stress of that metal and also its hardness. Thus the dispersion of fine particles throughout a metal will increase its strength and hardness. For example, 'tough-pitch' copper is strengthened and hardened by the dispersion of particles of copper oxide. Further, heat-treatable aluminium alloys are hardened by the precipitation of intermetallic particles that hinder the movement of dislocations. The alloying of metals can introduce alien atoms into the crystal lattice to form interstitial or substitutional solid solutions. Again, such point defects hinder the movement of dislocations, resulting in an increase in the yield stress and hardness of the metal.

8.2.6 *Stress relief and recrystallisation*

The stress relief and recrystallisation (annealing) processes for work-hardened metals have already been introduced in Volume 1. The processes so described will now be reconsidered in terms of dislocation theory. Reference to Fig. 8.14 shows that in the vicinity of an edge dislocation there are regions of tension and compression in the crystal lattice. At the point where the lattice is in tension the region will possess high strain energy. Further, high potential energy is associated with the dislocations congregating at the grain boundaries due to misorientation and to 'misfits' between adjacent crystals as previously discussed. These high-energy regions initiate seed crystals (nucleation) during the annealing processes.

When the temperature of cold-worked metals is raised, the first change to take place in the metal is stress relief. The increased vibration of the atoms associated with the rising temperature allows them to approach their equilibrium positions. There is no change in the distortion of the grain structure and no reduction in the hardness and tensile strength of the metal. In fact, in the case of α-brass, not only is the tensile strength of the metal slightly increased by stress-relief annealing, but the corrosion resistance of the metal is also improved. The intergranular corrosion of severely cold-worked metals due to the congregation of edge dislocations was discussed in Section 4.8.

At the relatively low temperatures associated with stress relief only 'glide' of the dislocations along the slip planes can occur. However, if the temperature is raised, edge dislocations actually start to move out of their slip planes by a mechanism known as *climb*. This is shown diagrammatically in Fig. 8.16, where it can be seen that the last row of atoms of an edge dislocation moves to a new location. This mechanism is known as *positive climb* and results in a reduction in strain energy in the region adjacent to the dislocation. If the last row of atoms of an edge dislocation is augmented, as shown in Fig. 8.16(c), then the mechanism is known as *negative climb*, resulting in an increase in the strain energy in the region adjacent to the dislocation. Whole rows of atoms do not move simultaneously during climb, any more than they move en bloc during slip. It is assumed that the atoms move singly or in small groups in a series of 'jogs' until the total movement is complete. When heated, the movement (diffusion) of interstitial atoms in an alloy can also give rise to positive and negative climb. When heating of a cold-worked metal results in stress relief, it can be assumed that positive climb is taking place since there is a reduction in locked-up internal strain energy.

Fig. 8.16 *Dislocation climb: (a) positive climb by vacancy diffusion; (b) positive climb (interstitial atom); (c) negative climb (interstitial atom)*

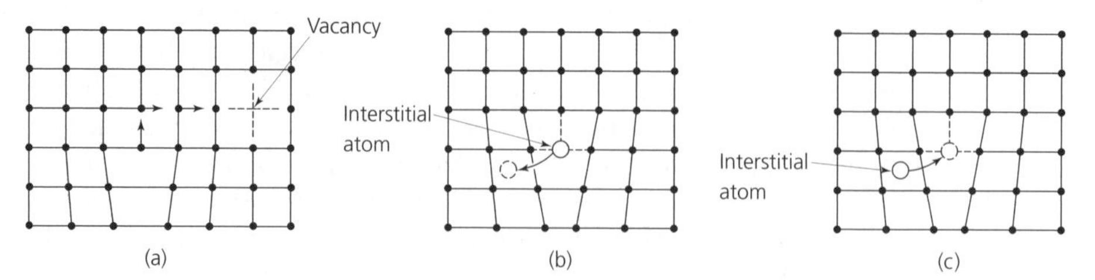

If the temperature is raised still higher to that required for annealing as described in *Engineering Materials*, Volume 1, the elastic strains are totally eliminated and recrystallisation takes place, with seed crystals being initiated at the most heavily deformed and stressed areas. If the annealing temperature is maintained, grain growth occurs by the merging of adjacent grains. The annealing reduces the strength and hardness of the metal but improves its ductility and malleability so that it can again be cold worked.

8.3 Deformation of materials (polymeric)

The molecular structures of polymeric materials and the properties of such materials have been discussed previously in *Engineering Materials*, Volume 1. The atoms in a polymer molecule are arranged in a more complex manner without the regular geometrical alignment of a metal crystal. Some polymeric materials, however, do show a degree of crystallinity in the form of crystallites and this has a marked effect on their properties.

Polymeric materials, which can be shaped by plastic deformation, will have molecular chains with little branching and few, if any, cross-linkages. Such materials can be considered as consisting of linear molecular chains which are intertwined in a random manner, as shown in Fig. 8.17(a). Single bonds in a polymer chain are capable of rotation, providing sufficient thermal energy is applied. When the temperature of the material is raised sufficiently to supply the necessary thermal energy, a 'writhing' movement of the molecules will take place continuously throughout the material, giving it great flexibility. Hence, polymeric materials are usually hot formed whilst the molecules are in this mobile condition. (Heat bending of rigid PVC sheets.)

There are two mechanisms by which linear polymers can be deformed. The applied stress may either straighten the molecules, as shown in Fig. 8.17(b), or it may cause molecular slip, as shown in Fig. 8.17(c). Elastomers are deformed by straightening their tightly curled molecules. When the applied stress is removed the molecules curl up again and the material returns to its original size and shape. When the changes take place instantaneously, the material is said to be an elastomer.

When the applied stress causes molecular slip, permanent plastic deformation takes place and the process is not reversible since little or no elastic strain occurs. In this instance, the applied stress must be sufficient to overcome the van der Waals forces between the molecules. The introduction of a plasticiser, which separates the molecular chains and weakens the van der Waals forces, acts as a lubricant, and aids the deformation of plastic materials. The molecular slip does not occur instantaneously, as in the straightening of elastomer molecules, but is time-dependent upon the viscosity of the material and the distortion is said to be *viscoelastic*.

It has previously been stated that, for plastic deformation to occur, molecular branching and linking should be on a limited scale or not present at all. Where branching and linking are present, or when thermoplastics are below their glass-transition temperature, all the atoms are held together either by chemical bonds or intermolecular forces. Under such conditions only limited elastic deformation and recovery are possible and, if the applied stress is too great, the material will shatter.

Fig. 8.17 *Deformation of polymeric materials: (a) no stress applied; (b) deformation by molecular straightening; (c) deformation by molecular slip*

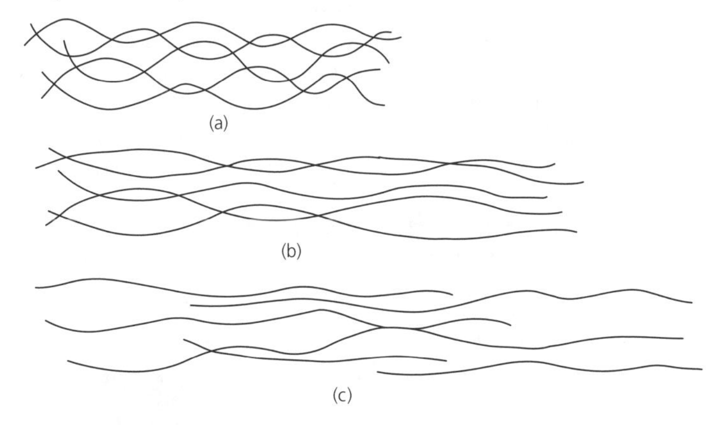

8.4 Failure of metals (fracture)

The failure of metals through fracture is a complex subject that can only be dealt with briefly in this section. Fracture is said to have occurred when there is cohesive failure of a metal resulting in the metal separating into two or more portions.

When an external force (load) is applied to a metal component, the metal is put into a state of stress, and a corresponding elastic and/or plastic strain occurs. If the stress is increased sufficiently the metal will eventually fracture. Fracture is described in various ways depending upon the behaviour of the material under stress or upon the mechanism of fracture, or even its appearance. For example, the fracture may be classified as *ductile* or as *brittle* depending upon whether or not plastic deformation precedes cohesive failure. The fracture may also be classified as *shear* if slip occurs in the crystal lattice immediately prior to failure. It may be described as *cleavage* if there is cohesive failure of the chemical bonds holding the planes of atoms together in the crystal lattice, as was shown in Fig. 5.3. Again, the fracture may be classified as *fibrous* from the dull fibrous appearance of the fractured surfaces when plastic deformation has preceded failure, or as *granular* from the shiny granular appearance of the fractured surfaces when cleavage has occurred. Thus, ductile fracture, shear fracture and fibrous fracture may all be considered as alternative names for failure preceded by plastic deformation. Brittle fracture, granular fracture and cleavage fracture may be considered as alternative names for failure that is *not preceded* by plastic deformation to any appreciable extent, or even at all.

8.5 Ductile fracture

Figure 8.18 shows a typical, ductile tensile specimen during and after testing. The 'necking' of the specimen which occurs before fracture and the typical 'cup and cone' of this type of fracture are clearly shown. Initially, only elastic deformation occurs but, as the applied load is increased, plastic deformation occurs and the 'neck' begins to form. The increased load, coupled with the reduction in cross-sectional area in the region of the neck, results in the specimen being subjected to increased stress. This stress rapidly reaches a magnitude where small internal cavities start to appear. These cavities 'nucleate' (form) more easily if impurities or a second alloy phase is present; thus pure metals tend to be more ductile than impure metals and alloys. The cavities join up to form a crack that is roughly normal to the axis of the testpiece. This weakens the metal still further and concentrates the stress, and simple shear fracture finally occurs at an angle of 45° to the axis of stress.

Fig. 8.18 *Tensile test specimen (ductile): (a) necking immediately prior to fracture; (b) typical ductile fracture showing cup and cone; (c) photograph of a cup-and-cone fracture of a mild steel tensile test specimen*

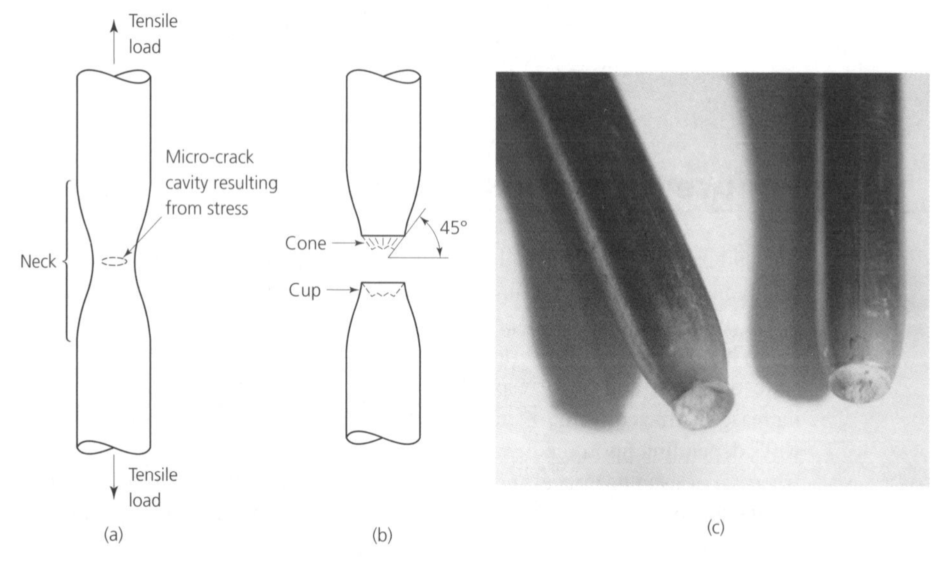

Since ductile fracture is preceded by plastic deformation, it is reasonable to assume that the normal mechanisms of dislocation and slip occur in the crystal structure of the metal. Figure 8.19 shows how dislocations 'pile up' when they are obstructed by a separate alloy phase or by an impurity. Such a 'pile up' in the metal nucleates micro-cracks which tend to join together (coalesce) and spread until the material is weakened sufficiently for fracture to occur.

Fig. 8.19 *A 'pile-up' of dislocations at an inclusion, leading to the formation of a fissure that will be propagated as the stress increases*

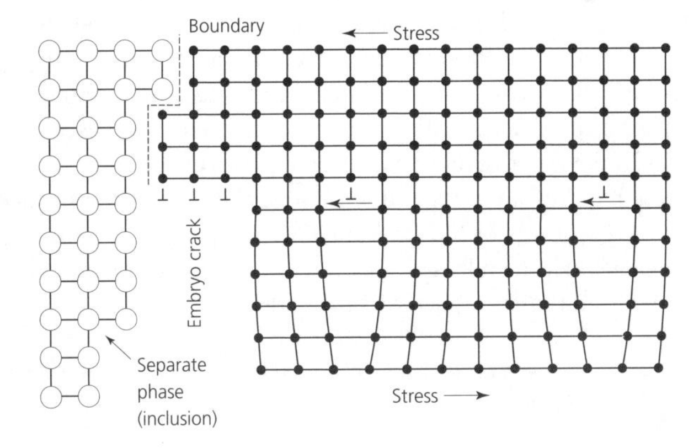

Ductile failure can also occur in compression, although this is not so common since the applied axial force tends to close any internal discontinuities. However, as the metal starts to bulge, axial surface cracks appear around the sides of the component. These cracks reduce the cross-sectional area of the remaining coherent metal and the stress builds up until failure occurs. Such cracks can be seen around the head of a rivet when it has been excessively cold worked or hot worked at too low a temperature.

8.6 Brittle fracture

This type of fracture is associated with non-metals such as glass, concrete and thermosetting plastics. In metals, brittle fracture occurs mainly when body-centred-cubic crystals and close-packed-hexagonal crystals are present, but rarely when face-centred-cubic crystals are present. Specimens fractured in this way show little or no plastic deformation. Figure 8.20 compares the stress–strain curves for a medium carbon steel when it is in the annealed and ductile condition and after it has been quench hardened and is brittle, lacking in ductility.

Fig. 8.20 *Stress–strain curves for a medium carbon steel: (a) in the annealed condition; (b) after quench hardening*

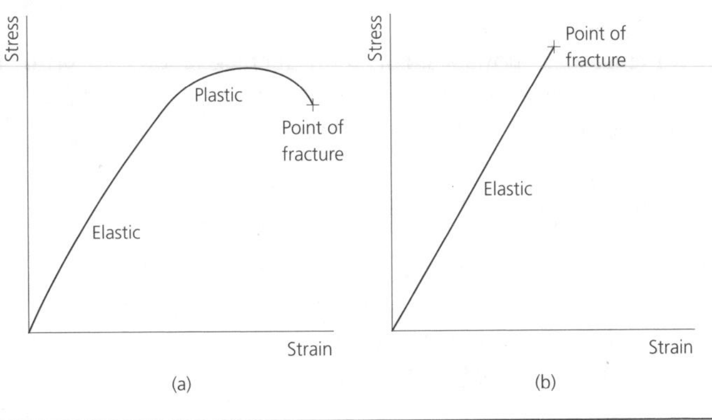

Figure 8.21 shows the crystal lattice for a brittle material in which cleavage failure is commencing. The failure mechanism can best be understood by adopting the following analogy. Assume that the bonds holding the atoms together are made from an elastic material, and that the applied force causes the bonds to become strained elastically (stretched). This results in strain energy being stored within the material. If the applied force is increased sufficiently the bonds will eventually snap with a release of strain energy. However, the crack that appears (Fig. 8.21(a)) creates new surfaces, but the creation of new surfaces requires an energy input. Therefore, it can be assumed that the energy released when the bond was broken is just sufficient to create the surfaces. Similarly, the crack will not propagate if the released elastic strain energy is too low. Figure 8.21(b) shows a broken hammer head that has failed by brittle fracture.

Fig. 8.21 *Crack propagation in a brittle material: (a) crack propagation mechanism; (b) hammer head that has suffered brittle fracture*

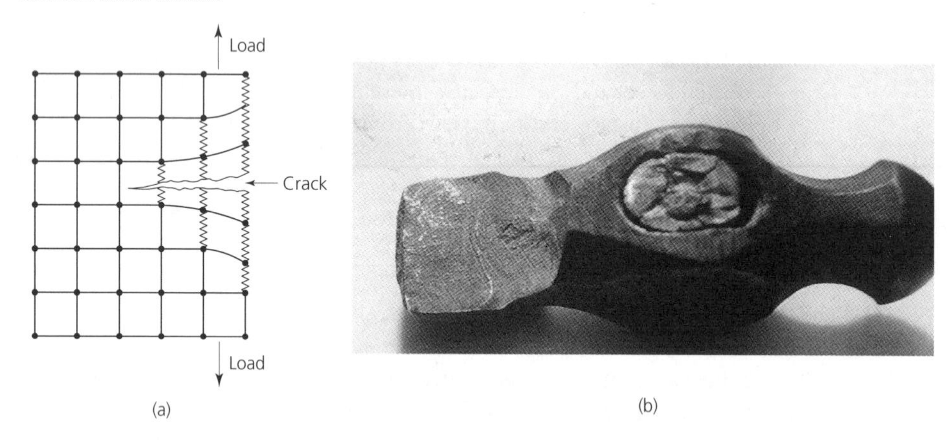

8.7 Griffith crack theory

According to A. A. Griffith, the energy required to fracture a brittle material is not uniformly distributed over the volume of the material since minute faults and cracks at the surface or within the material create regions of energy concentration. Thus the real strength of the material is substantially lower than the theoretical strength of the material, as derived from the inter-atomic forces.

Figure 8.22 shows a small elliptical crack in a rod of brittle non-metallic material. The relationship between the stress required to propagate the crack and the length of the crack can be determined from the expression:

$$\sigma = \sqrt{\frac{2\gamma E}{\pi a}}$$

where σ = crack propagation stress (N/m^2)
γ = surface energy per unit area (J/m^2)
E = tensile modulus (N/m^2)
a = half length for major axis of crack (m)

Note that the multiplying factor 2 is derived from the fact that the crack that is propagated has two surfaces.

When the applied stress reaches the value σ, the micro-cracks present can begin to propagate and coalesce so that a increases. It can be seen that as a increases, the value of σ decreases. Eventually, when fracture occurs and a spreads completely across the section so that the two portions separate, σ becomes zero. For any given situation all but σ and a are constants so, by transposing, we can derive the following proportionality:

$$a \propto 1/\sigma^2$$

Thus the larger the stress, the smaller will be the critical crack length to precipitate catastrophic failure.

Fig. 8.22 *Stress concentration due to a micro-crack*

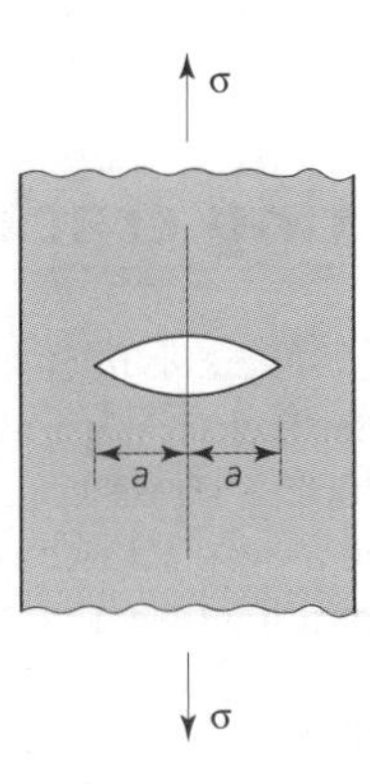

The Griffith relationship was derived for brittle materials such as concrete, glass, ceramics and thermosetting plastics. It is not valid for ductile metals or even for brittle metals where some plastic deformation, however small, always precedes fracture. When applied to metals, the relationship has to be modified to take into account the energy required for plastic deformation, as follows:

$$\sigma = \sqrt{\frac{2E(\gamma + \gamma_p)}{\pi a}}$$

where γ_p = the energy required for plastic deformation per unit area of crack.

Note: For most highly ductile materials (this includes most metals) γ is smaller than γ_p and may be ignored for ease of calculation (see Example 8.1).

EXAMPLE 8.1

A large sheet of glass is supporting a tensile stress of 35 MPa. Given that the surface energy per unit area and modulus of elasticity for this type of glass are 0.25 J/m² and 70 GPa respectively, calculate the maximum length of surface flaw that can be tolerated.

Solution

Given $\sigma = 35 \times 10^6$ Pa, $\gamma = 0.25$ J/m^2 and $E = 70 \times 10^9$ Pa

$$\sigma = \sqrt{\frac{2\gamma E}{\pi a}}$$

Rearranging and substituting in values:

$$a = \frac{2\gamma E}{\pi \sigma^2} = \frac{2 \times 0.25 \times 70 \times 10^9}{\pi \times (35 \times 10^6)^2}$$

$$= 9.1 \times 10^{-6}\ \text{m}$$

$$\mathbf{= 9\ \mu m}$$

8.8 Factors affecting crack formation

The effects of inclusions on crack nucleation and propagation in a material have already been discussed, but other factors must also be taken into account. Any sudden change in section that will produce a stress concentration can lead to cracking. Under highly stressed conditions even a deep scratch or a tooling mark can lead to crack propagation. Faults in design or manufacture that cause stress concentrations are referred to as *incipient cracks.*

The amount by which the stress is raised is proportional to the depth of the notch and inversely proportional to the radius at the root of the notch. Cracks in brittle materials have a very small radius, so the stress concentration at the root of the crack is very high. The propagation of the crack can be arrested by drilling a hole at its root. The effect of this is to increase its effective radius and thus reduce the stress concentration. Early attempts to produce ships with welded hulls led to disaster when cracks spread unhindered so that the ships broke up. Ships with riveted hulls did not suffer in this manner since any crack stopped at the next adjacent rivet hole. There is an approximate relationship between the applied stress and the stress at the root of the notch (Fig. 8.23) given by the expression:

$$\sigma_n = \sigma\left(1 + 2\sqrt{\frac{a}{\rho}}\right)$$

where σ_n = stress at root of notch
σ = applied stress
a = length of notch
ρ = root radius of notch.

Thus the increase in stress due to the notch is given by the expression:

$$\sigma_i = 2\sqrt{\frac{a}{\rho}}$$

where σ_i = increase in stress (See Example 8.2).

Fig. 8.23 *Dimensions of a typical notch*

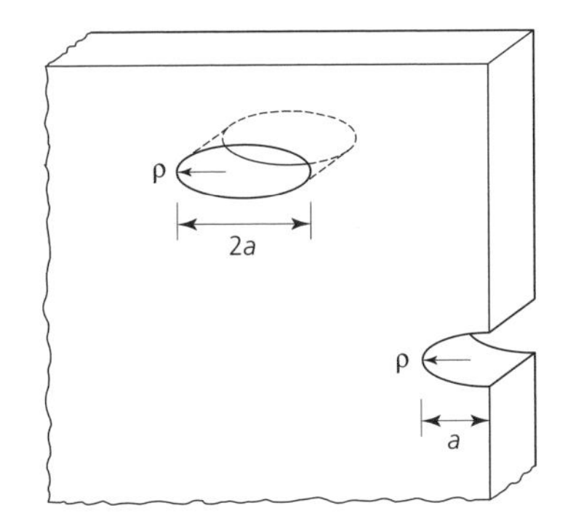

EXAMPLE 8.2

A steel plate has a slot milled for the attachment of a ladder support. Calculate the stress local to the slot if its dimensions are 25 mm long and 12 mm wide when an overall stress of 45 MPa is applied to the steel.

Solution

Given $2a = 0.025\,\text{m}$, $\rho = 0.006\,\text{m}$ and $\sigma_{app} = 45 \times 10^6\,\text{MPa}$

$$\sigma_n = \sigma_{app}\left(1 = 2\sqrt{\frac{a}{\rho}}\right)$$

Substituting in values

$$\sigma_n = 45 \times 10^6\left(1 + 2\sqrt{\frac{0.0125}{0.006}}\right)$$

$$= 174.9 \times 10^6\,\text{Pa}$$

$$= \mathbf{175\,MPa}$$

In ductile materials, the stress concentration at the root of the notch is less likely to lead to crack propagation and failure since much of the available energy is dissipated in plastic flow and insufficient energy may be available for the creation of new crack surfaces. Thus cracks in ductile metals are less likely to result in fracture than cracks in brittle metals.

The speed of loading can also affect the behaviour of the metal, and this was introduced when discussing material testing in *Engineering Materials*, Volume 1. This applied to tensile testing and, particularly, to Izod and Charpy impact testing where a notched testpiece is loaded by a sudden blow. If the loading is applied too rapidly to an otherwise ductile metal there may not be time for the dislocations associated with plastic deformation to take place and the metal will behave as though it were brittle.

Temperature can also have a profound effect on the behaviour of metals. As the temperature decreases, the movement of dislocations becomes more sluggish so that the internal stresses may exceed the shear strength for the metal. Therefore brittle fracture is more likely to occur at low temperatures and may even occur in otherwise ductile metals if the temperature is sufficiently low. This becomes an important design consideration when selecting materials for equipment operating under arctic conditions and for space vehicles. Metals with body-centred-cubic crystals and close-packed-hexagonal crystals such as beryllium, chromium, molybdenum, plain carbon steels (particularly the ferrite content), tungsten and zinc all suffer from low-temperature brittleness.

Figure 8.24 shows the effect of temperature on a low-carbon steel when it is subjected to impact loading. At low temperatures the metal has a BCC ferrite structure and, as stated above, this exhibits 100 per cent brittle fracture at low temperatures. As the temperature rises the dislocations become less sluggish so that some ductile fracture can occur even in BCC ferrite. At the transition temperature the failed surface will show 50 per cent brittle fracture and 50 per cent ductile fracture. As the temperature is raised still further, the failed surface will show 100 per cent ductile fracture. The presence of carbon, phosphorus and nitrogen raises the transition temperature for mild steel, and for this reason such impurities must be kept to a minimum when the steel has to operate at low temperatures. However, alloying elements such as nickel and manganese depress the transition temperature. Steel plates suitable for the welded hulls of ships, operating in the cold North Atlantic, will contain only 0.14 per cent carbon but as much as 1.3 per cent manganese.

Fig. 8.24 *Effect of temperature on the fracture of a low-carbon steel*

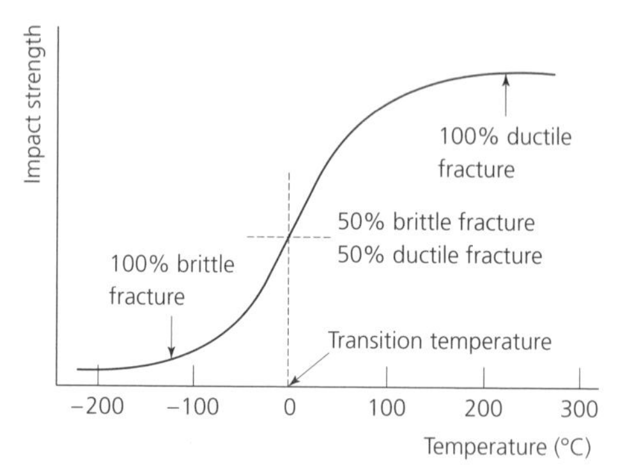

8.9 Fracture toughness

The factors affecting fracture have just been considered in Section 8.18, including the effect of temperature on impact toughness. Toughness is defined as the ability of a material to resist transverse impact loading without failure. In standard tests such as the Izod test and the Charpy test, which were described in *Engineering Materials*, Volume 1, a notched specimen is struck by a pendulum and the energy absorbed in bending and/or breaking the specimen is stated as a measure of the fracture toughness of the material.

It has been stated earlier that for a crack to propagate the critical elastic strain energy released at the root of the crack or notch must equal energy to create new surfaces. Thus *fracture toughness* can be defined in terms of the critical elastic strain release rate, G_c, measured in kJ/m^2. To determine G_c the Griffith relationship discussed earlier can be modified as follows:

$$\sigma = \sqrt{\frac{G_c E}{\pi a}}$$

where $G_c = 2\gamma$ (brittle materials)
$= 2(\gamma + \gamma_p)$ (ductile materials)

The greater the value of G_c, the greater is the fracture toughness of the material. Fracture toughness can also be defined in terms of the *stress intensity factor*, K, determined at the root of the crack or notch. Crack propagation occurs when the stress intensity factor reaches a critical value:

K_c = measured in $MN/m^{1.5}$, or $MPa\sqrt{m}$

where $K_c^2 = G_c E$ and $\sigma = \sqrt{\frac{K_c}{\pi a}}$

The greater the value of K_c, the greater the fracture toughness of the material. As well as the properties of the material, its thickness must also be taken into account. Figure 8.25(a) compares the fracture of thick and thin plates. Thin plates, subjected to axial tensile stress, σ, fail in shear at 45° to the axis of loading. However, thicker plates show a central, flat area where fracture is normal to the axis of loading. The thicker the plate the greater is this normal area and the greater becomes its influence on the stress intensity factor. Figure 8.25(b) shows how K_c has a high value for thin plate but that the value of K_c becomes less for thick plate, until the curve almost levels out at some value K_{1c} (pronounced kay-one-cee). This lower limiting value of the critical stress intensity factor is referred to as the *plane strain fracture toughness*. In addition to plate thickness, fracture toughness is also affected by such factors as:

- *Composition* The presence of alloying elements and/or impurities can have a significant effect upon the plane strain toughness factor. Impurities such as phosphorus and sulphur, even in small amounts, seriously reduce the fracture toughness of ferrous metals.
- *Heat treatment* Quench hardening plain carbon and alloy steels substantially reduces their fracture toughness. However, subsequent tempering progressively restores the fracture toughness – as the tempering temperature is raised, the fracture toughness is restored more rapidly than the hardness is reduced. The fracture toughness of heat-treatable aluminium alloys is also affected by solution treatment and precipitation hardening. The fracture toughness of other non-ferrous metals and alloys is affected by work hardening during processing and by stress relief annealing.
- *Service conditions* The fracture toughness value can also be affected by such factors as temperature, environment (corrosion) and cyclical loading (fatigue).

Fig. 8.25 *Effect of plate thickness on crack propagation: (a) effect of plate thickness on crack profile; (b) effect of plate thickness on stress intensity factor*

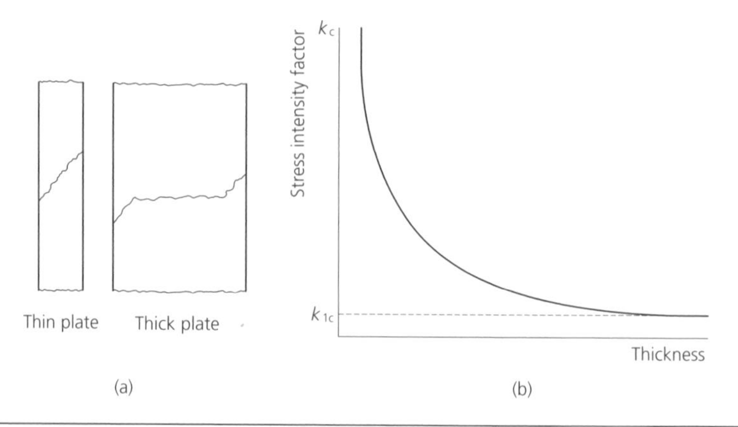

EXAMPLE 8.3

A plate made from a medium-carbon steel with a plane strain fracture toughness of 51 MPa√m is found on inspection to have a through thickness defect. Calculate the maximum tolerable size of this defect if the plate is to withstand a stress of 350 MPa.

Solution

Given $K_{1c} = 51 \times 10^6$ Pa and $\sigma = 350 \times 10^6$ Pa

$$\sigma = \frac{K_{1c}}{\sqrt{\pi a}}$$

rearranging and substituting in values:

$$a = \frac{1}{\pi}\left(\frac{K_{1c}}{\sigma}\right)^2 = \frac{1}{\pi}\left(\frac{51 \times 10^6}{350 \times 10^6}\right)^2$$

$$= 0.00676\,\text{m}$$

$$\mathbf{= 6.7\ \mu m}$$

SELF-ASSESSMENT TASK 8.2

1. Taking the expression for the stress at the root of a notch (page 208), put in the dimensions to the situation as if the notch is a circular hole and show that a hole drilled in a plate causes a three-fold local increase in stress.
2. Alumina (Al_2O_3) having a tensile strength of 210 MPa, contains a flaw of dimensions length 0.1 mm with a tip radius 5 μm. What is the maximum tensile stress that could be applied to the materials. (Warning: keep an eye on your units.)
3. A piece of mild steel at room temperature has a plastic deformation energy of 120 kJ/m^2. Given a modulus of elasticity of 200 GPa, calculate the size of the largest surface crack that can be tolerated at a stress of 200 MPa.

4. A piece of mild steel at subzero temperatures has a plastic deformation energy of $10\,J/m^2$. Given a modulus of elasticity of 200 GPa, calculate the size of the largest surface crack that can be tolerated at a stress of 200 MPa.
5. Discuss the difference between your answers to Questions 3 and 4.

8.10 Failure of metals (fatigue)

Since it is estimated that more than 75 per cent of failures in engineering components can be attributed to fatigue failure, and as the reliability and performance expected from engineering products rises, the need to understand this mode of failure becomes increasingly important. Fatigue failure, the factors affecting fatigue and fatigue testing for both metals and polymers were introduced in *Engineering Materials*, Volume 1. Let's now consider the effects of mean stress.

Figure 8.26 shows the conditions for fatigue loading. With a fluctuating load the mean stress is greater than the stress range; with a pulsating or repeated load the mean stress is equal to half the stress range; and with an alternating load the mean stress is zero. In each instance the stress amplitude is half the stress range. Figure 8.27 shows typical *S–N* diagrams, as considered in Volume 1, where the testpiece was subjected to an alternating load. However, this can be misleading if there is a steady-state component stress (mean stress). Both Goodman and Soderberg (see diagrams in Fig. 8.28) investigated the relationship between stress amplitude, mean stress and fatigue limit. When the mean stress

Fig. 8.26 *Conditions of fatigue loading: (a) fluctuating load; (b) pulsating or repeated load; (c) alternating load*

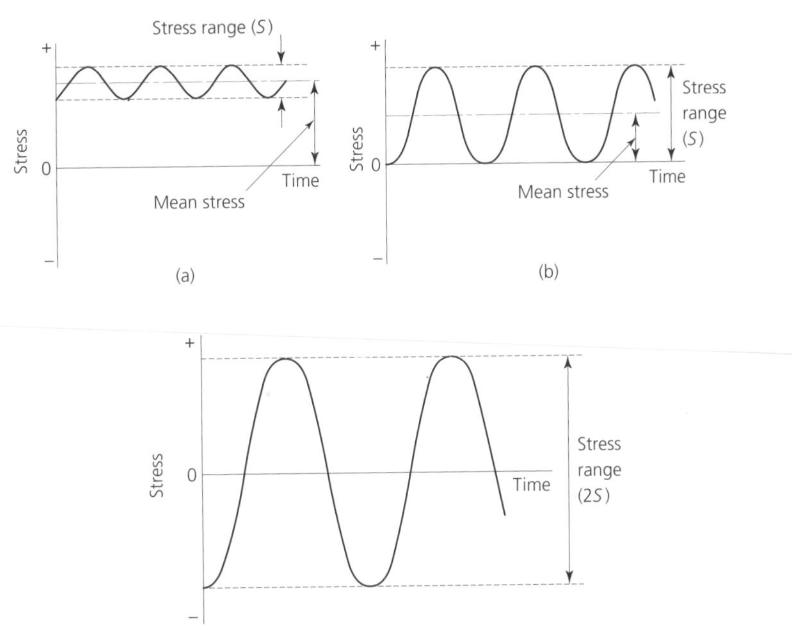

is zero (perfect alternation) the fatigue limit is at a maximum value before failure occurs. However, if a steady-state stress is superimposed upon the cyclical stress then this must also be taken into account. This steady-state stress is the mean stress previously shown in Fig. 8.26.

Fig. 8.27 *Stress reversal curves (S–N diagrams): (a) S–N diagram for a typical steel; (b) S–N diagram for a typical non-ferrous alloy*

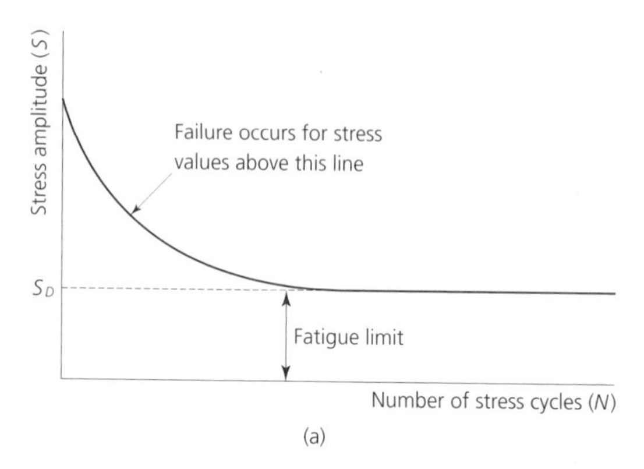

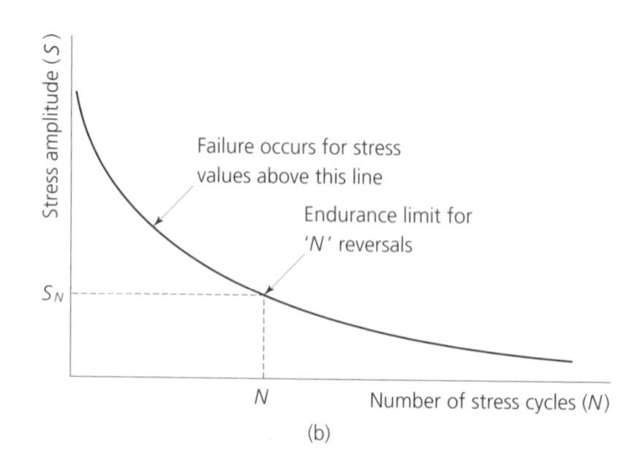

Fig. 8.28 *The effect of mean stress: (a) Goodman diagram; (b) Soderberg diagram*

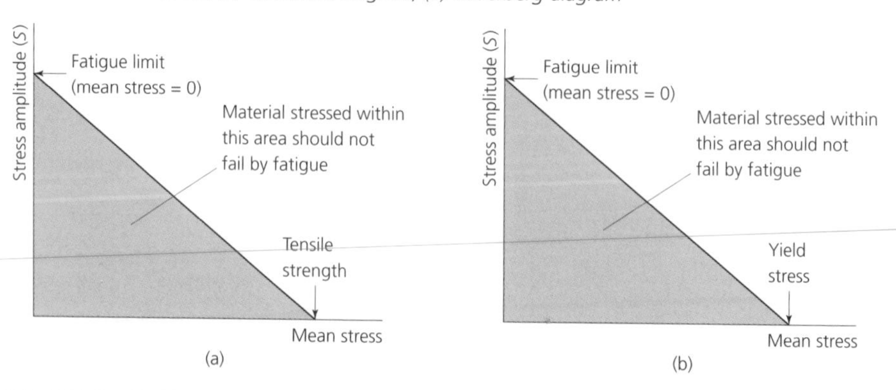

In the *Goodman* diagram the fatigue limit is zero when the mean stress is equal to the tensile strength of the material, since the material will fail at this value before any cyclical loading can commence. Therefore, if the point representing the stress amplitude and mean stress for any given set of conditions lies within the area bounded by the axes and the 'Goodman line', the shaded area, then according to the Goodman relationship the material should not fail by fatigue.

In the *Soderberg* diagram the fatigue limit is zero when the mean stress is equal to the yield stress of the material. Again the point representing the stress amplitude and mean stress for the material must lie within the shaded area bounded by the axes and the 'Soderberg line' if failure by fatigue is to be avoided.

Since perfect alternation (zero mean stress) rarely occurs in practice, *S–N* curves should not be used alone without consideration of the mean stress. Care must also be taken when using the Goodman or Soderberg diagrams since they tend to give a low value of fatigue limit for ductile materials and a high value of fatigue limit for brittle materials.

Cumulative fatigue damage must also be taken into account. So far prediction of fatigue failure has only been considered in terms of *S–N* curves and the Goodman and Soderberg diagrams. These are based on constant amplitude tests, but such conditions seldom apply in practice where the stress amplitude and any component of mean stress may constantly vary in an unpredictable manner. Such a spread of stress variables is referred to as 'spectrum loading'.

To predict the effects of spectrum loading, we make the assumption that a given stress amplitude and number of stress cycles will result in a certain amount of permanent fatigue damage. We also assume that subsequent operation at a different stress amplitude and number of stress cycles will produce additional fatigue damage. Therefore, a sequential accumulation of fatigue stress damage occurs until a critical level of damage is reached when the material fails.

A simple application of the above theory is *Miner's law*, also known as the *linear damage rule*. This assumes that n_1 cycles at a stress of σ_1, for which the average number of cycles to failure is N_1, causes an amount of damage equal to n_1/N_1 (where n/N is referred to as the 'damage fraction' or the 'cycle ratio'). Failure is predicted to occur when the sum of the damage fractions equates to unity, thus:

$$\frac{n_1}{N_1} + \frac{n_2}{N_2} + \frac{n_3}{N_3} + \frac{n_4}{N_4} + \ldots = 1$$

Despite its simplicity, Miner's law is easy to apply and gives a reasonable level of prediction reliability. However, the ratios n/N are not easy to determine and, with the availability of computer-modelling techniques, more accurate predictions can be made using simulated service conditions which approach close to reality.

The factors affecting fatigue failure have already been discussed in *Engineering Materials*, Volume 1, and can be summarised as:

- Design (e.g. effect of sharp corners and sudden changes of section).
- Surface finish (e.g. surface cracks and tooling marks).
- Ambient temperature.
- Residual stress.
- Corrosion.

8.11 Failure of metals (creep)

Creep can be defined as the gradual extension of material under a constant applied load over a prolonged period of time, particularly at elevated temperatures. Therefore it is a phenomenon that must be considered in the choice of materials that are required to work continuously at elevated temperatures – for example, the turbine blades of jet engines and gas turbines.

Figure 8.29 shows a typical creep curve for metallic materials. The three periods of creep have already been defined in *Engineering Materials*, Volume 1. The creep curve for any given metal will change with any variation in applied stress and/or temperature, and this is shown in Fig. 8.30.

Fig. 8.29 *Creep*

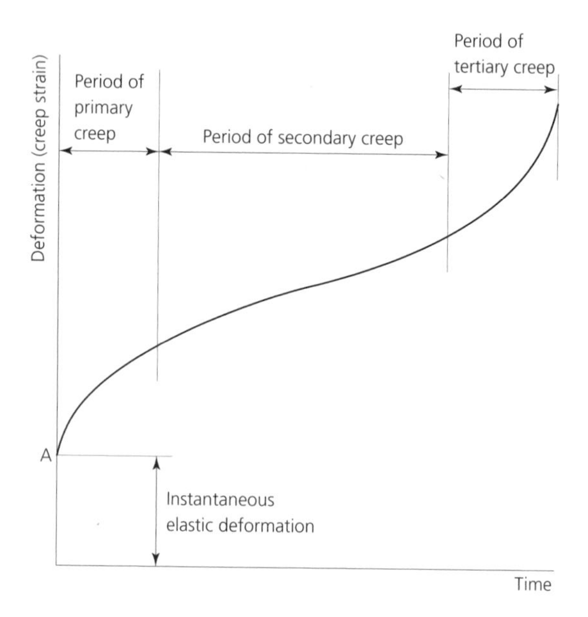

Fig. 8.30 *Effect of stress or temperature on creep rate*

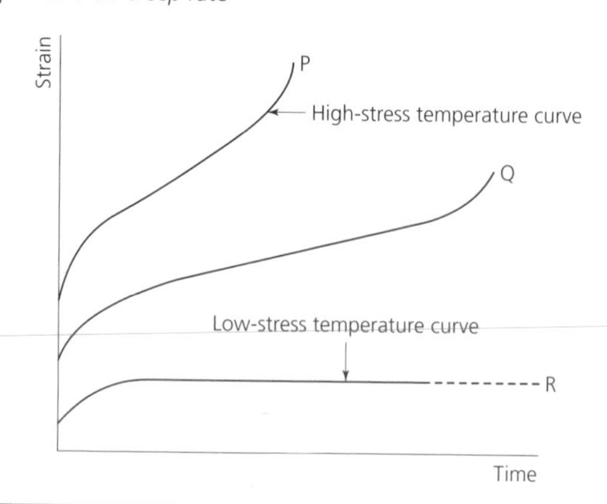

The variation in the curves is best understood by examining the mechanisms of creep. Creep is due to:

- Plastic deformation caused by normal dislocation along slip planes in crystalline materials (metals).
- Plastic deformation caused by viscous flow at the grain boundaries where misorientation occurs.

Consider curve P in Fig. 8.30. At first, plastic deformation resulting in primary creep is rapid as dislocation can take place relatively unhindered. This is because the thermal agitation, due to the elevated temperature of the metal, enables such barriers as solute atoms and dislocation pile-ups at the grain boundaries to be overcome. However, whilst the temperature of the metal and therefore the thermal agitation remains constant, the barriers to dislocation build up and the rate of creep decreases.

During secondary creep, plastic deformation continues more slowly as the work-hardening effect of the deformation increases but is offset, to some extent, by the recovery processes associated with the elevated temperature of the metal. This balance gives rise to a constant creep rate. Deformation by viscous flow as well as by dislocation occurs during secondary creep. A high level of stress has the same effect on the creep curve as a high temperature since a high level of stress can overcome the barriers to dislocation for a given level of thermal agitation and recovery.

During tertiary creep micro-cracks begin to appear at the grain boundaries as the barriers to dislocation movement become too great for the thermal agitation and the applied stress to overcome. These micro-cracks result in a rapid reduction in cross-section (necking) leading to a rapid increase in creep rate and fracture (creep failure).

Consider curve Q in Fig. 8.30. The creep rate is obviously lower, resulting from a lower applied stress and/or lower temperature. Not only does the lower stress reduce the ability of dislocations to overcome such barriers as solute atoms and dislocation pile-ups at the grain boundaries, but the movement of the dislocations, themselves, will be more sluggish because of the lower temperature.

Consider curve R in Fig. 8.30. Here the temperature is too low for recovery to occur and the work hardening which occurred during primary creep results in the stiffness of the metal becoming too great to be overcome by the limited applied stress and creep becomes negligible.

8.11.1 *Creep resistance (metals)*

It has already been shown that creep is dependent upon the movement of dislocations, therefore anything that can limit this movement and the formation of additional dislocations will reduce creep. Metals with close-packed crystal structures (FCC and CPH) are intrinsically resistant to creep, and their resistance may be improved still further by:

- Adding an alloying element to form a solid solution, providing the solute atoms have low mobility. Note that atoms that diffuse easily increase the susceptibility of the metal to creep.

- Adding an alloying element which will promote particle (dispersion) hardening by precipitation treatment. This is the technique used in the 'Nimonic' range of alloys developed for gas turbine blades.
- Increasing the grain size of the metal by heat treatment in order to reduce the total grain boundary area per unit volume. This retards creep at high temperatures where movement occurs mostly through viscous flow at the grain boundaries. At low temperatures a fine-grained structure is more creep resistant since the increased number of grain boundaries obstruct the spread of dislocations.

8.11.2 *Creep Limit*

It should now be apparent that conventional, short-duration, tensile testing does not provide adequate design information for components and assemblies where static and dynamic loads have to be sustained at high temperatures over prolonged periods of time. In creep testing, the specimen is subjected to a constant tensile load whilst it is maintained at a constant high temperature. Usually several specimens of the same metal are tested at the same temperature but with different applied loads. The maximum stress that can be applied with no measurable creep is called the *creep limit* or *creep stress.*

This is not a particularly satisfactory method of determining the creep resistance of a metal since the test may take several months to complete and the results are dependent upon the sensitivity and accuracy of the extensometers used as the extensions can be extremely small. A more useful measure of creep resistance is the rupture stress for a specified time and temperature – for example, at 400 °C the rupture stress for a low-carbon steel life of 10,000 hours is 220 MPa, but at 500 °C the rupture stress is limited to only 60 MPa.

Alternatively the *Larson–Miller parameter* may be used to estimate rupture time for a given stress, or the rupture stress for any given time. A curve, similar to that shown in Fig. 8.31, has to be prepared for each material using experimental data in conjunction with the relationship:

$$P = \theta(C + \log_e t)$$

Fig. 8.31 *Larson–Miller parameter curve*

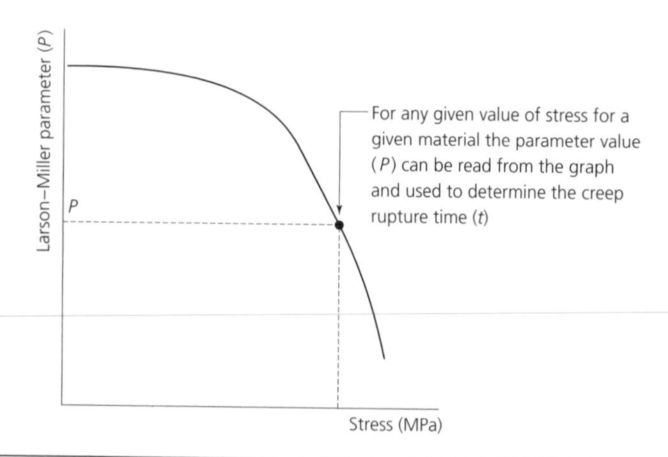

where P = the Larson–Miller parameter
θ = the service temperature (K)
C = constant (usually 20)
t = the time to rupture (hours)

To use this expression to estimate the time to rupture for a given stress, it is rearranged to make 't' the subject, thus:

$$\log_e t = \left(\frac{P}{\theta}\right) - C$$

8.12 Failure of polymers (creep)

Creep in metals usually occurs at high temperatures, whereas polymeric materials are subject to creep effects even under normal ambient conditions. Figure 8.32 shows typical creep curves for cellulose acetate at a temperature of 25 °C. It can be seen that the time to rupture is very much shorter than for a metal and that the degree of creep is very much greater than for a metal. These curves are typical for most thermoplastics.

As for metals, the creep properties of plastics are also largely dependent upon service temperature. Below the glass transition temperature, T_g, the material will be rigid and the creep rate will be low. As the temperature rises, the creep rate and the elongation will increase. Above the glass transition temperature, T_g, viscoelastic deformation occurs resulting in greater elongation, but at a lower creep rate, for a given applied stress.

Fig. 8.32 *Typical creep values (cellulose acetate)*

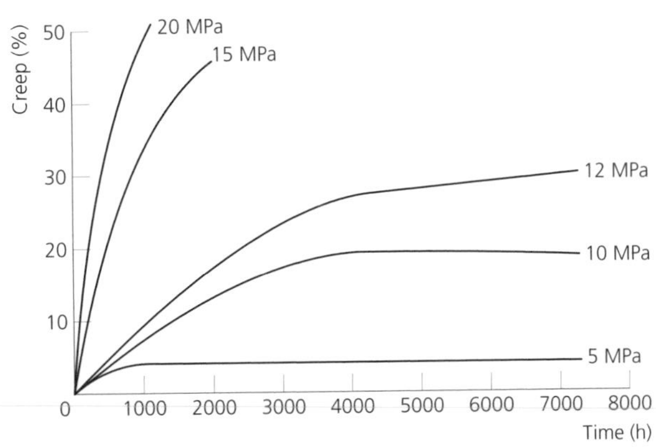

SELF-ASSESSMENT TASK 8.3

1. Regarding fatigue loading, sketch two graphs of stress against time clearly showing the difference between a pulsating load and an alternating load.
2. Describe how the formula $P = \theta(C + \log_e t)$ can be used together with a diagram such as Fig. 8.30 to work out the time to rupture, given the value of the rupture stress.

3. Work out the Larson–Miller parameter given the following:
   rupture time = 10,000 hours, temperature = 400 °C, $C = 20$.
4. Creep can cause failure in metals and polymers. Compare and contrast the behaviour of these materials in this mode of failure.

8.13 Material specification

The specification of a material for any given engineering application involves the consideration of three parameters:

- Properties.
- Processing.
- Commercial viability.

Inevitably the final choice will be a compromise between these parameters.

8.13.1 *Properties (general)*

Throughout this book, and in Volume 1, the general properties of the main groups of materials have been considered together with an understanding of how those properties are related to the physical and chemical structure of the materials, and how the properties may be modified by processing (e.g. plastic deformation, heat treatment, anti-corrosion finishing). From this general knowledge, and from a knowledge of the design and service requirements, it is possible to refer to British and International Standards and manufacturers' data sheets in order to draw up a 'short list' of potentially suitable materials for further consideration. To select the optimum material from the short list a number of techniques can be used, and some of the more important of these will now be considered.

8.13.2 *Critical properties*

For some applications a certain property or combination of properties may be paramount and may over-ride considerations of material cost and processing costs. An example would be the choice of the optimum material for a key aircraft component where failure could result in a major accident resulting in heavy loss of life, massive financial compensation claims and legal costs.

8.13.3 *Property limits*

There are three property limits to be considered:

- *Upper limit* Above this limit the component will be 'over-engineered'. This is not only wasteful of resources, but may unnecessarily increase the cost of processing, and can result in an excessively high unit material cost. Materials with properties above this upper limit can be excluded from consideration.

- *Lower limit* Below this limit the component will not function correctly, will not have an adequate service life, and may prove to be unsafe. Materials with properties below the lower limit must on no account be selected or even considered, however low in cost.
- *Target value* As its name implies, this is a property value or set of values that the chosen material should achieve within close limits set by the designer.

8.13.4 *Cost per unit property*

Materials are compared with a particular critical property or set of properties on a 'cost per unit property' basis. Thus, if the requirement of a material for a particular application in an aircraft or motor vehicle is its fatigue strength/density ratio, then a group of suitable materials with the required properties would be evaluated on a relative cost per unit property basis.

8.13.5 Processing

Unless the production run warrants investment in new plant, the materials chosen must be suitable for processing on the existing plant. Alternatively, some of the processing may have to be subcontracted to firms who have suitable facilities.

8.13.6 *Commercial viability*

This involves a number of compromises between:

- Cost, including unit material cost, processing cost, after sales servicing costs and warranty claims.
- Availability, use of standard materials and standard sizes wherever possible.
- Repeatability of quality and reliability in service.
- Customer satisfaction, customer loyalty and customer safety.

8.13.7 *Merit rating*

This is a property 'handicapping' exercise. Each property for each material is given a relative merit score. This score is then multiplied by a predetermined 'weighting factor' depending upon the relative importance of the property for the application under consideration. The merit rating is the sum of the individual weighted scores, and the material that is finally chosen is the one with the highest merit rating.

This procedure does not take into account the relative costs of the materials considered. A variation on the merit rating procedure is called the cost-modified merit rating. Comparable cost-modified merit rating factors can be determined from the expression:

$$\frac{\text{material unit cost}}{\text{maximum allowable cost}} \times \text{merit rating}$$

Any selection made on the basis of any of the above procedures can only, at best, be a draft specification. An initial batch of components should be made from the chosen material and rigorously tested by simulation and field trials. Such procedures are widely used when developing new aircraft and vehicles, and the results of such tests may influence the final material specification used in production. More problematical are major structures, such as bridges, which are manufactured on a 'once off' basis. Two procedures may be adopted when selecting materials for such projects.

- One is to examine the designs of previous, similar structures and follow accepted practice. Case studies of failures should be examined in order to avoid repetition of the faults in design, manufacture and material selection that caused such failures.
- Where new and attractive materials have been developed for which there are no precedents, the designer must proceed with caution. Large structures are usually made up of a number of smaller components, units or subassemblies, many of which may be used repetitively in relatively large quantities. In this case prototypes of these smaller components, units and subassemblies should be given simulation tests and field tests to destruction. Having evaluated the suitability of the new material in this manner, the results may be extrapolated using sophisticated computer-modelling techniques.

All successful companies will have an ongoing programme of testing and evaluation of the new materials that are being continually developed, and the assessment of the suitability of these new materials for their products.

EXERCISES

8.1 With the aid of diagrams, show how crystal orientation notation is derived from the Miller indices for the crystal.

8.2 With the aid of diagrams, describe how plastic flow in crystalline materials occurs by:
(a) block slip
(b) dislocation
(c) twinning

8.3 (a) Explain how interaction occurs between dislocations.
(b) Explain how the generation of dislocations occur in terms of:
(i) The Frank–Read source
(ii) misorientation at the grain boundaries

8.4 Discuss the interrelationship between dislocation, work hardening, and dispersion hardening.

8.5 (a) Discuss stress relief and recrystallisation in terms of dislocation theory.
(b) Explain the significance of positive and negative 'climb'.

8.6 Discuss the essential differences between the deformation of metallic materials and polymeric materials in terms of their relative structures.

8.7 Explain the essential differences between ductile and brittle fracture.

8.8 With reference to the Griffith crack theory, explain why the real strength of a material is substantially lower than the theoretical strength based upon inter-atomic forces.

8.9 Explain how the Griffith crack theory can be adapted to enable it to be applied to metallic materials.

8.10 Discuss the main factors affecting the nucleation and propagation of cracks and, with the aid of diagrams, show how component design can minimise the cracking.

8.11 Explain what is meant by the term 'fracture toughness' and state the factors, other than its inherent properties, which affect the fracture toughness of a material.

8.12 Explain why the *S–N* curve for a material can be misleading if there is a steady-state component force in addition to the alternating load.

8.13 Explain how Goodman diagrams and Soderberg diagrams are prepared and how these diagrams are used to determine the operational parameters for a material if failure by fatigue is to be avoided.

8.14 Explain what is meant by '*cumulative fatigue damage*', and describe how Miner's law is applied when predicting the fatigue life of a component under 'real life' service conditions.

8.15 Discuss the factors affecting the creep resistance of metals and describe what is meant by:

(a) creep limit (creep stress)
(b) rupture stress
(c) Larson–Miller parameter

8.16 Choose a component with which you are familiar, and select a suitable material for that component, taking into account the following factors;

(a) properties
(b) processing
(c) commercial viability

8.17 With reference to suitable examples of your own choice, explain the meaning of:

(a) critical properties
(b) property limits
(c) cost per unit property
(d) merit rating
(e) cost-modified merit rating

9 Bearings and bearing materials

The topic areas covered in this chapter are:

- Material surfaces.
- Sliding and rolling friction.
- Similar and dissimilar metals.
- Heating effects and lubrication.
- Hydrodynamic and hydrostatic bearings.
- Requirements of bearing materials.
- Materials for bearing applications.

9.1 Material surfaces

In order to understand the requirements of bearing materials and their selection, it is first necessary to examine the mechanics of the bearings themselves. Bearings can be classified mainly as:

- Plain bearings.
- Rolling bearings.
- Hydrodynamic and air bearings.

Let's start by considering what happens when movement takes place between the bearing surfaces under load conditions. A perfect plain surface is a geometrical plane devoid of irregularities. Such a surface never exists in practice. In reality, all material surfaces – no matter how smooth they may appear to the unaided eye – will have a texture similar to that shown in Fig. 9.1. *Real surfaces*, as defined in BS 1134, are the actual physical surfaces separating the component from surrounding space. Such surfaces have three main characteristics:

- *Waviness* That component of surface texture upon which roughness is superimposed. Waviness may be caused by such factors as machine or work deflections, vibrations, chatter, heat treatment, or warping strains.
- *Roughness* Irregularities of the surface texture that are inherent in the production process (tooling marks) but excluding waviness and errors of form.

- *Lay* The direction of the predominant surface pattern. Usually this is determined by the production method used.

All these characteristics will have a profound effect upon the performance of any two real surfaces that are in contact with each other to form a bearing. The life of these mating surfaces, when used for such applications as shafts and bearings, is dependent upon the surface texture as well as upon the physical characteristics of the materials from which the mating components are made and any lubricant used.

Fig. 9.1 *Surface characteristics and terminology*

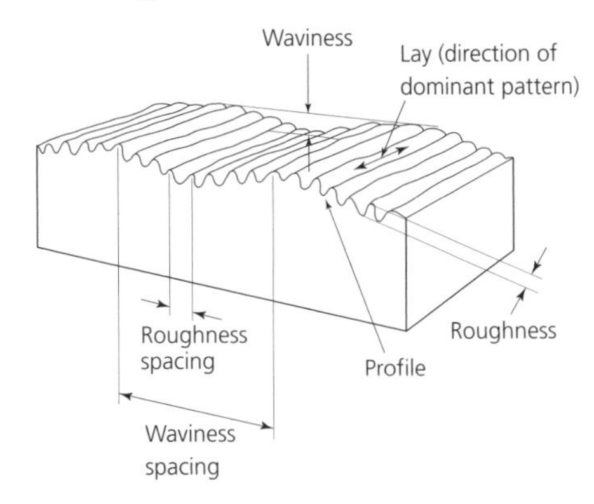

Fig. 9.2 *Effect of surface finish on choice of production process: (a) limits of size and process mismatched; (b) process suitable for limits of size*

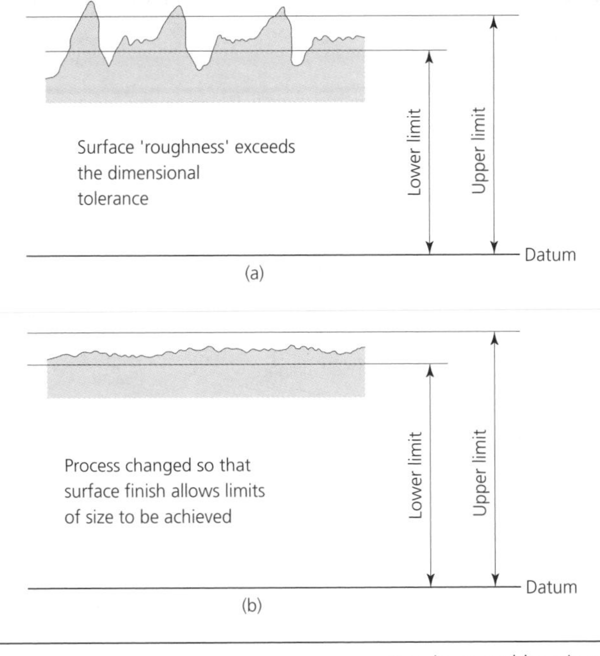

A rough surface with large peaks and valleys will have less contact area and will wear more quickly than a smooth surface. Unfortunately, the smoother and more geometrically accurate a surface becomes, the more costly it is to produce. Further, it is no use specifying a dimensional tolerance to a process whose inherent surface roughness lies outside the limits of that tolerance as shown in Fig. 9.2(a) and (b). Even two surfaces having the same roughness index can have different wearing characteristics, as shown in Fig. 9.3(a) and (b). They both have peaks of similar height and valleys of similar depth, with the same spacing, but it is obvious that the first surface will wear less quickly than the second surface under the same conditions of service.

Fig. 9.3 *Wear characteristics: (a) surface with low rate of wear characteristics; (b) surface with high rate of wear characteristics*

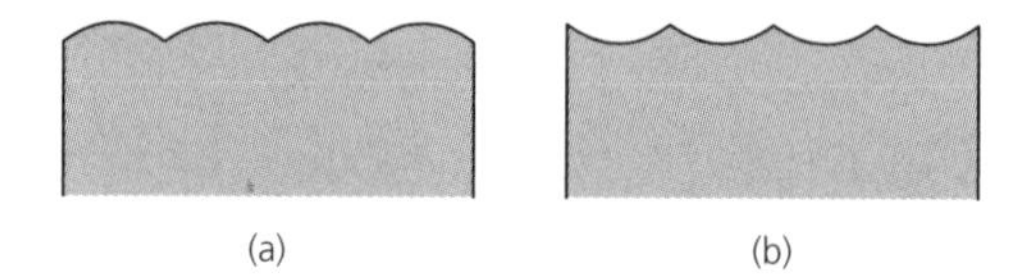

9.2 Wear

Figure 9.4 shows a magnified section through two mating surfaces devoid of any form of lubricant. The entire load on the surfaces is supported only on the high spots of the peaks. These high spots are called *asperities*. Since the contact surface area of the asperities may be very small, the contact pressure, even under light loads, will be very great and *plastic deformation* will occur in ductile materials. This will result in the contact area increasing, and the contact pressure decreasing, until the material can support the applied load.

Fig. 9.4 *Friction and wear*

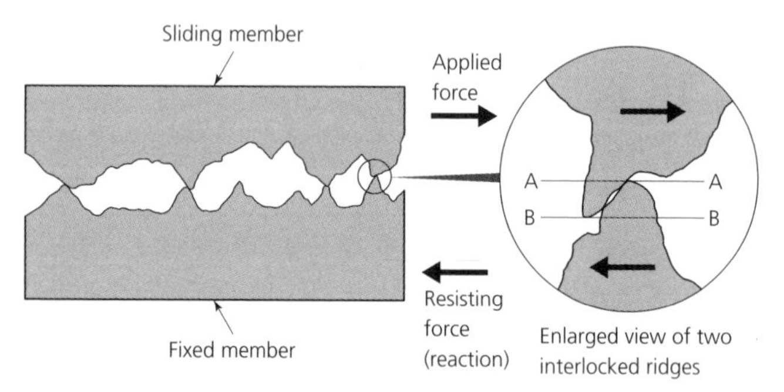

9.2.1 *Abrasive wear*

Figure 9.4 shows how the asperities become interlocked under load. Before the surfaces can slide over each other, the asprities have to shear along the planes AA or BB or both. The resisting (friction) force is the sum of these shear forces. The continual interlocking and shearing away of the ridges produces the *abrasive wear* that occurs between unlubricated (dry) surfaces. The particles of sheared asperities and foreign abrasive matter between the sliding surfaces greatly increases the friction and the rate of wear. Thus any lubrication system must have an efficient filter so that such abrasive matter is not recycled through the bearing surfaces. The dust that accumulates in the brake drums of vehicles are the sheared asperities from the surface of the brake drum and the brake shoes.

9.2.2 *Adhesive wear*

Certain metals have an affinity for each other and, when subject to local heating and intense pressure, will weld together. Such conditions exist in the dry bearing shown in Fig. 9.4. Two bearing surfaces that have become welded together in this way are said to have *seized*. The resisting forces created and the wear that takes place under such conditions, as the welded junctions shear, are very much greater than when the shearing forces are limited to the interlocking asperities. Repeated shearing of welded junctions causes *adhesive wear*.

In extreme cases the shaft may break in torsion or the bearing may be ripped from its housing before shearing of the junctions takes place. One of the advantages of a soft (white) metal bearing shell is that the heat generated under such conditions will result in the bearing shell melting before the shaft is broken. The replacement of the bearing shells is very much easier and cheaper than replacing the shaft.

9.2.3 *Corrosive wear*

When the bearing surfaces are attacked chemically or electrochemically, corrosive wear will take place. This may be caused by environmental attack, or by residual acids within the lubricant itself. Further, some extreme pressure additives such as active sulphur compounds can attack copper-bearing alloys such as phosphor bronze.

9.2.4 *Cavitation erosive wear*

When we discuss lubrication, you will see in Fig. 9.13 that the pressure distribution between a shaft and its bearing is not constant. The lubricant is subjected to areas of high and low pressure. Localised falls in lubricant pressure (cavitation) can result in the formation of vapour bubbles. These bubbles implode causing concentrated impact loadings at the bearing surface. These are sufficient to cause severe erosion unless the bearing material is designed to resist this effect. This effect is also referred to in Section 4.10.

9.2.5 *Surface fatigue*

The bearing surfaces considered so far have been *sliding*. Figure 9.5 shows what happens in rolling bearings. When a plain surface supports a perfect cylinder, only 'line contact' exists between the two surfaces, as is shown in Fig. 9.5(a). Since a line has no area, any load on the roller results in infinite contact pressure. In practice the roller, or the supporting surface, or both, will collapse slightly due to *elastic deformation* until the contact area is increased, as shown in Fig. 9.5(b). This results in a corresponding decrease in the contact pressure. The increase in elastic deformation will continue until equilibrium conditions are achieved and the bearing materials are able to sustain the load.

The constant flexing of the rolling surfaces due to elastic deformation can result in cyclical stresses that may exceed the fatigue resistance of the material. This can result in fatigue cracks below the surface of the material at a depth of 0.25 mm. These initial cracks usually propagate parallel to the surface, causing flaking (*spalling*) of the material from the surface. High surface hardness, and a highly polished surface free from scratches and fissures (hardening cracks and machining marks) will reduce the rate of fatigue wear.

Fig. 9.5 *Elastic deformation and flexure of rolling bearings: (a) theoretical rolling contact; (b) practical rolling contact*

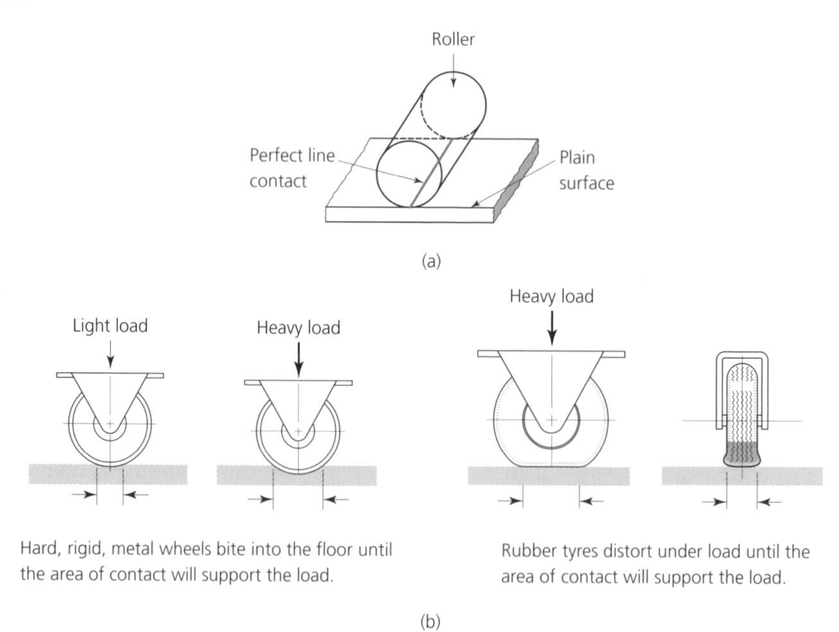

9.3 Sliding friction

Friction is defined as the *resistance that opposes the motion of one surface as it moves across another*. It has already been shown that, under dry conditions (no lubricant), this resistance to movement is caused by the interlocking and welding of the asperities of the mating surfaces. Let's consider Fig. 9.6. The true (real) area of contact (A_r) between the two surfaces is much smaller than the apparent (theoretical) contact area and can be approximated from the equation:

$$A_r = \frac{F_n}{Y_p}$$

where A_r = real contact area
F_n = force normal to the contact area (load)
Y_p = yield pressure causing plastic deformation of the contact areas.

Fig. 9.6 *Real contact during sliding*

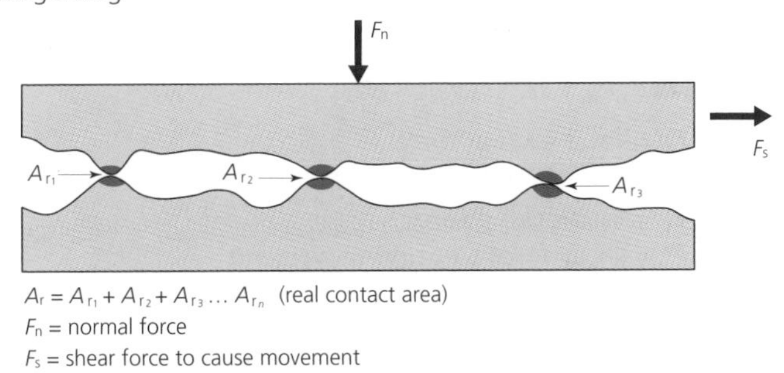

The force (F_s) that tends to cause relative sliding between the two surfaces is the force required to shear all the junctions. It is proportional to the shear strength of the material at the junctions and proportional to the total area in shear (which is also the real area of contact). Thus:

$$F_s = A_r \times S_u$$

where F_s = shear force to cause movement
A_r = total (real) area in shear
S_u = ultimate shear strength of the material at the junctions.

Dividing F_s by F_n:

$$\frac{F_s}{F_n} = \frac{A_r \times S_u}{A_r \times Y_p}$$

But from Fig. 9.7:

$$\frac{F_s}{F_n} = \mu$$

where μ = the coefficient of limiting friction. Therefore:

$$\mu = \frac{S_u}{Y_p}$$

Fig. 9.7 *Sliding friction*

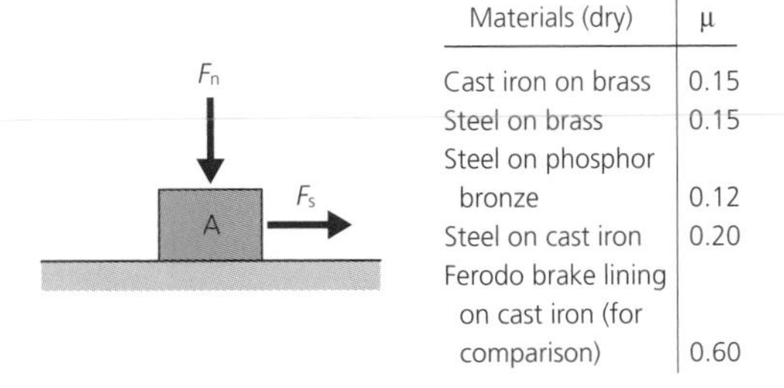

Materials (dry)	μ
Cast iron on brass	0.15
Steel on brass	0.15
Steel on phosphor bronze	0.12
Steel on cast iron	0.20
Ferodo brake lining on cast iron (for comparison)	0.60

That is, the coefficient of limiting friction μ is the total shear strength of the junctions divided by the yield pressure (hardness) of the softer material. In practice, the sliding friction may involve additional effects such as the ploughing of the softer material by the asperities of the harder material, and the interlocking of the surface irregularities. Therefore the total friction force will be equal to the sum of these three components:

$$F_t = F_s + F_p + F_i$$

where F_t = total friction force
F_p = ploughing force
F_i = interlocking force
F_s = shear force to cause movement.

The physicist Coulomb originally propounded the laws of sliding friction, hence sliding friction is often referred to as 'Coulomb friction'. These laws were based upon experimental data and are by no means exact since, in practice, it is impossible to obtain 'dry' surfaces free from oxidation products except under high vacuum conditions that were not available to Coulomb and, in any case, are beyond the scope of this book. However, these laws do form a useful basis for engineering calculations, and are satisfactory for the majority of general applications. The laws of 'dry' (no lubrication of the mating surfaces) friction may be summarised as follows:

- When an external force tends to cause one surface to slide over another surface, a reactionary frictional force is set up, acting tangentially to the surfaces so as to oppose the motion.
- There is a limiting value to the force of friction beyond which it cannot rise. If the externally applied force exceeds this value, sliding will commence.
- The force required to start sliding is greater than that to maintain it. Static friction is always greater than kinetic (moving) friction.
- The limiting value of the frictional force is independent of the area of contact.
- The limiting value of the frictional force maintains a constant ratio to the normal reaction between the surfaces. This ratio is called the *coefficient of limiting friction*, and is denoted by the Greek letter μ (mu).
- The coefficient of limiting friction depends upon the mating surfaces, their surface texture, surface contamination, and the physical properties of the materials. Figure 9.7 shows some typical values for μ for various combinations of materials. A typical calculation of the force required to overcome limiting friction is shown in Example 9.1.

EXAMPLE 9.1

Calculate the force to just move the block A *if the normal force is 32 N and the coefficient of friction μ is 0.125.*

Solution

$$\frac{F_s}{F_n} = \mu \quad \text{where} \quad F_s = \text{force to move } A$$
$$F_n = 32\,\text{N}$$
$$\mu = 0.125$$

$$F_s = \mu F_n$$
$$= 0.125 \times 32$$
$$= \mathbf{4.0\,N}$$

SELF-ASSESSMENT TASK 9.1

1. With the aid of diagrams, explain the difference between sliding and rolling bearings and give a typical example of each.
2. Discuss, in detail, the main causes of wear in a bearing.
3. Calculate the force to just move block *A* in Fig. 9.7 if the normal force is 60 N and the coefficient of friction is 0.3.

9.4 Rolling friction

Rolling friction occurs when a cylinder or sphere rolls over a plain surface. The load is assumed to act through the centre of the roller or sphere. Lubrication is necessary because, as has previously been stated, some elastic deformation occurs at the point of contact so that true rolling does not actually occur and a small amount of slip is present. However, this is very slight and rolling friction is very much less than sliding friction. Rolling friction is proportional to the applied load and inversely proportional to the diameter of the roller or sphere (see Fig. 9.8). The coefficient for rolling friction is the same as for static and kinetic friction, whereas for sliding friction, the coefficient for static friction is very much greater than the coefficient for kinetic friction. In rolling bearings most of the friction losses and heating effects are due to the *elastic hysteresis* losses within the component materials.

Fig. 9.8 *Rolling friction*

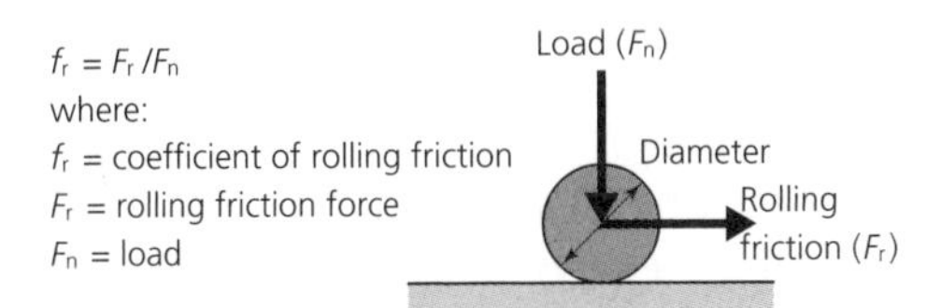

9.5 Surface contamination

When metals are exposed to air they are always covered by a film of metal oxide, together with a film of moisture and, frequently, with surface layers of other chemical compounds and absorbed gases. These greatly affect the adhesion and friction between the moving surfaces and, as long as these films are present, the coefficient of limiting friction rarely exceeds 1.5. However, for perfectly clean metal surfaces prepared under high vacuum conditions the coefficient of friction may reach very high values. Under such conditions adhesion will be very strong and a metallic bond may form between the surface asperities, resulting in cold welding and seizure. The coefficient of friction for diamond under normal atmospheric conditions is 0.05, but when it is 'outgased' in a vacuum its coefficient of friction rises ten-fold to 0.5.

Another example is graphite. This material has a low coefficient of friction under normal atmospheric conditions, not only because of its *laminar structure* but also because of its surface film of water and dissolved gases. Under such conditions the coefficient of friction for a graphite-to-graphite surface is 0.1, but when dried and 'outgased' in vacuum its coefficient of friction rises to between 0.5 and 0.8 and it rapidly wears away to a fine dust.

9.6 Similar and dissimilar materials

Amonton's laws of friction states that the coefficient of friction is independent of load and that the real contact area increases directly with an increasing load. This increase occurs during sliding between the surfaces because the area of the junctions increases as the result of the combined effect of plastic deformation and shearing. This applies particularly to pure metals. Further, the interface between two surfaces of the same or similar metals shows a greater coefficient of friction and a greater adhesion than an interface between two dissimilar metals. Again, homogeneous materials tend to show greater adhesion than heterogeneous materials. For example, cast iron on steel has a coefficient of friction of only 0.3, whereas nickel, pure iron (ferrite) and austenitic stainless steel (all highly homogeneous) show coefficients of friction within the range 1.2–1.5. Note that the very low value for cast iron is, to some extent, brought about by the lubricating properties of the flake graphite in this material.

Soft and ductile metal surfaces showing high levels of plastic deformation also show a correspondingly high tendency to adhesion, whereas hard metal surfaces showing only elastic deformation have a much lower tendency to adhesion, and seizure is less likely to occur. On the other hand, the ploughing and shearing resistance will be higher for hard materials.

Polymeric bearing materials behave very much like metal. Deformation occurs at the junctions and strong adhesion may take place. However, unlike metals, this is viscoelastic deformation and depends upon the surface texture, the load, and the duration of loading. Thus there is a lower level of junction growth and this accounts for the fact that, unlike metals, the friction of polymeric materials does not increase linearly with the load and that the coefficient of friction actually decreases with increasing load.

Adhesion tends to be strong between polymer materials and their coefficients of friction tend to be fairly high, ranging from 0.3 to 0.5. The exception is polytetrafluoroethylene (PTFE). This material has the lowest coefficient of friction of any known substance. A typical value under normal atmospheric conditions is 0.04.

9.7 Heating effects

When unlubricated surfaces slide over each other the mechanical energy required to overcome frictional resistance is converted into heat energy. This is not uniformly distributed over the apparent contact area but is concentrated mainly at the asperities where it is generated. This may raise their temperatures above their melting points and accounts for the formation of welded junctions. The surface temperature depends upon a number of factors such as the relative load and speed, thermal conductivity of the surface materials, the mass of metal available to conduct the heat away, and the coefficient of friction. Polymer materials present a particular problem as they are very good heat insulators and thermal conductivity through the bearing is low. Further, they usually have low softening and melting temperatures.

Frictional heat can produce very smooth surface finishes if the mating surfaces are properly 'run in'. Surface material is transferred from the peaks to the valleys, and 'hot spots' result in softened or molten metal being spread over the surface where it cools to form a characteristic layer of a glassy, amorphous appearance called the *Beilby layer*.

9.8 Lubrication

The friction between two surfaces has already been shown to result from the interlocking, shearing and welding of the surface asperities. However, separating the contact surfaces of the bearing by the *use of a lubricant* results in the friction and wear becoming negligible. For example, it is more difficult to drag a small boat over the shingle of the beach than to move it once it is floating, where a layer of water acts as a lubricant between the boat and the shingle. Further, the wear on the bottom of the boat will be negligible once it is in the water compared to the damage that will be done dragging it over the dry shingle beach.

There are two mechanisms of lubrication: *fluid lubrication* (as described above) and *boundary layer lubrication*. Full fluid lubrication occurs when the bearing surfaces are completely separated by a thick film of lubricant capable of supporting the loads applied to the bearing. Under these conditions there is no direct contact between the material surfaces, as shown in Fig. 9.9(a), and the coefficient of friction becomes very low (0.001–0.01) being independent of the bearing materials and is determined only by the viscosity of the lubricant.

Fig. 9.9 *Lubrication: (a) full fluid (hydrodynamic) lubrication; (b) boundary lubrication (adsorbed layers only)*

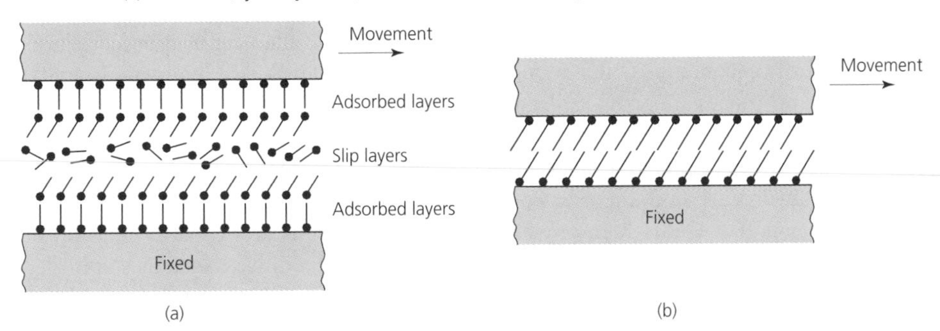

Boundary layer lubrication occurs when the layer of lubricant adhering to the bearing surfaces is only a few molecules thick, as shown in Fig. 9.9(b). This is often only comparable with the valleys and asperities of the bearing surfaces. Lubrication will be minimal and there will be some contact between the asperities resulting in high rates of wear. This occurs during the starting and stopping of machines and engines when oil flow is negligible or is non-existent and only a residual film is present.

Between these two extremes there is the condition when oil is commencing to flow into the bearing and the bearing surfaces are partly in contact and partly separated by the lubricant. This is referred to as *partial fluid lubrication*. Figure 9.10 shows how the coefficient of friction can change with these different conditions.

Fig. 9.10 *Coefficient of friction changes with service conditions*

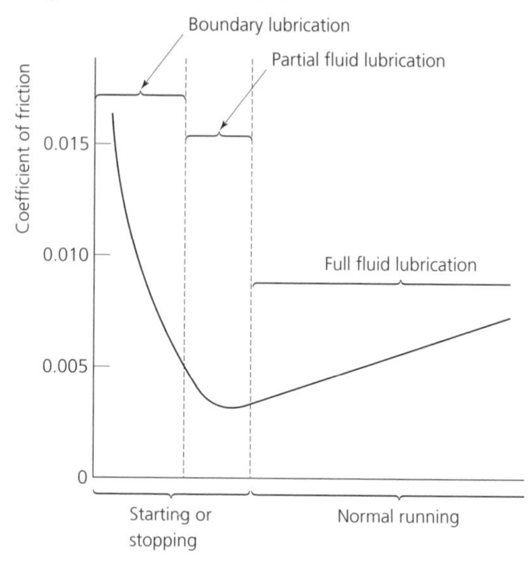

Changes in temperature results in changes in the viscosity of a lubricant. Therefore, a compromise has to be achieved between an oil that is 'thin' enough at low (starting-up) temperatures to avoid excessive drag, yet does not become so thin under the higher temperatures of operating and overload conditions that it cannot protect the bearing. This has resulted in the development of 'multi-grade' oils.

Mechanisms are usually operated under full fluid lubrication conditions since, under these conditions, the friction is very low and wear is minimal. Boundary lubrication properties are provided by adding fatty acids such as carboxyl to a lubricating oil that has been refined from a petroleum crude. The reaction of the fatty polar groups with the metal surface results in a monomolecular layer that adheres strongly to the surface with a long hydrocarbon chain oriented outwards, perpendicular to the surface, as previously shown in Fig. 9.9. Under high pressures and speeds there is an excessive build up of heat and the boundary film melts and breaks down. Special additives, such as sulphides, chlorides and phosphides, that will form films on the metal surface are used. Such films have sufficient shearing strength and high-temperature resistance to withstand severe service conditions. These are called *extreme pressure additives* and lubricants that contain them are called *extreme pressure lubricants*. They are ineffective on inert metal surfaces such as chromium, but are widely used where sliding and rolling occur at the same time under heavy load conditions, as in the hypoid bevel gears of road vehicle final drives. As fast as the boundary film is rubbed away, it is replaced by the chemical reaction between the additive and the bearing surfaces.

Improved boundary lubrication and reduced wear may also be provided by the addition of inorganic materials such as *graphite* or *molybdenum disulphide* to the lubricant. The laminar structure of these substances gives them a high compression strength to resist the bearing load, whilst ensuring a low shearing strength to reduce friction. Because the platelets of these materials are relatively coarse, care must be taken to ensure that they are not removed by any filtration components in the lubrication system. Filters must be designed to accept them.

9.9 The hydrodynamic lubrication of plain bearings

In a plain bearing, the shaft is separated from the bearing by a lubricant that cools and lubricates the members of the bearing. The lubricant must also locate the shaft in the bearing since there must be a finite gap between the shaft and the bush if the shaft is to rotate. Lubrication and location is provided by a wedge of oil, referred to as the *hydrodynamic wedge*. This occurs during full film lubrication when the conditions shown in Fig. 9.11 are applied.

Fig. 9.11 *The hydrodynamic wedge*

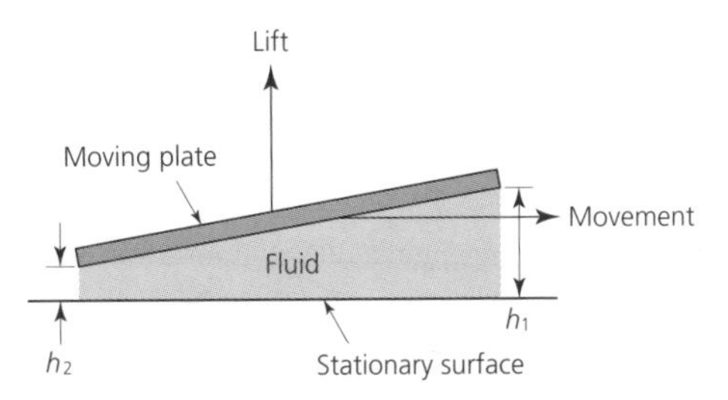

Osborne Reynolds evolved the theory of the hydrodynamic wedge in 1886. The load-carrying capacity of the top plate is dependent upon the following factors:

- Oil viscosity (thickness).
- Plate area.
- The ratio h_1/h_2.
- Plate velocity.

Unfortunately, those factors which increase the load-carrying capacity of the bearing also increase the fluid friction (drag). However, Reynolds evolved two important mathematical equations to support his theory, and these showed that whilst the one for friction was a first-order (linear) equation, the one for load-carrying capacity was a second-order (quadratic) equation. It is therefore possible to 'play off' these two equations against each other and obtain an optimum set of conditions that will give maximum lift for minimum drag. The pressure distribution curve is shown in Fig. 9.12, and it can be seen that maximum lift does not occur at the point of minimum gap, as is often assumed, but at a point along the plate.

Fig. 9.12 *Plain bearing pressure distribution*

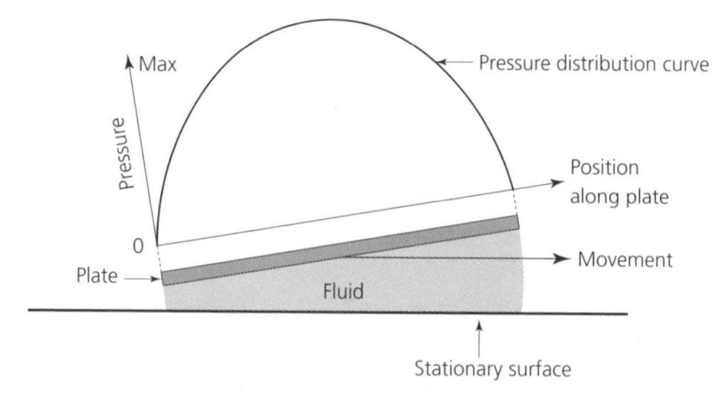

In practice, it is the plain journal bearing that makes use of Reynolds' converging films theory. Figure 9.13(a) shows the position of the shaft centre in a plain cylindrical bearing under 'rest', 'start' and 'run' conditions. It can be seen that the shaft centre moves about as a result of the changing and uneven distribution of lubricant pressure. Therefore, this simple bearing is unsuitable for applications in which the axis of the shaft must remain constant as in a machine tool spindle. The introduction of oil grooves upsets the pressure distribution still further, but without these grooves uniform lubrication of the bearing is not possible. For precision bearing applications *constant centre-line bearings* have been

Fig. 9.13 *Plain journal bearing: (a) simple journal bearing; (b) constant centre-line bearing. The pressure system created by the rotating shaft causes the tilting pads to 'float' until the pressure lobes are uniformly distributed. Any disturbing force on the shaft upsets the balance of the system causing local pressure increases that oppose the disturbance. Thus the shaft is kept centred in the bearing*

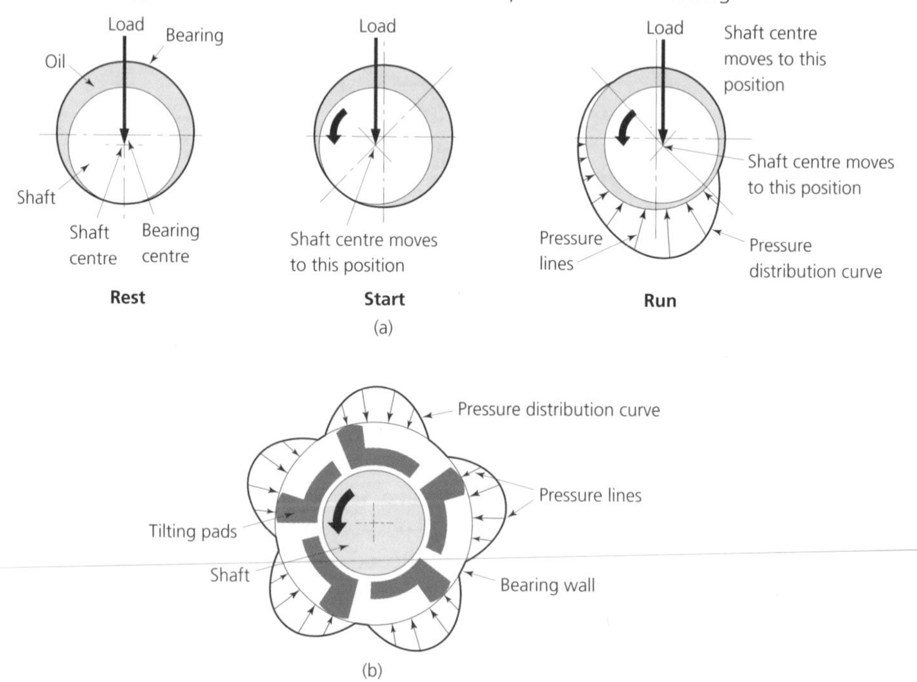

developed. The principle of one such bearing is shown in Fig. 9.13(b). The pressure system created by the rotating shaft causes the tilting pads to 'float' until the pressure lobes are uniformly distributed. Any disturbing force on the shaft upsets the balance of the system, causing local pressure increases that oppose the disturbance. Thus the shaft is maintained in a central position in the bearing.

9.10 Hydrostatic bearings

In this type of bearing the shell is perforated with fine holes through which air or oil is forced under high pressure. There is an appreciable gap between the shaft and the bearing shell and the shaft floats on a cushion of air or oil in a similar manner to a hovercraft. The principle of the bearing is shown in Fig. 9.14.

Fig. 9.14 *The hydrostatic bearing*

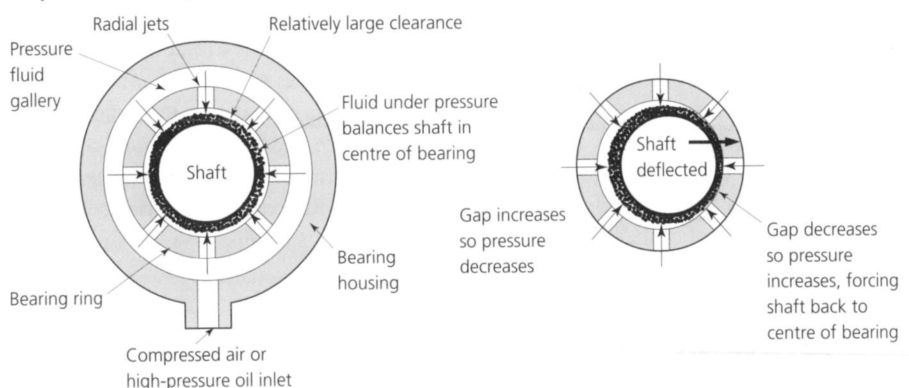

9.11 Requirements of bearing materials (sliding)

Having considered the mechanics of bearings we must now consider the requirements of materials used in bearings. There are essentially two types of bearing: those where the moving parts *slide over each other*, as in plain bearings and machine-tool slideways, and those where the moving parts *roll over each other*, as in ball and roller bearings.

Sliding surface bearings work under full fluid film lubrication conditions that separates the moving surfaces. However, some direct surface-to-surface contact occurs from time to time when the lubrication film breaks down under overload conditions or when sliding starts or stops. Materials for sliding bearings require the following properties:

- *Coefficient of friction* (μ) This should be kept as low as possible to avoid wear, wasting energy and generating excessive heat in the bearing.
- *Strength* The bearing must have sufficient compressive strength to support the shaft and any load that may be applied to it in service.

- *Wear resistance* The bearing material must resist wear but, at the same time, it is invariably better for the bearing to eventually wear out rather than the journal on the shaft. It is easier and cheaper to replace the bearing shell.
- *Plasticity* It is impossible to obtain perfect alignment between a shaft and its associated bearing. Therefore, a bearing material should be capable of slightly distorting and 'bedding in' to ensure as perfect alignment as possible. White metals and leaded bronzes are better in this respect than the harder and more rigid phosphor bronzes.
- *Surface texture* A perfectly smooth surface would be a poor bearing surface as there would be no provision for the retention of pockets of lubricant. An ideal bearing material consists of hard facets of wear resistant, antifriction material dispersed through a soft matrix. The matrix wears away between the facets to leave pockets for retention of the lubricant. The soft matrix also has sufficient 'give' to assist alignment.
- *Corrosion resistance* The bearing material should resist corrosion by impurities in the lubricant or in the working environment. It should also be resistant to attack by any additives in the lubricant intended to give it greater lubricity (extreme pressure additives).
- *Thermal conductivity* Since the best combinations of bearing materials and lubricants will offer some degree of friction, there will always be some energy loss in the bearing and some corresponding rise in temperature when the shaft is rotating. To keep the temperature rise to a minimum it is necessary to dissipate the heat energy as quickly as possible. This heat energy can only be dissipated through the lubricant and by conduction through the walls of the bearing. If the heat energy generated in the bearing is not conducted away quickly enough, the temperature rise could reach the melting point of the bearing material with disastrous results.

9.12 Requirements of bearing materials (rolling)

The requirements for materials suitable for rolling bearings are somewhat different to those just described for sliding bearings. To appreciate the requirements of the materials used in rolling bearings it is necessary to consider the construction and use of such bearings. Rolling or 'antifriction' bearings consist of hardened balls or rollers arranged to roll between hardened steel rings called 'raceways'. The balls and rollers are kept equally spaced around the bearing by means of a 'cage'.

9.12.1. *Ball bearings*

Ball bearings consist of four main components:

- The inner race, which is a light press fit on the shaft.
- The precision ground balls.
- The cage that keeps the balls equally spaced.
- The outer cage, which is a light press fit in the bearing housing.

Theoretically, the balls should roll on the races without the occurrence of slip. Also, since sliding should not take place, lubrication should not be necessary. However, because of the elastic deformation of the bearing elements under load this is not the case, and true rolling does not occur. Therefore, lubrication is essential, and it is also necessary to lubricate the balls in their cage where sliding friction occurs.

Ball bearings are normally selected for high-speed applications that are only lightly loaded (theoretically such bearings only offer *point support*). They are less critical in their installation than roller bearings and tend to be self-aligning. Figure 9.15 shows some typical ball bearings.

Fig. 9.15 *Typical ball bearings: (a) journal bearing; (b) thrust bearing; (c) angular contact bearing (combined journal and thrust)*

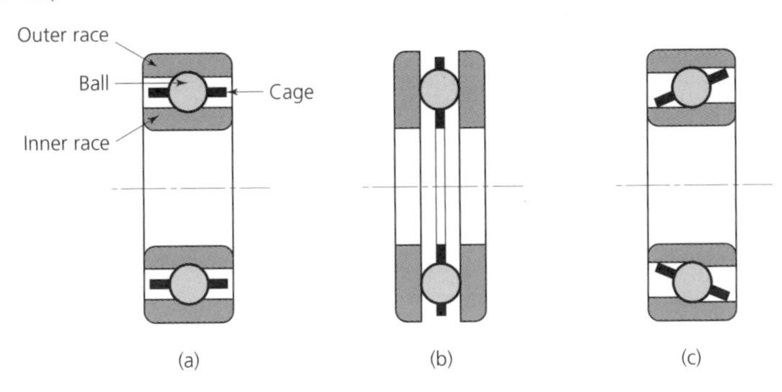

9.12.2 *Roller bearings*

These are built up in a similar manner to ball bearings – the exception being needle roller bearings (these are used where space is limited) where the rollers bear directly onto the shaft journal which is hardened and ground and there is no inner race. A typical needle roller bearing is shown in Fig. 9.16. Since roller bearings offer line support they are capable of handling heavier loads than ball bearings. They are widely used in supporting machine-tool

Fig. 9.16 *Needle roller bearings*

spindles subject to heavy cutting forces. In this connection, tapered roller bearings are used, in opposed pairs, to provide axial as well as radial restraint. Figure 9.17(a) shows the geometry for a tapered roller bearing where all axes and surfaces meet at a common point when they are projected. Figure 9.17(b) shows typical applications of cylindrical and taper roller bearings as applied to a machine-tool spindle. The opposed taper roller bearings adjacent to the spindle nose resist the axial thrust. The cylindrical roller bearings at the tail of the spindle allow for axial expansion as the spindle warms up. Thus the plane of the spindle nose remains constant irrespective of any change of length of the spindle due to heating or cooling.

Fig. 9.17 *Roller bearing applications: (a) geometry of a taper roller bearing; (b) a machine-tool spindle*

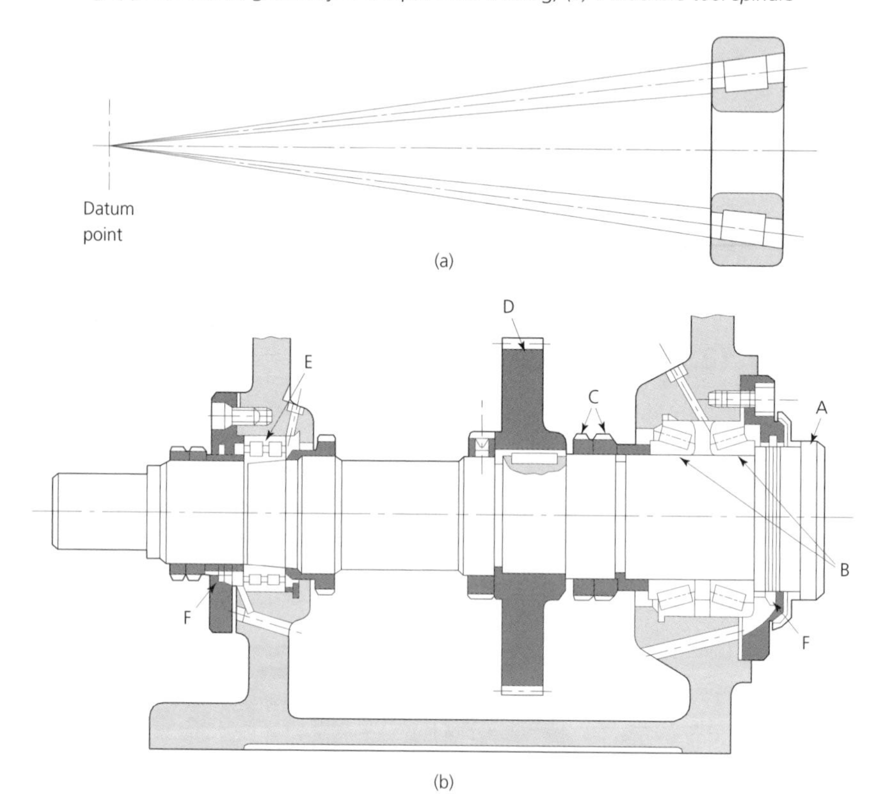

9.12.3 *Material requirements*

The material requirements for rolling bearings are as follows:

- *Wear prevention* For wear prevention in rolling bearing materials it is important that they have a high degree of hardness and are finished to a high degree of smoothness.
- *Fatigue resistance* The elements of rolling bearings are constantly flexing at the point of contact (rather like a tyre on the road) and this leads to high levels of fatigue stress. Therefore, a material with high fatigue resistance is essential and it must be finished

without surface imperfections that might accelerate fatigue failure. (See Sections 9.2 (surface fatigue) and 8.10.)

- *Strength* At one time case-hardened alloys were used exclusively for rolling bearings as it was thought that this would provide the necessary hardness to prevent wear and, at the same time, the toughness and strength necessary to support the load and provide a degree of deformation to assist alignment. In practice, the constant flexing (elastic deformation) caused the hard case to flake away from the more flexible core. Modern practice for high-quality bearings favours the use of high carbon–chromium steel alloys, hardened to give a martensitic structure throughout, except where heavy shock loads are encountered in which case case-hardening alloys are still preferred.

SELF-ASSESSMENT TASK 9.2

1. State the essential properties of bearing materials:
   (a) for sliding bearings
   (b) for rolling bearings
2. Explain the importance of lubrication in bearings and discuss the essential properties of a lubricant for medium-duty sliding bearings.
3. Compare and contrast the properties and uses of metallic and non-metallic bearing materials.

9.13 Bearing materials (metallic)

Traditionally, sliding bearing materials were alloys of tin, lead and antimony, referred to as 'white-metal' alloys or 'Babbitt-metal'. Copper-based alloys, aluminium-based alloys and polymeric materials are also used nowadays for sliding bearings.

9.13.1 *White-metal alloys*

These alloys represent the majority of bearing metals in which small particles of a hard phase are embedded in a ductile solid solution matrix. The hard particles are usually intermetallic compounds that have good wear resistant and antifriction properties. The ductile matrix provides toughness (shock resistance) and sufficient ductility to allow the bearing to 'bed in' during the running-in period.

White-metal alloys may be either tin based or lead based. High-quality Babbitt metals may contain up to 90 per cent tin and up to 10 per cent antimony. Typical alloys are listed in Table 9.1. Tin and antimony form cuboid crystals of the tin–antimony intermetallic compound SbSn. These cuboids of the intermetallic compound constitute the hard, antifriction phase. In a lead-free Babbitt metal these hard cuboids are dispersed through a soft, ductile matrix of a solid solution of antimony in tin. White-metal bearings are only suitable for lightly loaded applications and are no longer used in modern internal combustion engines.

Table 9.1 ***Compositions and properties of common bearing materials***

Category	*Composition (%)*					*Properties and applications*
	Sn	Sb	Cu	Pb	P	
White metal	93	3.5	3.5	—	—	Big-end bearings for light- and medium-duty, high-speed internal combustion engines
	86	10.5	3.5	—	—	Main bearings for light- and medium-duty, high-speed internal combustion engines
	80	11.0	3.0	6.0	—	General-purpose, heavy-duty bearings. Lead improves plasticity where alignment is a problem
	60	10.0	28.5	1.5	—	Heavy-duty marine reciprocating engines, electrical machines
	40	10.0	1.5	48.5	—	Low-cost, general-purpose, medium-duty, bearing alloy
Bronze	10.5	—	89	—	0.5	Good anti-friction properties, suitable for heavy loads, rigid
	10.0	—	79.9	10	0.1	Good anti-friction properties, lubrication not critical, lead content reduces rigidity and helps alignment
	3	—	74	23	—	Leaded (plastic) bronze, excellent self-alignment properties due to high lead content. For duty intermediate between white metal and phosphor bronze
	Fe	C	Si	Mn	S/P	
Cast Iron	94	3.3	1.3	1.0	0.1/0.3	The flakes of graphite (carbon) in grey cast iron gives it self-lubricating properties. Suitable for heavy-duty, low-speed applications where lubrication is difficult, e.g. machine tool slideways
Plastic	Polytetrafluoroethylene					Teflon: can withstand much higher temperatures than most plastics. Very expensive anti-friction coating – very low coefficient of friction. Does not require lubrication
	Polyamide					Nylon: Can be moulded into bushes and gears. Does not require lubrication. Use for office and food-processing machinery
	High-density polyethylene					Low-cost bearings. Does not require lubrication. Cannot support such high loads as Nylon or Teflon

Sn = Tin, Sb = Antimony, Cu = Copper, Pb = Lead, P = Phosphorus, Fe = Iron, C = Carbon, Si = Silicon, Mn = Manganese, S = Sulphur

Lead is added to white metals to reduce the cost and to increase the plasticity of the material to ease alignment. In such alloys there is a tendency for the intermetallic tin–antimony cuboids to float to the surface of the molten metal. The molten metal eventually solidifies to form a eutectic of two solid solutions. One of these will be lead-rich and one will be tin-rich. To prevent the segregation of the tin–antimony cuboids, copper is added. The copper forms an intermetallic compound with the tin that separates out before the tin–antimony cuboids. These remain trapped in the mesh of the copper–tin matrix thus preventing segregation.

9.13.2 *Copper–lead alloys*

These alloys produce a bearing material with a hard-phase matrix within which there is a softer phase network. The antifriction properties of these alloys result from the smearing of a thin film of lead over the surface of the harder copper during the 'running-in' period. Shearing of the asperities occurs in this lead film, whereas in Babbitt metals it occurs in the soft matrix. An example of a copper–lead alloy is given in Table 9.1.

Copper–lead alloys provide superior strength-bearing materials for highly loaded applications. Since copper-based materials are relatively hard, they have to be plated with an overlay to provide adequate soft-phase properties. This also improves their corrosion resistance. Lead–indium alloy is the preferred overlay material as it combines maximum fatigue strength with optimum soft phase properties. The combination of cast lead bronze and lead–indium overlay is widely accepted as the design standard for heavy diesel engines and high-performance racing car engines.

9.13.3 *Aluminium-based bearing materials*

Aluminium bimetals have become the industry standard for medium load engine applications. These materials consist of tin or tin and silicon in a matrix of aluminium. Tin gives the material good soft-phase properties whilst the addition of silicon gives the material the property of conditioning cast iron crankshaft journals. This group of materials also has excellent corrosion resistance.

They were not considered as a commercial proposition until the introduction of roll-bonding techniques for making bimetallic strips. The aluminium–silicon–tin alloy is roll-bonded onto a steel shell with an aluminium foil interlayer. The aluminium–silicon alloy usually contains 7 per cent tin to prevent galling and to improve the antifriction properties. Small amounts of nickel and copper may also be present.

9.13.4 *Copper–tin bronze alloys*

Tin bronzes, such as gunmetal, phosphor bronze and aluminium bronzes are widely used for heavy-duty bearing materials. They are stronger than the materials so far described and are self-supporting. These bearing materials are very rigid, and careful machining and alignment are essential since very little bedding-in can take place. Lead is added (leaded-bronze) to improve the plasticity and antifriction properties of the alloy, but with some loss of strength. Typical alloys, their composition and some typical applications are listed in Table 9.1.

9.13.5 *Porous metal bearings*

Porous metal bearing materials are produced by the compaction and sintering of powdered metals. Sintered particulate manufacturing processes are discussed fully in *Manufacturing Technology*, Volume 2. The bronze-bearing metal alloy in powder form is compressed into the required shape and this compact is then sintered at a high temperature in a reducing atmosphere to produce a micro-porous structure. (A metal 'sponge'.) Typical structures are shown in Fig. 9.18. Bearings produced in this manner are impregnated with a suitable lubricant that normally lasts the life of the bearing. They are used for applications such as domestic appliances where regular maintenance and lubrication cannot be relied upon, or for applications where an automatic lubrication system would be inconvenient or excessively expensive. Alternatively, provision may be provided for periodical replenishment of the lubricant when the use of a porous bearing increases the working period between servicing.

Fig. 9.18 *Microstructures: (a) low-density, high-porosity metal; (b) high-density, low-porosity metal*

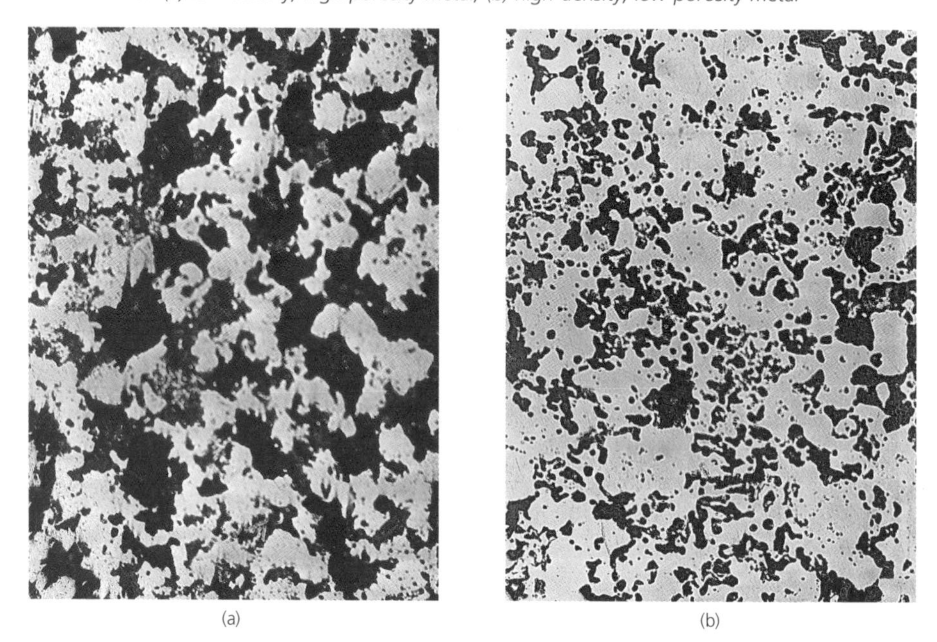

(a) (b)

9.13.6 *Overlay materials*

The use of overlays was introduced in the section on white-metal alloys. Some bimetals, depending upon the application, are overlay plated to improve their soft-phase properties. An overlay-plated bimetal bearing is referred to as a *trimetal* bearing. An overlay consists of a thin surface layer (up to 0.033 mm thick) that is precision electroplated onto the bimetal surface after finish boring. The overlay is supplied in a range of thicknesses in order to achieve the desired balance of properties.

- Thinner overlays promote a higher load-carrying capacity.
- Thicker overlays provide greater conformability and embeddability.

Cast lead bronze is overlay plated with either lead–indium or lead–tin–copper, while sintered lead bronze is plated with lead–tin. Let's now look at these overlays in greater detail.

9.13.7 *Lead–indium*

Cast lead bronze is overlay plated with lead–indium by successively electroplating with lead and then with indium. The indium is then diffused through the lead and into the lead pools of the bronze substrate so that corrosion protection provided by the indium is retained should the overlay be worn away in service.

9.13.8 *Lead–tin–copper*

Lead–tin–copper is an alternative overlay to lead–indium. It lacks the fatigue strength of lead–indium but by increasing the copper content it can be made harder, thus improving the wear resistance. The overlay is electrolytically deposited onto cast lead bronze but requires a nickel interlayer to prevent the tin diffusing into the lead bronze. Such diffusion would reduce the corrosion resistance of the overlay.

9.13.9 *Lead–tin*

Lead–tin overlay is co-deposited onto *sintered* lead bronze and offers a good compromise of properties.

9.13.10 *Polymer composite-bearing materials*

Polymer-based materials have been developed for a wide range of applications including compressors, pumps, suspension struts, shock absorbers, etc. They include materials designed to operate in non-lubricated, marginally lubricated and grease lubricated environments. The polytetrafluoroethylene (PTFE) based materials are mainly used in dry-running applications, whilst acetal and polyether ether ketone (PEEK) based compositions are designed for marginally lubricated and grease lubricated conditions. Most of the materials are produced by bonding polymer composites (in either powder or tape form) onto a steel-backed bronze or sintered base.

9.14 Bearing materials (non-metallic)

Polymeric materials such as teflon, nylon and high-density polystyrene can all be used for sliding bearings when lightly loaded and when the speed is not excessive so that temperature rise is minimal. They have the advantage that normal atmospheric moisture is sufficient lubricant and no oil or similar lubricant is required. This makes them ideal for use in office machinery and food-processing equipment. They are also used where corrosive environments preclude the use of metallic-bearing materials.

For heavier duty applications reinforced polymer materials such as 'Tufnol' may be used successfully. For example, this material has largely ousted the use of lignum vitae wooden inserts for the propeller-shaft stern bearings of large ships. The presence of sea water in the bearing prevents the use of metallic materials because of corrosion. A further advantage in using Tufnol for this application is that the water present in the bearing provides adequate lubrication.

Carbon and graphite bearings (e.g. some clutch thrust bearings in road vehicles) are used where lubrication and environmental conditions are difficult. They may or may not be lubricated with water, but usually there is sufficient moisture present in the atmosphere to provide adequate lubrication. Some typical examples of non-metallic bearing materials are listed in Table 9.1.

9.15 Surface coatings

Surface hardening has already been considered in Section 5.13 of *Engineering Materials*, Volume 1, including case hardening, flame hardening, induction hardening, and nitriding. In addition, the wear resistance, antifriction, and embeddability properties of bearing surfaces can be improved by the use of surface coatings.

9.15.1 *Wear resistance*

In addition to the surface-hardening processes described above, hard coatings may be applied to the bearing surface by a variety of processes.

- *Hard-chrome plating*, which must not be confused with decorative chrome plating, is used to electroplate a relatively thick deposit of pure chromium onto the surface being treated. Compared to the decorative coating which may only be a few microns thick (1 micron = 0.001 mm), a hard chrome deposit may be up to 0.4 mm thick. After finishing by grinding and superfinishing (e.g. honing) such a surface has high hardness, very low friction and high corrosion resistance. The slight porosity of the surface can be carefully controlled to improve its lubrication properties.
- *Hard metals*, such as stellite – a cobalt-based alloy which is so hard that it can only be machined by grinding – can be deposited on the bearing surface by welding using an oxyacetylene torch with a 'reducing' flame setting. A considerable thickness of metal can be deposited and such a surface not only has considerable wear resistance, but can operate continuously at high temperatures.
- *Ceramics and metal carbides* can be deposited onto the bearing surfaces by various processes in order to provide wear-resistant bearing surfaces. Ceramics such as alumina or zirconia may be sprayed onto the surface using a 'gun' in which the powder is melted by an oxyacetylene flame and the molten material is sprayed onto the surface by compressed air. Hard carbides, such as tungsten carbide, may be applied in a similar manner or they may be preformed and brazed onto the bearing surface. For sliding bearings a material with hard carbides dispersed through a softer and tougher matrix is to be preferred, for example tungsten carbide particles in a titanium carbide matrix.

9.15.2 *Embeddability*

The development of harder, heavy-duty bearing materials, such as aluminium–silicon alloys and the lead bronzes, presents problems resulting from their low embeddability properties. That is, they lack the ability to absorb small, abrasive particles of foreign matter and this can result in damage to the shaft journals. The problem is overcome by surface coating these bearing materials with an 'overlay' of lead–indium or lead–tin alloy. This overlay is plated onto the bearing metal and is just thick enough to absorb the dirt particles, but not so thick as to reduce the strength of the bearing. Such an overlay is shown in Fig. 9.19.

Fig. 9.19 *Embeddability: the bearing material is cast leaded bronze (designated VP2) and is electroplated with a lead–indium overlay; the dark, central debris particle is composed of silica, while the two light-coloured particles are ferrous (i.e. either steel or cast iron)*

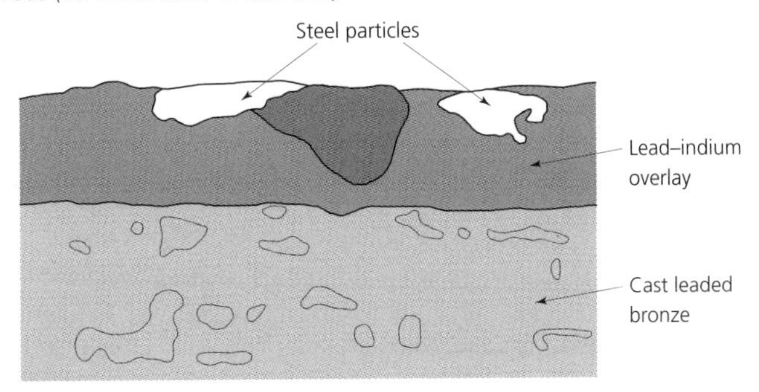

9.15.3 *Friction*

The use of 'Teflon' (PTFE) for solid plain bearings has already been considered. It is also used as a surface coating on machine-tool slideways where friction and the problems associated with 'stick-slip' has to be reduced to a minimum – for example, in computer-controlled machine tools where excessive 'stick-slip' could adversely affect accurate positioning of the worktable. Teflon has the lowest coefficient of friction of any known bearing material.

Phosphide coatings also have a low coefficient of friction and are frequently used to coat steel sheet and steel wire before extreme cold-drawing processes.

9.16 Strength of sliding bearing materials

An important factor that must be taken into account when assessing the suitability of a sliding bearing material is that the effective strength of a thin layer of material increases rapidly as its thickness decreases. This is shown for a typical Babbitt metal in Fig. 9.20. For example, micro-Babbitt (Babbitt metal bearings less than 0.01 mm thick) layers are substantially stronger than the more normal thickness of 0.4 mm. Again, plated overlays of very soft alloys such as lead–indium and lead–tin can be used over a copper–lead intermediate strata without reducing the strength of the bearing. The plating thickness, in this instance, has to be a compromise between strength (which requires a thin film) and embeddability (which requires a thicker film). Table 9.2 lists some of the thin-walled materials in current use and the load/projected area they can be expected to use in, say, a medium duty internal combustion engine.

Fig. 9.20 *Relationship between effective strength and thickness for a typical thin-walled Babbit-type bearing material*

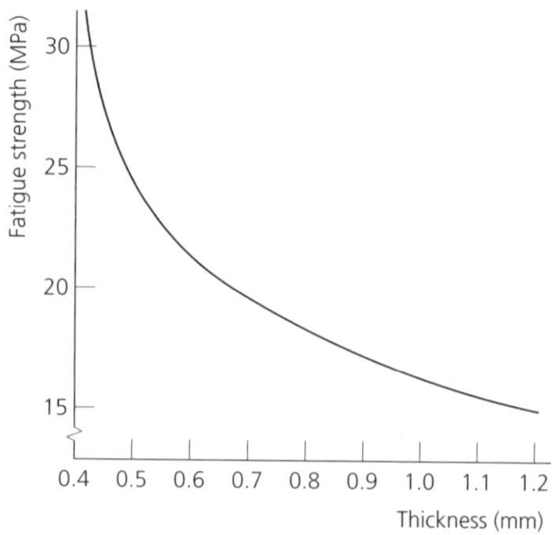

Table 9.2 *Safe loads for some typical thin-walled bearings*

Bearing material (thin walled)	*Specific load over the projected area (MPa)*
Lead–tin based Babbitt metal	12.50
Micro-Babbitt metal	17.50
Sintered copper–lead alloy (with plated overlay)	60.00
Recticular 20% tin, aluminium–tin alloy	42.00
Cast 23% lead, copper–lead alloy, 0.02 mm overlay	38.00
Cast 17% lead–bronze alloy, 0.02 mm overlay	60.00
Cast 10% lead–bronze alloy, 0.02 mm overlay	69.00

Finally, the bearing material must also be compatible with the shaft material with which it is used. This is a complex subject beyond the scope of this book. However, data are available from bearing manufacturers for combinations of materials that have been proved to be satisfactory by prolonged experimental research and practical experience.

SELF-ASSESSMENT TASK 9.3

1. Compare and contrast the properties and applications of 'white' bearing metals and 'tin-bronze' bearing metals.
2. (a) Discuss the significance of surface contamination in bearing materials.
   (b) Discuss the combination of similar and of dissimilar materials in bearings.
3. Discuss the significance of surface coatings on bearing materials in terms of wear resistance and embeddability.

9.17 Some commercially available bearing materials

Having considered the mechanical aspects of bearings together with a survey of some bearing materials, let's now consider some commercially available materials, their manufacture and their applications, before considering rolling bearings. Flow diagrams of the manufacturing processes for a range of plain bearing materials are given in Fig. 9.21. Note that the edge preparation of the steel strip for cast copper-based materials refers to turning up the edges of the strip to form a shallow 'tray' into which the molten copper alloy can be cast. The blanks for bearing shells are stamped from this strip and pressed to shape.

9.17.1 *Aluminium-based materials*

Glacier Vandervell AS15

This bearing material has a composition of:

- Tin (Sn) 20 per cent.
- Copper (Cu) 1 per cent.
- Aluminium (Al) remainder.

This was the first commercially available aluminium – 20 per cent tin-bearing material. This percentage of tin gives the alloy optimum seizure resistance whilst maintaining the necessary mechanical properties without the need for an overlay (unlike the 'hard' bronzes). Superior tensile strength is achieved by working and annealing treatments that give continuous tin along the aluminium grain edges but not along the grain faces. This strong alloy matrix is known as a reticular tin structure. This bearing material is used for a wide range of applications for both camshaft and crankshaft bearings, ranging from single-cylinder compressors to commercial vehicle diesel engines.

Glacier Vandervell AS1241

This bearing material has a composition of:

- Tin (Sn) 12 per cent.
- Silicon (Si) 4 per cent.
- Copper (Cu) 1 per cent.
- Aluminium (Al) remainder.

This bearing material has a similar reticular tin structure to Glacier AS15 but with silicon associated with the tin network. Copper is present within the aluminium phase as a solid solution hardener. This bearing material is used principally in connecting rod bearings against nodular cast-iron crankshafts where the silicon content helps to 'condition' the cast-iron bearing surfaces. This material does not require an overlay.

Glacier Vandervell AS11

This bearing material has a composition of:

- Tin (Sn) 6 per cent.
- Nickel (Ni) 1 per cent.

Fig. 9.21 *Manufacturing processes of plain bearing materials (reproduced courtesy of Glacier Vandervell Ltd)*

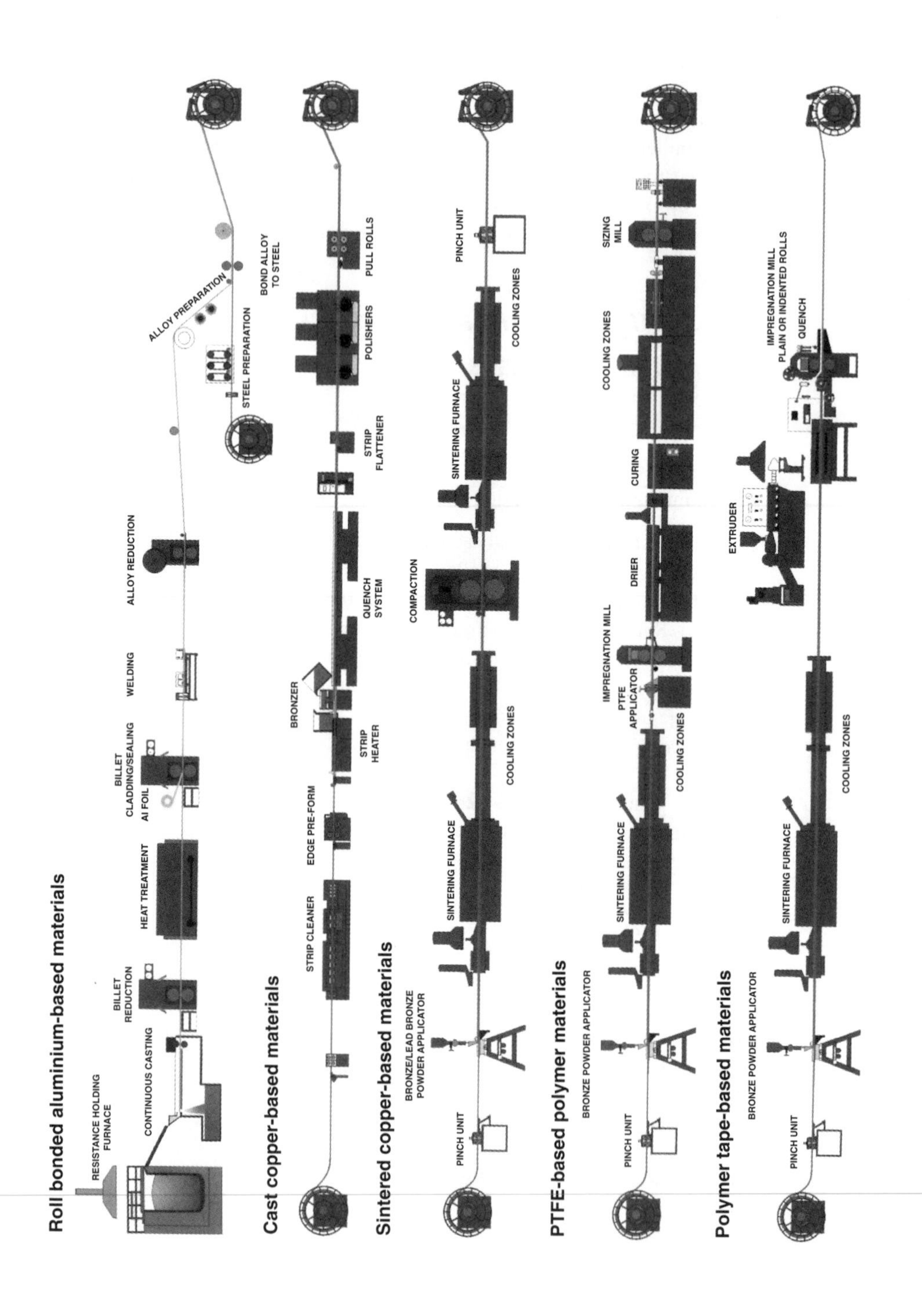

- Copper (Cu) 1 per cent.
- Aluminium (Al) remainder.

This bearing material has a reduced tin content compared with other aluminium-based bimetals and is often chosen as a thrust washer material. Its surface properties can be enhanced for use as a high strength, corrosion resistant, crankshaft material by use of an overlay.

9.17.2 *Cast copper-based materials*

Glacier Vandervell VP2

This bearing material has a composition of:

- Lead (Pb) 23 per cent.
- Tin (Sn) 1.5 per cent.
- Copper (Cu) remainder.

This is a lead-bronze material continuously cast onto steel strip. Casting and quenching conditions are controlled in order to produce a dendritic microstructure with vertically oriented bronze columns. This gives the bearing material its superior strength by increasing its compressive load capacity. This bearing material is overlay plated with either lead–indium or lead–tin–copper and is specified in many applications with high specific loads, such as highly rated diesel engines and high-performance engines.

Glacier Vandervell VP1

This bearing material has a composition of:

- Lead (Pb) 17 per cent.
- Tin (Sn) 5 per cent.
- Copper (Cu) remainder.

The reduced lead and increased tin content of this bearing material increases its load-carrying capacity. It is used unplated for bushes and when plated with an overlay it is used for highly loaded half-bearing applications where cavitation corrosion resistance is required.

Glacier Vandervell VP10

This bearing material has a composition of:

- Lead (Pb) 10 per cent.
- Tin (Sn) 10 per cent.
- Copper (Cu) remainder.

This bearing material is a very high-strength bronze. Its load-carrying capacity and shock resistance means that it is ideal for the small end bushes of the connecting rods used in diesel and high-performance engines.

9.17.3 *Sintered copper-based materials*

Glacier Vandervell SP

This bearing material has a composition of:

- Lead (Pb) 26 per cent.
- Tin (Sn) 1.5 per cent.
- Copper (Cu) remainder.

This bearing material is produced by spreading a lead-bronze powder onto a steel strip. The strip is then heated in a sintering furnace, compacted by re-rolling and finally resintered. The sinter process allows a wide range of material compositions to be produced even where the materials are normally immiscible in the molten condition. When plated with a lead–tin overlay it can be used in similar but slightly less arduous conditions to the VP2 bearing material. Sintered lead-bronze bearing materials manufactured in a similar manner are also available with properties comparable with the cast lead-bronze bearing materials such as VP1 and VP10.

9.17.4 *Trimetal engine technology*

Figure 9.22 shows a trimetal bearing. This has significantly improved wear resistance and a longer life than bearings that have conventional electroplated overlays. This is achieved by the inclusion of alumina particles in a lead-tin overlay of standard thickness. Trimetal bearings have improved fatigue strength and excellent cavitation corrosion resistance at an affordable price.

Fig. 9.22 *A trimetal engine bearing (reproduced courtesy of Glacier Vandervell Ltd)*

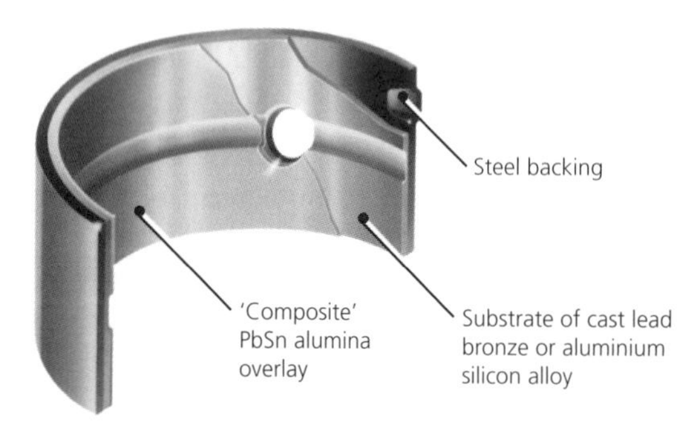

9.17.5 *White metals*

Glacier Vandervell B1 and B99

B1 bearing material is tin based and has a composition of:

- Antimony (Sb) 7.5 per cent.
- Copper (Cu) 3 per cent.
- Tin (Sn) remainder.

B99 bearing material is lead based and has a composition of:

- Antimony (Sb) 10 per cent.
- Tin (Sn) 6 per cent.
- Lead (Pb) remainder.

The use of white metals is declining but they are used in some internal combustion engine camshaft bushes.

9.17.6 *Polymer-based composite materials*

Glacier DU

This is a polymer-based bearing designed to operate without lubricant, or in environments where the working fluid is not a conventional lubricant. It incorporates PTFE and lead impregnated into a layer of porous tin-bronze on steel, and will operate from −200 °C to +280 °C. This material resists most solvents and many harsh environments, has negligible 'stick-slip' and is tolerant of dusty environments. It is available in bush, flanged bush, thrust washer or strip form and has applications in many automotive components from door hinges to shock absorbers and hydraulic pumps.

Glacier DX

This bearing material is designed to operate with marginal lubrication or in an environment where continual lubrication is not available, and is particularly suited to oscillating, frequent stop/start and high/low speed conditions. It has a similar structure to Glacier DU, consisting of a porous bronze substrate but impregnated with an acetal copolymer. The bearing surface is supplied indented for grease lubrication or non-indented where fluid lubrication is available. Glacier DX can operate between −40 °C and +130 °C and is available in bush, thrust washer or strip form. It is used in many automotive applications such as starter motor bushes and steering control arm bushes, and also in a diverse range of industrial application.

Glacier Hi-ex

This bearing material can operate under similar conditions to Glacier DX but has a surface layer consisting of PEEK (polyether ether ketone) combined with various fillers including PTFE and graphite. It can operate from −150 °C to +250 °C and its fatigue resistance under fluid film lubrication conditions approaches that of aluminium bimetals. Typical applications are compressor main bearings and diesel fuel pump bushes.

Glacier DP3

This material has been developed for suspension applications to withstand more demanding strut operating conditions. It is designed to withstand the increased side loads and rod flexure that has made conventional strut guide materials susceptible to wear and cavitation erosion. This material achieves a strengthened PTFE matrix by incorporating finely divided calcium fluoride into its structure. Figure 9.23 shows a schematic diagram of the structure of a polymer-based bearing. The polymer-based bearing materials described above are impregnated into a layer of porous bronze on a steel backing shell.

Fig. 9.23 *Polymer-bearing structure (reproduced courtesy of Glacier Vandervell Ltd)*

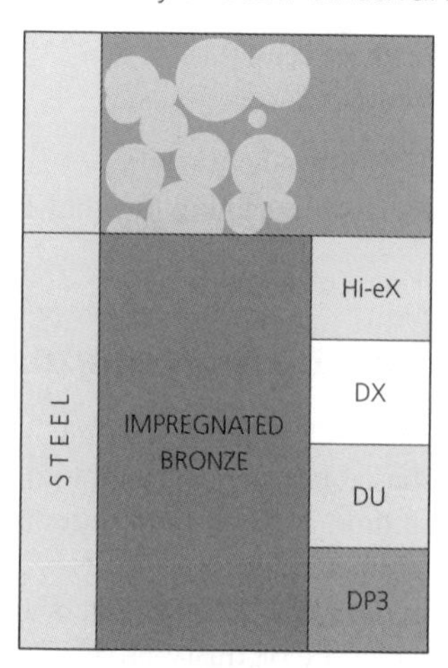

9.18 Materials for rolling bearings

Steels used for bearing rings or washers and the rolling elements must be capable of being adequately hardened and must have high fatigue strength and wear resistance. The structural and dimensional stability of the bearing components must be satisfactory at the operating temperatures that can be expected. In many cases the choice of a particular steel is dictated by the manufacturing techniques used, e.g. deep drawing for the drawn cups of certain types of needle roller bearings.

9.18.1 *Through-hardening steels*

The most common through-hardening steels for rolling bearings is a carbon–chromium steel containing approximately 1.0 per cent carbon and 1.5 per cent chromium. A typical alloy is BS 970: 534A99. Its composition is 1.0 per cent carbon, 0.45 per cent manganese, 1.40 per cent chromium, and the remainder iron. It is through hardened by oil quenching from 810 °C and tempering at 150 °C. For bearing components having large cross-sections, steels alloyed with manganese and molybdenum are used because of their superior through-hardening properties (increased) ruling section.

9.18.2 *Case-hardening steels*

Case-hardening steels are used for bearings that have to operate under severe conditions where they will be subjected to heavy shock loads. Chromium–nickel and manganese–chromium alloyed steels with a carbon content of approximately 0.15 per cent are those case-hardening steels most commonly used for rolling bearings.

In the majority of applications there is virtually no difference in behaviour between bearings made from through-hardened steels or case-hardened steels. This fact has been acknowledged by ISO in that no distinction is made between steel types in the life calculations of bearings. Steel cleanliness and proper manufacturing methods, as well as bearing design, are the decisive factors. However, there are applications where a particular type of steel has certain advantages. Special steels and heat-treatment procedures are required for temperature-resistant bearings and for corrosion-resistant bearings.

Recently, the fatigue resistance of rolling bearings (usually the limiting factor in the duration of their working life) has been improved 2.5 times by lowering the inclusions in the steel, particularly by vacuum de-gasing. This is an expensive process and is only used where service conditions warrant it, as in aircraft and marine gas turbines and jet engines. A cheaper process that gives similar results is *electro-slag-refining* of the steel (e.s.r.), which is also known as *electro-flux-refining* (e.f.r.). Special steels have been used for corrosive environments and high-temperature environments. These tend to be less hard and the life of the bearing suffers accordingly. The normal maximum operating temperature for a rolling bearing is 150 °C. In any case, lubrication problems occur above this temperature.

9.18.3 *Materials for cages*

The main purpose of the cage is to keep the rolling elements at an appropriate distance from each other and to prevent immediate contact between two neighbouring rolling elements in order to keep friction, and thus heat generation in the bearing, to a minimum.

Pressed steel cages are stamped out of sheet steel and are standard for many deep-groove ball bearings, spherical roller bearings and most taper roller bearings. Machined steel cages are used for large-size bearings or where season cracking of brass cages would occur. To improve sliding and wear resistance properties some machined steel cages are surface hardened by carbonitriding.

Pressed brass cages are used for some small and medium-sized bearings, but most brass cages are machined from cast or wrought material. Brass cages should not be used above 300 °C. They are unaffected by most commonly used bearing lubricants, including synthetic oils and greases, and can be cleaned by using normal organic solvents. The use of alkaline cleaning agents is not recommended. Ammonia, as used in some refrigeration plants, causes *season cracking* in brass cages and machined steel cages must be used instead under such operating conditions.

Polyamide cages, made from glass fibre-reinforced polyamide 66, are used in some bearings. This material is characterised by a favourable combination of strength and elasticity. The good sliding properties of the plastic on lubricated steel surfaces, and the smoothness of the cage surfaces in contact with the rolling elements, means that little friction is generated by the cage so that heat generation and wear in the bearing are at a minimum. The low density of the material means that the inertia of the cage is small. The excellent running properties of polyamide cages under lubricant starvation conditions permits continued operation for a time without risk of seizure and secondary damage.

The organic solvents – such as white spirits or trichloroethane – normally used to clean rolling bearings do not affect cage properties, nor do dilute alkali cleaners at room temperature providing the contact period is short. Ammonia, the chlorofluorocarbons and their more environmentally friendly substitutes now used in refrigeration equipment do not attack polyamide. In vacuum, however, polyamide cages become brittle due to dehydration.

EXERCISES

9.1 Discuss the causes of energy losses in sliding and rolling bearings under normal operating conditions.

9.2 Discuss the reasons why a shaft and a bearing are normally made from dissimilar metals.

9.3 Explain the essential differences between full fluid lubrication and boundary layer lubrication.

9.4 State the essential difference between oil and grease, and explain with examples where each should be used.

9.5 Research manufacturers' literature and select a suitable material for each of the following bearings and indicate any surface treatment that may be an advantage:

(a) a drill bush
(b) a main bearing shell for a car engine
(c) the balls and races of an antifriction bearing
(d) a water pump impeller bearing
(e) the 'brasses' for a plumber block type journal bearing

9.6 State which of the following properties are hard phase and which are soft phase. Write brief notes on the importance of each property.

(a) load-carrying capacity
(b) compatibility
(c) wear resistance
(d) cavitation erosion resistance
(e) conformability
(f) embeddability

9.7 With the aid of diagrams, explain the difference between bimetal and trimetal bearing materials.

9.8 Discuss the advantages and limitations of polymer-based composite bearing materials.

9.9 Discuss the advantages and limitations of the following materials for manufacturing the cages of rolling bearings:

(a) steel
(b) brass
(c) glass-reinforced polyamide

9.10 Research manufacturers' literature and discuss the materials used for the load-bearing components of rolling bearings and their heat treatment.

10 Tool materials

The topic areas covered in this chapter are:

- Requirements of cutting-tool materials.
- Requirements of flow forming tool materials.
- Requirements of electrode materials for ECM and EDM processes.
- Metallic tool materials.
- Non-metallic and composite tool materials.
- Selection of tool materials.

10.1 Introduction

The choice of a cutting-tool material for a particular application is determined principally by technical and economic requirements. The technical requirements are dependent upon the process, the material being processed, the rate of processing, the equipment being used and the condition of that equipment. The economic requirements are related to such factors as: initial tool cost; tool life; refurbishment costs versus 'throw-away' and replacement costs; rate of processing, and batch size.

Tool materials must have sufficient strength to resist the forces acting upon the tool; sufficient hardness to resist wear; and, in the case of cutting tools, the ability to maintain a sharp cutting edge. Tool materials must also be resistant to softening at the elevated temperatures that often accompany the processing of engineering materials. Such properties may be *intrinsic* as in carbides and diamonds that are naturally hard and heat resistant, or they may be *conferred* as in metal alloys where heat treatment is required to increase their hardness and strength. In the case of carbides, wear resistance may be increased by the application of a hard 'coating'. The problems relating to the selection of a suitable tool material for a particular process can be complex and require detailed knowledge and wide experience. The following, simple example describes some of the factors that need to be taken into account when selecting suitable tooling and tool materials for the process to be used.

10.1.1 *Production by hand*

Consider the component shown in Fig. 10.1. There are several ways by which the step can be produced. If only one or two components are to be produced, then the flat could be filed on by hand.

Engineers' files are usually made from a plain carbon steel with a high-carbon content (1.2 per cent carbon). In order that the teeth of the file will be hard enough to cut metal, the steel from which the file is made will be hardened and tempered. For a hand tool this is a very suitable material. It takes a keen cutting edge that reduces the manual effort required and insufficient heat is generated whilst filing to *draw the temper* of the tool. It is also readily available and has a relatively low initial cost.

Fig. 10.1 *Component with flat; material BS970.080M40 (dimensions in millimetres)*

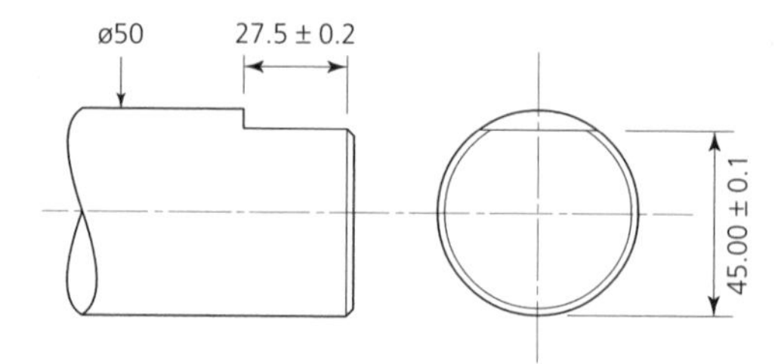

10.1.2 *Production by machining*

If larger batches of the component are to be made, it would not be economical to produce the flat by hand filing and a milling process would most likely be used. The choice then lies between the use of *high-speed steel* or *cemented carbide* cutting tools. The economic rate of production (productivity) expected from a machining process precludes the use of high-carbon steel tools since the heat generated by cutting would quickly soften the cutting edge. For small and medium batch sizes, a conventional end-milling cutter made from high-speed steel would be suitable. The cost of the cutter is reasonable; it would have an adequate life between regrinds and could be re-sharpened relatively easily.

However, if the rate of production had to be increased to satisfy the demands of the customer, and if the batch size warranted the financial outlay, then a cutter with cemented carbide teeth would be used. This choice assumes that the milling machine is sufficiently powerful to exploit the advantages of cutting with carbides. The choice then has to be made between a milling cutter in which the carbide inserts are brazed permanently in place and a milling cutter in which disposable inserts are clamped in place. The former type of cutter has a lower initial cost but refurbishment by grinding the cutting edges is relatively costly, whilst the latter type will have indexable inserts so that refurbishment merely involves indexing the inserts so as to present a fresh and sharp cutting edge. When all the cutting edges have become dulled, the inserts are rejected and new, standard inserts are clamped quickly and easily in place.

This example shows that the choice of tool material and its application involves some quite complex technical and economic decisions even for a simple operation. The tooling engineer would solve this relatively simple operation by personal experience of previous, similar jobs but, for more complex machining operations, mathematical modelling would have to be resorted to in order to arrive at the correct choice.

One simple technique is the use of a break-even diagram. Figure 10.2 shows such a diagram comparing the batch size and relative costs of using a high-speed steel drill and a cemented carbide drill. It can be seen that the initial cost of the carbide drill is greater than that for the high-speed steel drill. However, the unit production costs with the carbide drill is lower since it can be used at higher cutting speeds. The lines cross at the *break-even* point. To the left of this point it is more economical to use the high-speed steel drill, but to the right of this point it is more economical to use the carbide drill, despite its greater initial cost.

Fig. 10.2 *Comparison of HSS and carbide drills*

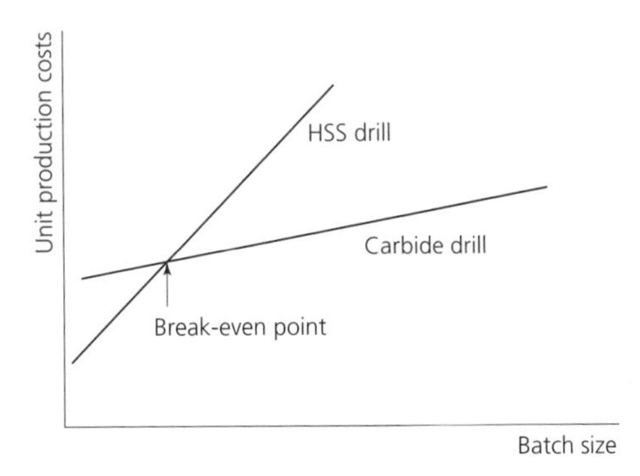

Some of the more common areas of application for tool materials are:

- Cutting tools (machining).
- Press tools (cutting and forming).
- Plastic moulds.
- Forging dies.
- Drawing and extrusion dies.
- Electrodes for ECM and EDM machining.

Let's now consider the properties required by tool materials for these various applications in detail.

SELF-ASSESSMENT TASK 10.1

1. List the main requirements of a typical cutting-tool material.
2. Explain briefly why high-carbon content, plain carbon steels are widely used for hand tools but are unsuitable for tools used on metal-cutting machines under modern production conditions.

10.2 Requirements for cutting tools (machining)

Cutting tool materials for the machining of metallic and non-metallic materials require the following general properties.

10.2.1 *Strength*

The tool material will require sufficient strength to resist the cutting forces acting upon the tool. The nature and magnitude of these forces, in turn, depend upon a number of factors. Figure 10.3 shows the difference between positive and negative rake cutting. With positive rake geometry, the tip of the tool is in *shear*, whilst with negative rake geometry the tip of the tool is in *compression*. This is why brittle tool materials such as tungsten carbide are generally used with a zero or negative rake-cutting geometry.

Fig. 10.3 *Effect of tool geometry on tangential cutting force: (a) positive rake; (b) negative rake*

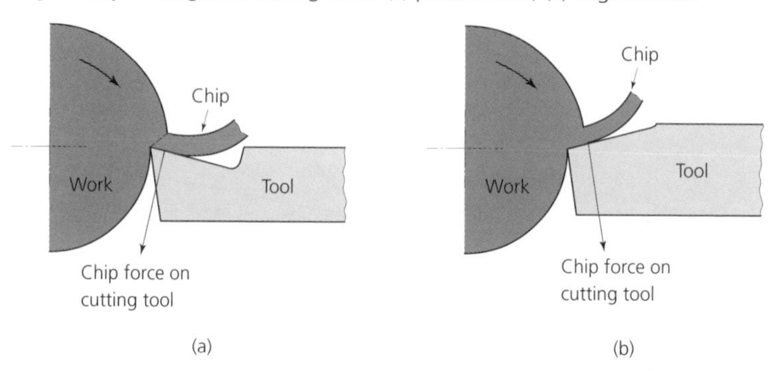

The magnitude of the cutting forces depends upon the properties of the material being cut and the area of cut but is independent of the cutting speed. Increasing the cutting speed increases the power required but not the tangential cutting force on the tool. This increased power (increased rate of doing work or using energy) results in greater heat energy being generated in the cutting zone and a corresponding rise in temperature. This, in turn, leads to softening of the cutting edge and a reduction in the life of the tool or even its destruction.

10.2.2 *Toughness*

Where intermittent cutting takes place or when cast or forged components with rough and uneven surfaces are being rough machined, the tool material requires the property of toughness as well as strength. Unfortunately toughness is usually achieved at the expense of hardness.

10.2.3 *Hardness*

In order to retain a sharp cutting edge over a reasonable tool life, a cutting tool must be substantially harder than the material being cut. Unfortunately many very hard materials are also brittle and relatively weak. Thus very hard materials, that will retain a keen cutting edge, are generally used for light, high-speed finishing cuts, whilst for roughing and general machining less hard but tougher cutting-tool materials are used.

A cutting-tool material must be capable of retaining its hardness at the high temperatures encountered in the cutting zone (Section 10.9). Although high-carbon steels can achieve very high hardness values, they start to soften at relatively low temperatures (just above the boiling point of water) and, therefore, they are only suitable for hand tools.

Cutting-tool materials must also be resistant to thermal shock – that is, they must not crack when alternately heated and cooled, as occurs when the coolant supply is inadequate or applied manually.

10.2.4 Abrasion resistance

This property, which is a function of hardness, is also required in a cutting-tool material to withstand the scouring action of the chip as it flows over the rake face of the tool. Not only is abrasion resistance required to prevent 'cratering' just behind the cutting edge, it is also required to ensure that the rake face of the tool retains a smooth, low-friction surface. Cratering is the wearing of a hollow in the tool just behind the cutting edge, as shown in Fig. 10.4. This weakens the tool, resulting in reduced tool life caused by failure of the tool. Roughening of the rake face of the tool increases the friction between the chip and the tool. This increased friction leads to increased heating of the tool and a reduction in tool life. It also leads to chip welding and the formation of a 'built-up edge', as shown in Fig. 10.5. This

Fig. 10.4 *Cratering of a cutting tool*

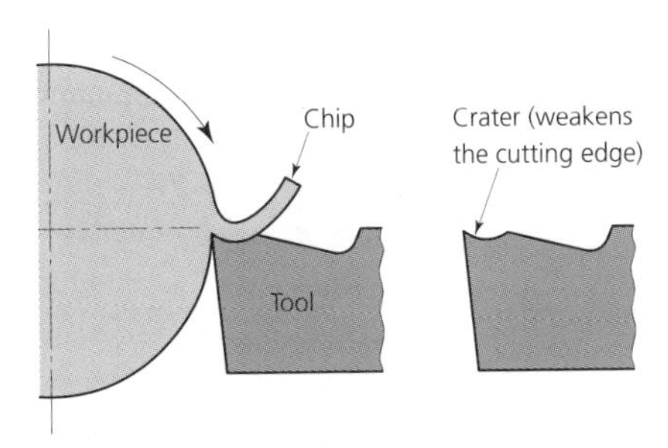

Fig. 10.5 *Chip welding (built-up edge): (a) layers of chip material form on the rake face of the tool; (b) excessive chip welding produces an unstable built-up edge; particles of built-up edge material flake away and adhere to the workpiece, making the machined surface rough; they also adhere to the chip, making it jagged and dangerous; the result is a poor surface finish*

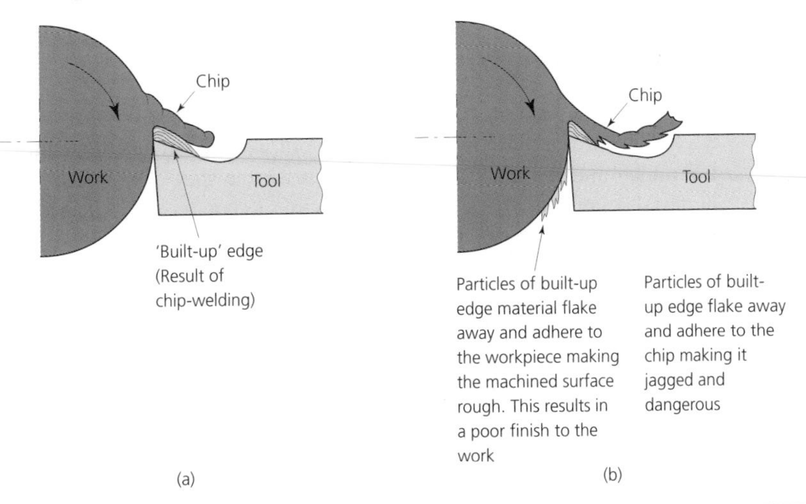

results in a further lowering of the cutting efficiency of the tool and a poor finish on the work. Additional information on metal cutting and the requirements of metal-cutting tools can be found in *Manufacturing Technology*, Volumes 1 and 2.

10.2.5 *Compatibility*

Cutting-tool materials must not react chemically with the materials being cut under normal process conditions. Mutually reactive materials may either cause corrosion of the rake face of the tool (this would aggravate mechanical erosion) or they could aggravate any tendency towards chip welding if they have an affinity for each other.

10.3 Requirements for press tools (cutting)

The requirements for press tools depend upon whether the pressing operation involves cutting (e.g. blanking or piercing) or forming (e.g. bending or cupping). Figure 10.6 shows a section through a blanking tool, and it can be seen that the conditions are different to those for cutting metal on a machine tool.

Fig. 10.6 *Blanking tool*

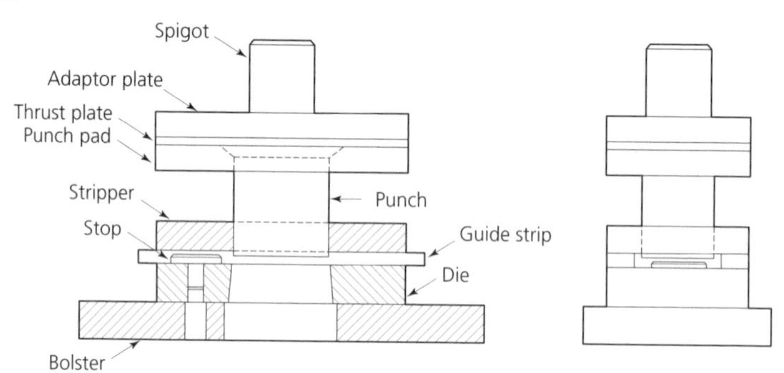

10.3.1 *Strength*

It is apparent from Fig. 10.6 that the punch and die are in compression close to their cutting edges. Therefore, material with a high compressive strength is required. At the same time the punch and die are subjected to considerable shock loads each time they close on the material being cut. Therefore, the punch and die materials also need to be tough and shock resistant.

10.3.2 *Hardness*

As for any cutting tool, the punch and die must retain sharp cutting edges and have an economical tool life. Therefore, the cutting-tool material must again be substantially harder than the material being cut. Unlike machining, the cutting action for press tools is intermittent and any heat generated is dissipated between each blow of the tool. Further, there is a greater mass of metal behind the cutting edges to conduct the heat away. This is reflected in the alloy steels used in making press tools. The quantity of the refractory metals such as tungsten and cobalt is much reduced compared with machining tool alloys and may even be omitted altogether. The exception is the tooling used in high-speed 'dieing' machines that produce small components at very high rates of production. Here considerable heating occurs and cemented carbides and high-speed steels frequently have to be used.

10.3.3 *Abrasion resistance*

Considerable abrasion occurs where the blank passes through the throat of the die just behind the cutting edge. Therefore, the die steel must be highly abrasion resistance or die wear will occur at this point causing 'bell-mouthing' of the die orifice. This will not only reduce the cutting efficiency, resulting in a heavy 'burr' round the edge of the blank, but may also result in the blanks not dropping clear and becoming jammed in the tool. When this happens, the tools and the press can suffer severe damage.

10.4 Requirements for press tools (forming)

There are many types of forming tools used in presswork. Figure 10.7 shows a cupping tool. As for all forming tools used in presswork, the properties required are somewhat different to those for blanking and piercing (cutting) tools.

Fig. 10.7 *Cupping tool*

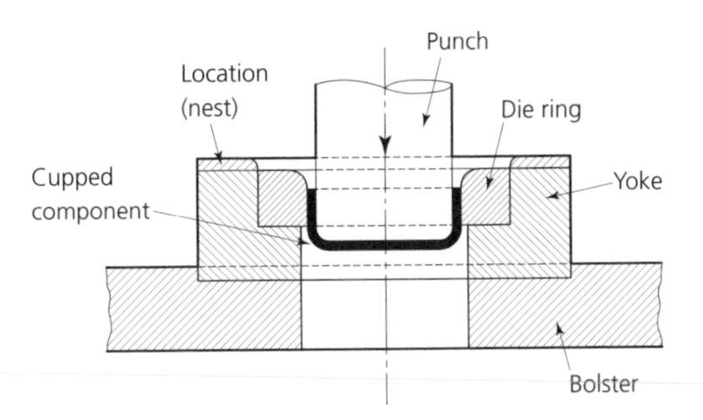

10.4.1 *Strength*

In press forming tools the working surfaces will be in compression whist forming is taking place. For the example shown, the die ring of the cupping tool will also be subjected to severe hoop stresses. Therefore, it is usual to use die cheeks of a hard, wear-resistant material set in a yoke of low- or medium-carbon steel. The yoke has a high tensile strength and prevents the tool bursting. As for all press tools, forming tools must also have a high resistance to impact loading.

10.4.2 *Hardness*

The wearing surfaces are usually hardened, ground and polished to reduce friction and to prevent marking the surfaces of the material being formed. 'Through hardening' is preferable to 'case hardening' as it prevents the die cheeks from collapsing under load and introducing inaccuracies in the formed component.

10.4.3 *Abrasion resistance*

Most of the wear on forming tools is caused by abrasion as the blank bends or flows to its finished shape. Tools where the load per unit area is high are made from high-carbon–high chromium (HCCr) die steels. For large forming tools used for shaping relatively thin materials, as in car body panels where the load per unit area is quite low, forming and deep-drawing tools are frequently made from high-grade grey cast iron as this material has excellent antifriction and wear-resistance properties.

A lubricant should be used to reduce die wear and special lubricants are available that will withstand the high point pressures that often occur in forming process. These pressures are very much greater than those found in bearings, and ordinary mineral lubricating oils are only suitable for the most undemanding forming processes. In extreme cases it is not unusual to subject the blanks to pickling in order to ensure a chemically clean surface. The pickled blanks are then treated with a phosphate bonding agent and lubricated with a metallic stearate soap. The phosphate coating ensures that the lubricant adheres to the surface of the blank and also provides additional, extreme pressure lubrication in its own right.

10.5 Requirements for moulds and dies

Figure 10.8 shows a section through a typical plastic mould suitable for positive moulding, whilst Fig. 10.9 shows a section through a typical injection mould. Die-casting dies are similar to the injection mould in principle but they have to operate at much higher

Fig. 10.8 *Positive mould for thermosetting plastic materials*

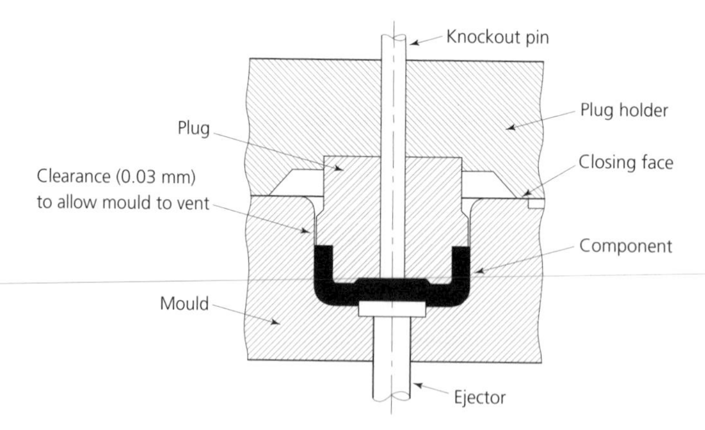

temperatures and pressures. Therefore, plastic moulding tools and die-casting dies require similar properties to each other. The tool materials, from which both of them are made, have to operate at elevated temperatures without their properties being impaired and they are both subjected to high bursting forces.

Fig. 10.9 *Injection mould for thermoplastic materials*

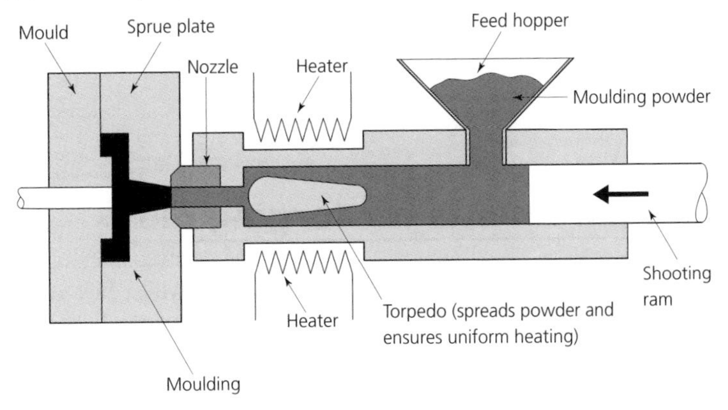

10.5.1 *Strength*

The working surfaces of plastic moulds and die-casting dies are subjected to very high compressive forces from the entrapped or injected workpiece material. It is essential that the metal does not collapse under these forces otherwise the mould or die cavities will become misshapen and enlarged, resulting in inaccurate components. The moulds and dies are also subjected to very high bursting forces. Therefore the die material must combine high tensile and high compressive strengths. It is reinforced additionally by being set in a yoke of medium-carbon steel. Die-casting dies are generally subjected to considerably greater forces than plastic moulds.

10.5.2 *Hardness*

Both plastic mouldings and die castings are expected to have high levels of accuracy and surface finish so that finishing costs are reduced to a minimum. They may also contain a considerable amount of fine detail. Since moulds and dies are very costly to produce, it follows that they must have a long working life and be capable of producing many thousands of components before replacement. Therefore their working surfaces must be extremely hard and wear resistant. Because of the high quality of lustre and finish expected on plastic mouldings, the mould cavity is usually hard chromium plated prior to finishing to size and polishing. Hard chromium plating, which provides a relatively thick, wear-resistant surface, must not be confused with decorative chromium plating which is only a few microns thick. The moulds and dies are constantly heating up and cooling down. This thermal shock must not cause surface cracks in the mould or die that could lead to the surface flaking away.

10.5.3 *Abrasion resistance*

The moulding powder and granules used in the manufacture of plastic components and the scouring action of the molten metal in die casting both cause abrasive wear on the walls of the mould and die cavities. The mould and die materials have to be resistant to this abrasion, and must also be compatible with the materials being moulded or cast. At the elevated temperatures involved these materials tend to become chemically reactive with the mould and die and materials and this chemical erosion can cause early destruction of the mould and die cavity surface finish.

10.6 Requirements for forging dies

Figure 10.10 shows a typical set of forging dies, whilst Fig. 10.11 shows the temperature ranges for a variety of ferrous and non-ferrous metals and alloys. You can see that forging operations are carried out at very much higher temperatures than any of the processes described so far. However, because of the high temperatures involved, oxidation of the surface of the workpiece occurs and, generally, accuracy and finish are of a much lower order than for the processes described so far.

Fig. 10.10 *Examples of drop forging dies*

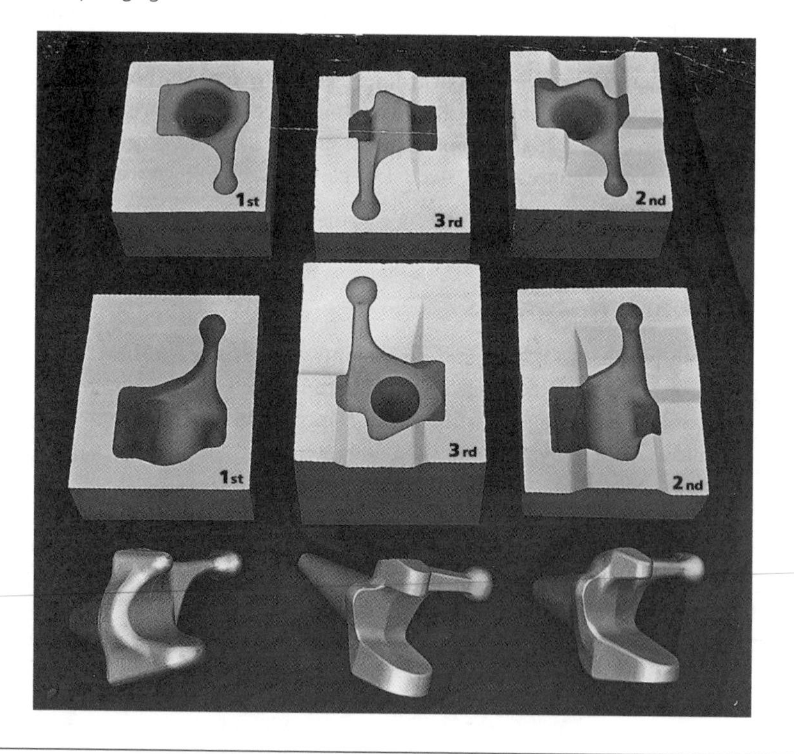

Fig. 10.11 *Typical forging temperatures*

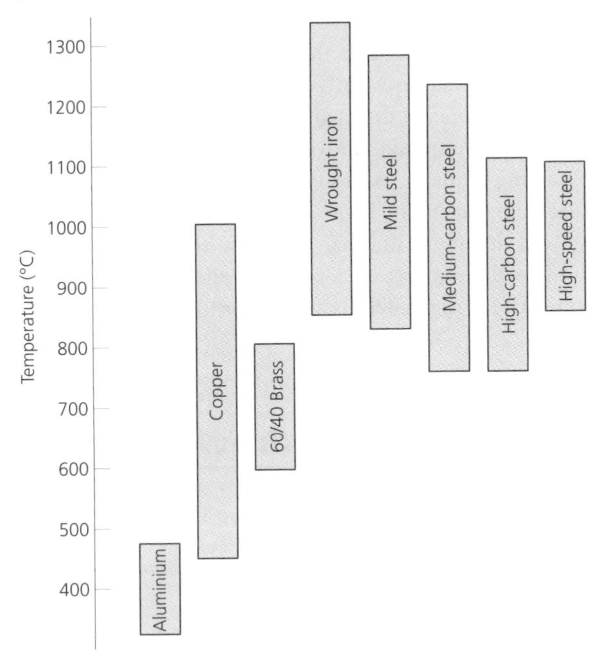

10.6.1 *Strength*

As the cavity walls of forging dies are subjected to high compressive loading, the die material must have high compressive strength. At the same time the die material must have a high resistance to impact loading.

10.6.2 *Hardness*

The walls of the cavities of forging dies are subjected to considerable abrasive wear, not only from the flow of the metal in the dies but also from the scale that exists on the surface of the heated workpiece material. Thus forging dies must be hard as well as strong and tough. This hardness must be maintained at the high temperatures at which the dies have to operate. However, because of the mass of the dies and the fact that they are only in intermittent contact with the workpiece materials, their working temperature is not as high as Fig. 10.11 would imply. The intermittent heating and cooling of the dies leads to thermal shock that can cause surface cracking of the die cavity walls and this can lead to flaking of the walls of the cavity and early failure of the dies.

10.6.3 *Abrasion resistance*

As has already been stated, the high temperatures involved results in the workpiece material becoming covered in a highly abrasive oxide film called 'scale'. This scale lacks plasticity, flakes off the workpiece material during the forging operation and becomes trapped in the die cavity where it causes abrasive wear. The die material must, therefore, be resistant to this wear. Scale that has flaked from the hot workpiece during the forging process is removed between operations by the application of a high-pressure spray of release-agent/lubricant directed into the die cavities.

Brass and aluminium alloys are often used to produce hot stampings, particularly for the plumbing industry. Although similar in principle to forging, the process is designed to produce components of high accuracy and good surface finish. Because hot stamping is carried out at only a dull-red heat, due to the lower melting temperatures of the alloys involved, scaling is less of a problem and can easily be removed by acid pickling. The properties of the die materials are similar to those required for forging.

10.7 Requirements for drawing and extrusion dies

Figure 10.12 shows the principles of wire and tube drawing whilst Fig. 10.13 shows the principle of hot extrusion. These processes produce wires, rods, tubes and sections with high length/diameter ratios.

Fig. 10.12 *Principles of metal drawing: (a) wire drawing; (b) tube drawing*

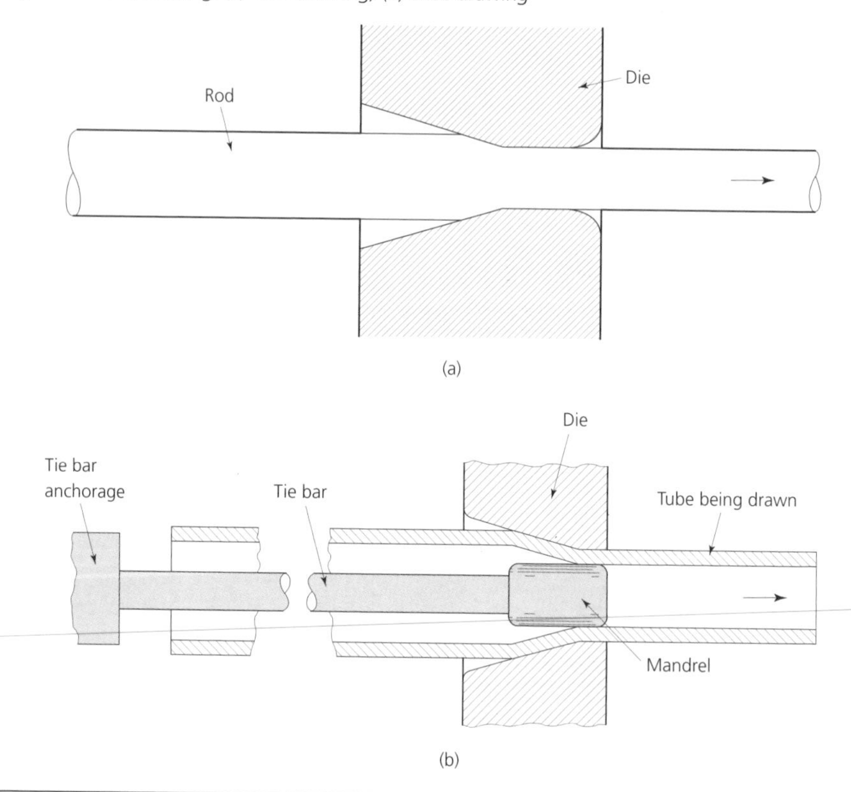

Fig. 10.13 *Extrusion: (a) commencement of extrusion stroke; (b) completion of extrusion stroke*

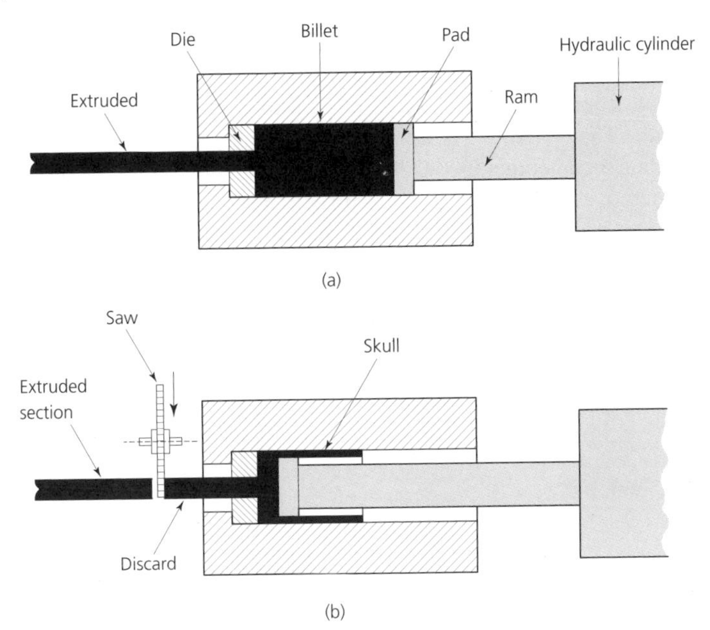

10.7.1 *Strength*

Drawing and extrusion dies are subjected to high levels of hoop stress. At the same time they are subjected to very high levels of surface wear. Therefore such dies are usually made from hard and wear-resistant materials surrounded by a high-tensile steel bolster or yoke that resists the hoop stresses.

10.7.2 *Hardness*

It has already been stated that drawing and extrusion dies need to be hard and wear resistant. This is because the metal being processed is in constant and moving contact with the working surfaces throughout the processing cycle and very considerable contact pressures are involved. For fine wire drawing the die is often made from industrial diamonds inserted in a suitable steel bolster. Extrusion dies are usually made from alloy steel that is heat resistant since the metal being extruded is at a dull-red heat.

10.8 Requirements for electrodes for ECM and EDM process

Electrode materials for electrochemical machining (ECM) and electric discharge machining (EDM) vary according to the technique being used. Generally, electrode materials must be good conductors of electricity, capable of being readily formed to shape, and non-reactive with the electrolyte being used.

10.8.1 *Electro-chemical machining*

The principles of this process are shown in Fig. 10.14. This process is the reverse of electroplating. Instead of adding metal to the surface of the workpiece, metal is stripped away from it. The tool never comes into contact with the workpiece and no wear takes place. The electrolyte is usually a solution of sodium chloride, but a solution of sodium nitrate is sometimes used. The electrolyte is pumped through the system at a high flow rate to remove the reaction products (dross). These are filtered out and the electrolyte is reused. The electrodes are usually made from brass, copper or bronze by any appropriate forming process. These materials have good electrical conductivity, adequate rigidity, are easily machined to shape, and do not react with the electrolyte.

Fig. 10.14 *Electrochemical machining (ECM)*

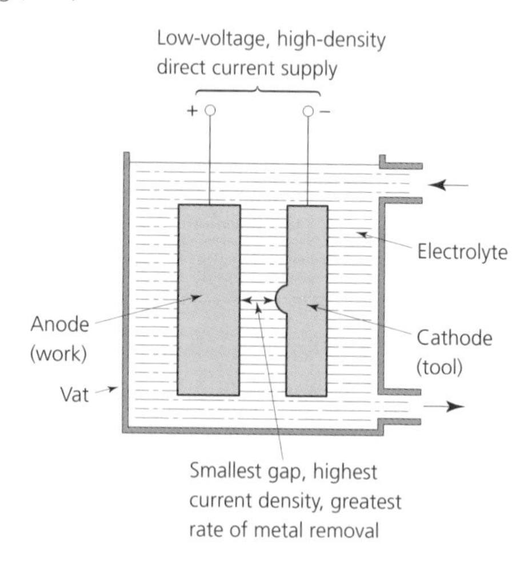

Electrochemical grinding is a variation on the above process. The electrode is the grinding wheel and, unlike conventional grinding wheels, the abrasive (aluminium oxide or diamond) particles are metal bonded. The abrasive particles take no part in the forming process, acting merely as insulators to space the wheel at a constant distance from the workpiece and also to help to remove the reaction products.

10.8.2 *Electric discharge machining*

The principles of this process are shown in Fig. 10.15. The surface of the workpiece material is eroded away by electric spark discharges. This process is also known as 'spark erosion'. The dielectric acts as an insulator to control the spark, as a coolant, and it also carries away the debris from the cutting zone. Tools for EDM are usually made from brass, copper, copper–tungsten alloy, or graphite. Tool wear is important since it affects the tolerances and the shape produced in the workpiece. Separate electrodes for roughing and finishing are often used. Tool wear is related to the melting temperatures of the tool materials used; hence graphite has the highest wear resistance, but has to be handled carefully since it is soft and relatively weak. Care must also be taken when machining graphite as the dust produced can cause serious illness if inhaled. Tungsten wire is used for machining fine holes in hard materials, particularly if the holes have a high length/diameter ratio.

Fig. 10.15 *Electrical discharge machining (EDM)*

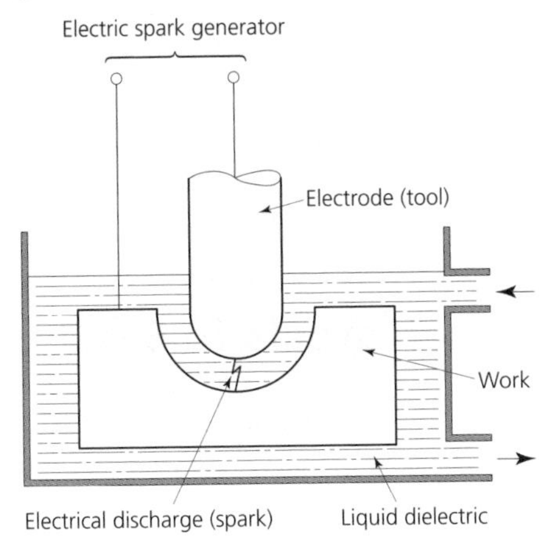

A variation of this process is continuous wire (travelling wire) EDM. This process is used for cutting out shaped holes in the workpiece, as shown in Fig. 10.16. The wire is continuously fed through the cut and, if subjected to sufficient wear, is used only once, which adds to the cost of the process, particularly if tungsten wire is used. Brass, copper or tungsten wire can be used. Again, the higher the melting temperature of the wire, the longer will be its life and the more accurate the cut made. No matter which technique is used the workpiece material must be an electrical conductor.

Fig. 10.16 *Continuous wire EDM*

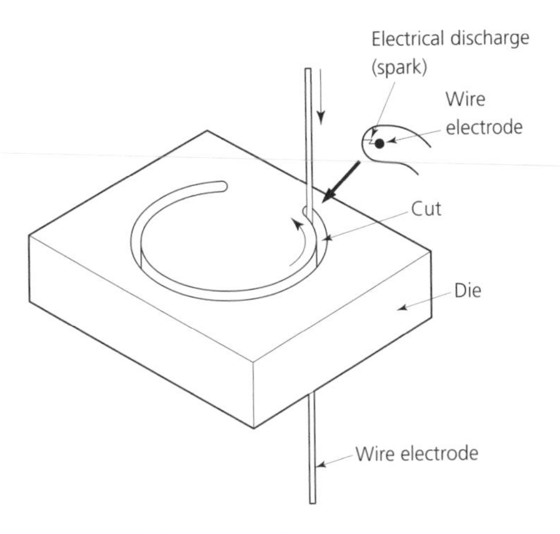

SELF-ASSESSMENT TASK 10.2

1. Research manufacturers' literature and select suitable tool materials for the following applications:
   (a) a press-blanking tool for tamping out discs 100 mm diameter from 1.0 mm thick sheet steel; batch size 25,000 blanks between regrinds
   (b) a pair of closed forging dies for a motor car engine connecting rod
   (c) an EDM electrode for sinking the impression in a plastic compression mould
2. Briefly explain the main difference in the properties required by a cutting-tool material compared to a press-tool material with a die steel for a mould.

10.9 Plain carbon steels (0.8–1.2 per cent)

Figure 10.17 compares the hardness of some typical cutting-tool materials at various temperatures. It can be seen that plain carbon steels with a high-carbon content have a high initial hardness but that this hardness is rapidly reduced as the temperature increases (i.e. as the temper is easily 'drawn').

Fig. 10.17 *Hardness–temperature curves for some cutting-tool materials*

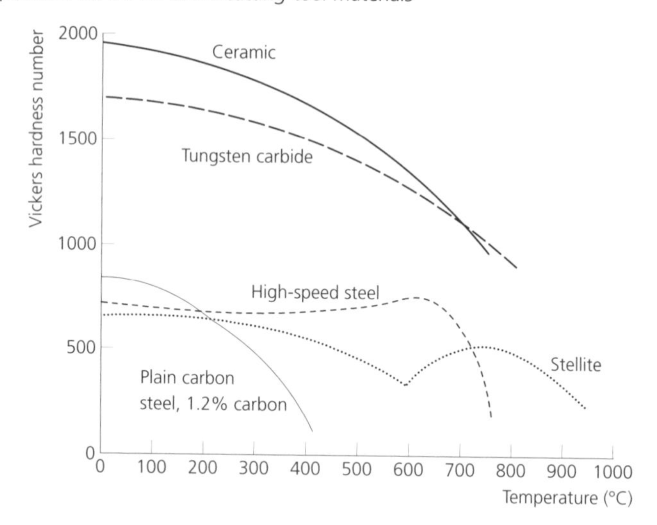

Steels in the lower carbon range are used where toughness rather than hardness is required, and some typical applications are blacksmiths' forging tools such as hot and cold sets, swages, fullers punches and drifts. Such steels are also used for making screwdriver blades, hammer heads, drifts, cold chisels, low-grade spanners and similar hand tools.

Steels in the higher carbon range are used where a keen cutting edge is required. It can be seen from Fig. 10.17 that such steels are harder than alloy steels at room temperature and this makes them very suitable for such applications as carpenters' chisels, knives, wood drills and engineers files, where maximum sharpness is required but where the cutting temperature is low. However, Fig. 10.17 also shows that the hardness of plain carbon steels falls off rapidly as the temperature increases and this makes them unsuitable for machining operations under present-day production conditions.

The hardening and tempering of plain carbon steels has already been considered in detail in *Engineering Materials*, Volume 1, and also in Chapter 2 of this text. The tempered martensitic structure of high-carbon steel tools is a two-phase structure of ferrite and iron carbide that differs from the lamellar structure of pearlite in that tempered martensite is particulate. The extremely fine particles are uniformly distributed through the mass of the steel on a submicroscopic scale and this greatly interferes with dislocation along the slip planes. Therefore tempered martensite is almost as hard as the martensite initially produced by quenching. Since the particles are entrapped in a tough ferrite matrix, any crack that may develop has difficulty in spreading so that tempered martensite is tougher than both martensite and pearlite.

10.10 Carbon–chromium steels

It was stated in Chapter 1 that when chromium is alloyed with carbon steel complex carbides are formed during heat treatment and the steel become much harder, but at the expense of ductility. The hardness is retained at higher temperatures than is possible with plain carbon steels. Unfortunately high-chromium content steels tend to suffer from grain growth during heat treatment and this can lead to loss of strength. Therefore, in high-chromium content steels the manganese content is also increased as this element reduces the tendency for grain growth and increases the toughness of the alloy.

10.11 Vanadium tool steels

The addition of the alloying element, vanadium, to carbon–manganese steels and carbon–chromium steels restrains the onset of grain growth, increases the toughness of the steel and improves its 'hot hardness'.

10.12 Low-tungsten steels

The addition of the relatively expensive metal, tungsten, results in the formation of hard and stable carbides when heat treated. These carbides remain hard and stable at elevated temperatures, therefore tools made from steels containing tungsten can operate continuously at high temperatures without loss of hardness. Such alloys are used for hot-working dies and moulds. For metal-machining applications a greater tungsten content is required as in the high-speed steels to be considered next.

10.13 High-speed steels

These are tool steel alloys designed to operate continuously at elevated temperatures. Their name derives from the fact that they can operate at very much higher cutting speed than plain carbon and low-tungsten steels. There are three groups of alloys:

- Tungsten–chromium alloys, known as 'high-speed steels'.
- Tungsten–chromium–cobalt alloys, known as 'super high-speed steels'.
- Molybdenum alloys, known as 'economy high-speed steels' since the cheaper alloying element, molybdenum, is substituted for much of the tungsten and cobalt content.

The heat treatment of high-speed steels differs from other alloy tool steels inasmuch that they are hardened and then *secondary hardened* instead of being hardened and tempered. This secondary hardening actually increases the hardness as well as increasing the toughness of the steels, whereas tempering increases the toughness at the expense of some loss of hardness.

High-speed steel tools can be given a very thin titanium nitride coating by chemical vapour deposition (CVD). Titanium nitride (TiN) is a hard and heat-resistant ceramic material. Twist drills are frequently treated in this manner to reduce wear of the flutes and the radial lands. Drills so treated have a characteristic golden colour.

10.14 Application of British Standard specifications to tool steels

Table 10.1 is derived from BS 4569 and shows some typical examples of the tool steels described in Section 10.13. Unlike BS 970 where the identification code is based upon the composition and properties of the steel, BS 4569 applies the number and letter code in a more arbitrary manner. The coding is based upon that of the American Iron and Steel Institute (AISI) and the only difference is that the British code is prefixed by the letter B, as shown in Table 10.2. Where a number follows the letter code, the number denotes a specific alloy. The number is applied arbitrarily and in no way indicates the composition or properties of the alloy.

Table 10.1 *Heat treatment and applications of die and tool steels*

(Derived from BS 4596)

Type of steel	*Composition (%)*								*Heat treatment*	*Applications*
	C	*Mn*	*Cr*	*Mo*	*V*	*Si*	*W*	*Co*		
BD3	2.0	0.3	12.5	—	—	—	—	—	Heat slowly to 750–800 °C and heat quickly to 960–990 °C and quench in oil. Temper at 150–400 °C for 30 to 60 minutes depending upon application	*High-carbon–high-chromium (HCCR) die steel.* Widely used for high-quality blanking and forming punches and dies for sheet metal pressings. Gauges, and thread rolling dies. Low shrinkage and distortion during heat treatment
BH12	0.35	—	5.00	1.50	0.40	1.00	1.35	—	Heat slowly to 800 °C, soak and heat quickly to 1020–1050 °C. Air cool, and temper at 540–620 °C for up to $1\frac{1}{2}$ hours depending upon section and application	*Hot-working die steel*, suitable for extrusion dies, hot-stamping and forging dies. Hot-pressing dies for copper and aluminium alloys
BT1	0.75	—	4.25	—	1.20	—	18.0	—	Heat slowly to 900 °C, soak, and heat quickly to 1290–1310 °C. Quench in oil or air blast. Double temper 565 °C for 1 hour	*18% tungsten high-speed steel* for general-purpose cutting tools for lathes, shapers, planers, milling machines etc. Threading taps, hacksaw blades, master gauges etc.
BT6	0.80	—	4.75	0.50	1.50	—	22.0	12.0	Heat slowly to 900 °C, soak, and heat quickly to 1300–1350 °C. Quench in oil or air blast. Double temper at 565 °C for 1 hour	*'Super' high-speed steel.* Cutting tools for machining hard and tough materials such as high-duty alloy steels and cast irons
BM2	0.85	—	4.25	5.00	1.90	—	0.5	—	Heat slowly to 900 °C, soak, and heat quickly to 1200–1250 °C. Quench in oil or air blast. Double temper at 565 °C for 1 hour	'Economy' high-speed steel, developed during World War II when tungsten was in short supply. Similar in characteristics to BT1. Still used where shock resistance is required as it is less brittle than BT1. Useful for roughing and heavily scaled forgings and rustings

Table 10.2 ***Coding of tool and die steels***

AISI Code	*BS 4659 Code*	*Description*
A	BA	Medium-alloy steel; hardenable; but only suitable for processes which do not involve appreciable rise in temperature
D	BD	High-carbon, high-chromium die steel. This has better wear resistance than type BA, but can only be used for processes which do not involve appreciable rise in temperature
H	BH	Chromium- or tungsten-based steel for hot-working processes
L	BL	Low-alloy tool steels for special applications
M	BM	Molybdenum-based steel for hot-working processes
O	BO	Oil-hardening tool steels; only suitable for processes which do not involve appreciable rise in temperature
T	BT	Tungsten-based high-speed steels
W	BW	Water-hardening tool steels

10.15 Stellite

This is a cobalt-based alloy, containing little or no iron; it can only be cast to shape or deposited as a hard facing material using an oxyacetylene torch. It requires no heat treatment to make it hard, and can only be machined by grinding processes since it cannot be softened. Although slightly softer than high-speed steel, it retains its hardness even when the heat generated by the cutting process causes it to glow red-hot. It is considerably more expensive than high-speed and super high-speed steels because of the large amounts of tungsten and cobalt present in the alloy. Stellite can be deposited, by welding, onto plain carbon and alloy steels to build up hard facings. It is also available in the form of ground tool 'bits' to fit standard tool holders.

Stellite has sufficient strength and toughness to withstand positive rake cutting with the tool 'bit' supported as a cantilever. A typical composition is:

- Cobalt, 50 per cent.
- Tungsten, 33 per cent.
- Carbon, 3 per cent.
- Various, 14 per cent.

10.16 Cemented carbides

These are *cermet* composite materials and were introduced in Section 5.27. They consist of hard ceramic particles in a tough metallic matrix. Preformed tool tips made from metallic carbides are produced by a technique known as *sintering*. The production of sintered powder-metal compacts is discussed in *Manufacturing Technology*, Volume 2. Such tool tips are not only much harder than stellite and high-speed steels, they also retain their hardness at very much higher cutting temperatures. Carbide cutting tools fall into four categories:

- Tungsten carbide.
- Mixed carbides (tungsten and titanium carbides).
- Titanium carbides.
- Coated carbides.

The ISO classification of carbides for cutting tools is given in Table 10.3.

Table 10.3 *Carbide grades for metal-cutting tools*

ISO code	*ISO grades*	*General applications*
P (Blue)	P01 → P50	Ductile materials such as plain carbon and low-alloy steels, stainless steel, long-chipping malleable cast iron, ductile non-ferrous metals and alloys
M (Yellow)	M10 → M40	Tough and 'difficult' materials such as: high-carbon steels and high-duty alloy steels, manganese steels, cast steels, alloy cast irons, austenitic stainless steel castings, malleable cast iron, heat-resistant alloys
K (Red)	K01 → K30	Materials lacking in ductility and components which cause intermittent cutting. Cast iron, chill-cast iron, short-chipping malleable cast iron, hardened steel, non-ferrous free-cutting alloys, free-cutting steels, plastics, wood, titanium alloys

Cutting properties

P01 ◄———————— K30
Increasing hardness and ability to withstand wear – high cutting speeds and fine feeds

P01 ————————► K30
Increasing toughness and ability to withstand interrupted cutting with coarse feeds – rough machining high-strength materials

10.16.1 *Tungsten carbide*

Tungsten carbide cutting-tool tips are made from particles of tungsten carbide bonded together in a matrix of metallic cobalt. They are very hard and brittle and are used mainly for cutting such materials as cast irons and cast bronzes that have a relatively low tensile strength, which form a discontinuous chip, and which have a highly abrasive skin. Because of its brittleness and low strength, 'straight' tungsten carbide is not suitable for roughing cuts on ductile, high-stength materials that form a continuous chip, although it can be used for light, finishing cuts using a high cutting speed and fine feed. Nor is it advisable to use it for interrupted cutting. Further, 'straight' tungsten carbide cutting-tool materials tend to be porous and particles of the metal being cut can become embedded in the matrix. Although the metal being cut will not chip weld directly to the tungsten carbide, it will chip weld to the embedded particles of the metal being cut to form an undesirable built-up edge on the tool. Workpiece materials such as cast iron, cast bronze and free-cutting brass that do not form a continuous chip and are, therefore, less susceptible to chip-welding can be successfully cut using tungsten carbide tools.

10.16.2 *Mixed carbides*

Mixed carbides are mixtures of titanium, tungsten, molybdenum and tantalum carbides in a matrix of cobalt. As well as being tougher and less susceptible to chipping, they have improved crater-wear characteristics and better hot-hardness than 'straight' tungsten carbide, and are suitable for taking heavy roughing cuts on such materials as high-strength alloy steels. They are also suitable for interrupted cutting. Mixed carbide tool tips are less porous than 'straight' tungsten carbide with the result that there is less tendency for particles of the workpiece material to become embedded in the matrix and, therefore, less tendency for chip welding to occur.

10.16.3 *Titanium carbide*

Titanium carbide-based tool tips have a matrix of nickel–molybdenum alloy. They have a higher wear resistance than tungsten carbide tool tips and mixed carbide tool tips but are not as tough. They are less susceptible to chip welding and are suitable for cutting hard materials at high speeds.

10.16.4 *Coated carbides*

Coated carbides are more expensive than tungsten carbide and mixed carbides but have the advantage of being able to operate at significantly higher cutting speeds. After compaction and sintering, the tips are given a very thin coating of titanium carbide, titanium nitride, or aluminium oxide by chemical vapour deposition (CVD). The coating thickness is usually 5–9 μm. Alloy tool and die steels may also be coated by this process to increase their wear resistance. Multi-layer coatings are also available.

Titanium carbide coatings improve the wear resistance of the tool, whilst titanium nitride coatings reduce adhesion (chip welding) and galling of the rake face. The use of titanium nitride greatly increases tool life. However, these coatings do not perform well at low cutting speeds as the coating tends to be chipped and removed by adhesion. Lubrication of the chip/tool interface is therefore important Aluminium oxide coatings have a low thermal conductivity and, therefore, form a good thermal barrier. Such coatings prevent the substrate of the tip becoming overheated when machining at very high cutting speeds. As any attempt to sharpen the coated tips by grinding will remove the coating, they are only disposable tooling systems.

10.17 Ceramics (oxides)

Ceramic tool tips are even harder than those made from metallic carbides and can also withstand higher operating temperatures without loss of hardness. Unfortunately, they are also more brittle. The ceramic material most widely used for cutting tools is finely ground aluminium oxide (Al_2O_3) particles together with chromium oxide, titanium oxide and titanium oxide bonded together by sintering. The aluminium oxide forms at least 70 per cent of the mix. Ceramic tips have a low transverse strength and are susceptible to edge chipping. Therefore they are only used for high speed finishing operations where a low feed rate is used and cutting is uninterrupted. Unlike carbide tool tips, ceramic tips cannot be brazed to the tool shank but must be clamped in place. Figure 10.18 shows a typical toolholder suitable for both carbide and ceramic tips. The triangular tip is not usually resharpened but can be indexed to provide three alternative cutting edges before it is discarded.

Fig. 10.18 *Tool holder for carbide and ceramic tips*

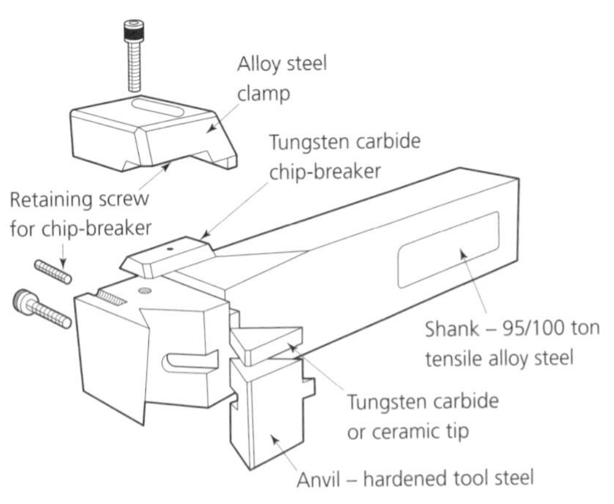

To reduce thermal shock, cutting fluids should not be used. The cutting-tool tip should have negative rake geometry and the machine tool should be very rigid as vibration soon leads to tool failure.

10.18 Cubic boron nitride

This is the hardest material presently available except for diamond. Tool tips are made by bonding a layer of polycrystalline cubic boron nitride (CBN) to a substrate of cemented carbide. The boron nitride provides very high wear resistance and edge strength whilst the carbide provides the shock resistance and support. CBN is chemically inert to iron and nickel and is also resistant to oxidation at high temperatures. It is particularly suitable for machining hardened ferrous and high-temperature alloys.

10.19 Diamond

Diamond – a crystalline allotrope of carbon – is the hardest substance known at present. In addition to its extreme hardness diamond has a high wear resistance, low coefficient of friction, low coefficient of thermal expansion, and high thermal conductivity. Because diamond is brittle, tool shape and mounting is important. Crystal orientation is also important to obtain optimum performance. The abrasive wear resistance of diamond can vary by at least ten-fold depending upon crystal orientation. Wear in diamond tooling usually takes place by micro-chipping due to thermal stressing, and transformation to amorphous carbon due to frictional heating.

Diamond tools can be used satisfactorily at almost any speed, but are principally used for light uninterrupted cuts with high cutting speeds and low feed rates. Diamond tools are widely used for finish machining non-ferrous metals and alloys and abrasive non-metals to high dimensional accuracy and high surface finish. For example, the finish machining of motor car engine pistons where the aluminium alloy used cannot be satisfactorily ground. Unfortunately, diamond cannot be used for machining ferrous metals and alloys and nickel alloys as it has a high chemical affinity for these metals.

Gem stones are not used as they are too costly and chips of brown- and black-burt stones are generally used. However, there is now an increasing tendency to use *polycrystalline diamond.* Small synthetic crystals are fused together by a high-pressure, high-temperature process, and bonded to a hard carbide substrate. The random orientation of the polycrystalline diamond prevents the propagation of cracks through the structure. The carbide substrate provides the toughness and support that the diamond layer requires. Diamonds are also used for manufacturing drawing dies for fine wires and tubes (hypodermic needles), and for dressing and shaping grinding wheels.

10.20 Abrasives

Abrasives are generally made from hard crystalline materials such as:

- Aluminium oxide (emery).
- Silicon carbide (corundum).
- Cubic boron nitride.
- Diamond.

Abrasive wheels are made from large numbers of crystalline abrasive particles, called grains, held together by a bond to form a multi-tooth cutter whose cutting action is similar to a milling cutter. Since an abrasive wheel has many more 'teeth' than a milling cutter, and because they are arranged in a random pattern, a ground surface usually has a very high standard of surface finish. However, the reduced chip clearance between the grains of an abrasive wheel, compared with the space between the teeth of a milling cutter, results in a lower rate of material removal. Thus grinding is a finishing process.

Figure 10.19 shows the dross from a grinding operation and it can be seen that this consists of a mixture of metal chips and blunt grains. The chips are remarkably similar to the chips produced by a milling cutter – proof that abrasive wheels cut metal from the workpiece and do not rub or burnish the surface. The blunt, spent grains are torn from the bond to expose new, sharp, active grains. When the abrasive grains become dull, the cutting forces acting upon them increase and the crystal particles shatter to expose new, sharp cutting edges, or the grains are ripped out of the bond wholesale to expose completely new, active grains. Figure 10.20 shows a comparison between the cutting actions of an abrasive grain and a milling cutter tooth.

Fig. 10.19 *Grinding wheel dross*

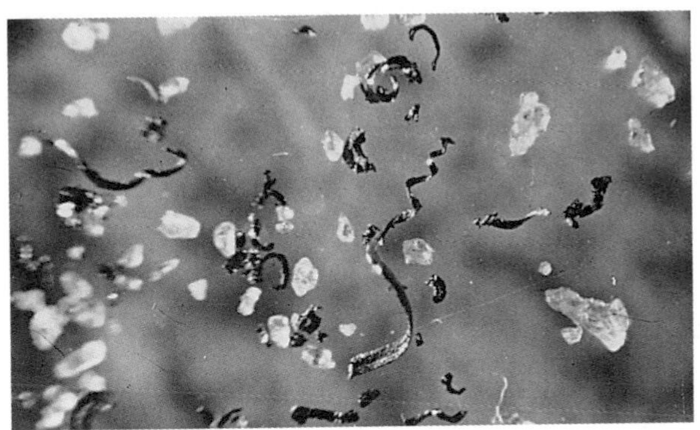

Fig. 10.20 *Cutting action of abrasive wheel grains*

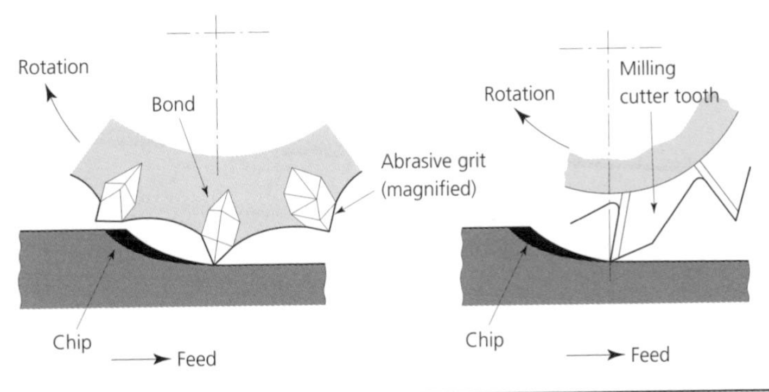

SELF-ASSESSMENT TASK 10.3

1. Select a suitable plain carbon steel for the following applications, giving reasons for your choice:
   (a) hammer head
   (b) cold chisel
   (c) engineer's file
2. Select suitable ISO grades of carbide tool tips for the following machining applications, giving reasons for your choice:
   (a) grey iron casting
   (b) free-cutting mild steel rod
   (c) a high-strength alloy steel forging

3. Discuss the advantages and limitations of coated carbide cutting tools compared with uncoated carbide cutting tools.
4. Research manufacturer's literature and select suitable abrasive grits for grinding:
   (a) mild steel
   (b) hardened and tempered high-carbon steel
   (c) high-speed steel tool bits
   (d) tungsten carbide tool tips

10.21 Selection and application of cutting-tool materials

Let's now consider the properties, selection and application of cutting-tool materials.

10.21.1 *Resistance to thermal softening*

The mechanical energy required for cutting and forming engineering materials is invariably converted into heat energy which, in turn, results in an increase in temperature at the chip/tool interface. This may range from a gentle warming during manual cutting and forming operations, to increasing the temperature of the chip to red heat when machining with carbide- and ceramic-tipped tools. Therefore all tool materials require some degree of 'hot hardness'.

10.21.2 *Coefficient of friction*

Since forming operations and cutting operations on ductile materials that form a continuous chip result in the workpiece material sliding over the tool material, a low coefficient of friction is desirable. This not only reduces wear, but also reduces the forces required to perform the forming or cutting operation. In the case of metal-cutting tools, a low friction drag factor at the chip/tool interface can reduce the possibility of chip welding occurring.

10.21.3 *Wear resistance*

Forming and cutting tools are costly, especially the large press tools used for car body panels and large plastic moulding dies, and the cost needs to be spread over a very large number of components. Therefore, the tool materials must have a high resistance to wear, particularly if a high standard of product dimensional accuracy and surface finish is required.

10.21.4 *Resistance to chemical reaction*

Care must be taken to ensure that chemical reaction does not take place between the workpiece material and the tool material. For example, diamond reacts with ferrous metals under cutting conditions and its tool life is lower than less hard materials such as coated carbides, or ceramics. Again molten aluminium alloys react with some die steels and care is required in the choice of materials for pressure die-casting dies.

10.21.5 *Electrical conductivity*

Electrodes for electric discharge machining (EDM), and electrochemical machining (ECM) require a high conductivity (low resistance) so that the flow of the electric current is not impeded by them.

10.21.6 *Dimensional stability*

The accuracy and surface finish of the workpiece is dependent upon the accuracy of the tooling, it is essential that the tooling material has a high dimensional stability. This applies particularly to large forming tools made from iron castings, forging dies, large die-casting dies and plastic moulds. After rough machining the tool materials should be stabilised by 'weathering' or by a normalising heat treatment to remove internal stresses. The slow release of these stresses under working conditions can cause distortion and dimensional inaccuracy of the tools. This particularly applies to hot-forming processes such as forging, die casting and plastic moulding where the repeated heating and cooling of the tool material can cause creep and distortion to occur. This is aggravated if the heating and cooling is not uniform. For hot-forming processes the tool materials should have a low coefficient of expansion.

10.21.7 *Machinability*

This applies only to metallic tooling materials. Most of these materials can be softened and machined to shape using conventional techniques. They can then be finished to size by grinding after hardening by heat treatment. The exception is Stellite alloy that can only be machined by grinding. Carbides and ceramics cannot be machined except by grinding. Tool tips made from these materials are preformed by compaction of the powdered metallic carbide and the powdered metallic matrix into the required shape, followed by sintering, prior to finish grinding.

10.21.8 *Mechanical properties*

The material properties required by various typical tooling applications were discussed in Sections 10.2 to 10.8 inclusive. By analysing the stresses involved, and choosing suitable materials to match these stresses, considerable cost savings can be achieved together with longer tool life and lower maintenance and replacement costs. The use, for example, of wear pads of expensive die steel set into a yoke of low-cost plain carbon steel is more cost effective than making a forming tool for pressed steel components from solid die steel. Not only will the tough yoke resist the working stresses of the tool better than the hardened die steel but, as wear takes place, the individual wear pads (die cheeks) can be replaced more easily and cheaply than having to replace the whole tool.

10.21.9 *Size availability*

Tool designers should always use standard components and standard size materials. This not only reduces the cost, it also reduces the lead time in getting the tools into production by using components and materials that are readily available. Further, tool designers must also consider the 'ruling section' for the materials selected where heat treatment is involved if the required properties are to be achieved (see Section 2.5).

10.21.10 *Cost*

The tooling is only a means to an end; the main requirement is the production of components in the quantity, quality and time specified. Materials should be selected to keep the cost of the tooling to a minimum commensurate with adequate tool life between replacement. When choosing tool materials, the cost of manufacturing the tool should be considered as well as the cost of the individual materials. The cost of the tooling has to be shared out over the number of components to be produced (unit tooling cost). Simple, low-cost tooling may increase the production cost unduly, whilst sophisticated tooling may increase the unit tooling cost beyond the saving in production cost. Usually a compromise has to be achieved between these two extremes and other economic parameters by the use of appropriate mathematical modelling.

In this chapter we have reviewed many types of tool materials and some of the processes for which they are suitable. As a summary, Fig. 10.21 compares the properties of strength and toughness with the hardness and abrasion resistance of the main categories of cutting-tool materials.

Fig. 10.21 *Range of properties for various groups of tool materials (adapted from Kalpakjian, Manufacturing Engineering and Technology, Addison Wesley)*

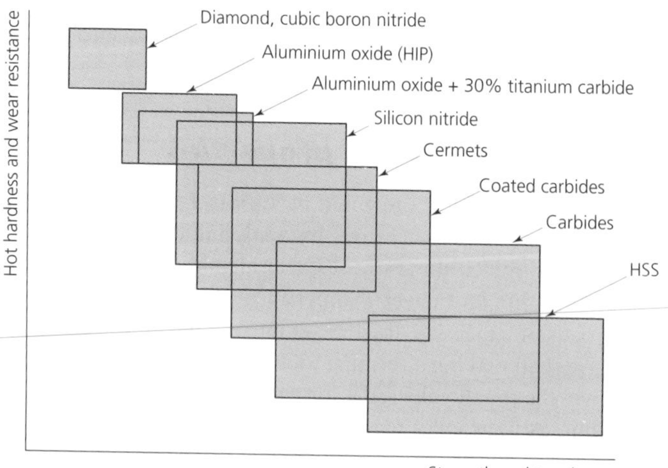

EXERCISES

10.1 Discuss the essential requirements of tool and die steels and explain how these requirements are met by the addition of suitable alloying elements.

10.2 Explain, with the aid of examples, what is meant by intrinsic properties, and what is meant by conferred properties as applied to cutting-tool materials.

10.3 Compare and contrast the properties and applications of the following tool materials: high-carbon steel, high-speed steel, high-carbon–chromium die steel, and tungsten carbide.

10.4 State the composition, properties and **two** typical applications of 'Stellite'.

10.5 Explain the essential differences between the following cutting-tool materials and state a typical application for each material giving reasons for your choice.

(a) tungsten carbide
(b) mixed carbides
(c) titanium carbide
(d) coated carbides

10.6 Describe typical applications of cubic boron nitride and diamond as cutting-tool materials and explain the precautions that must be taken in their use.

10.7 List the common abrasives used for grinding and honing operations, giving typical applications for each material with reasons for your choice.

10.8 (a) Describe the main factors which must be considered when selecting a cutting-tool material for a given application.
(b) For a cutting or forming operation of your choice, present a reasoned argument for the selection of a suitable tool material.

10.9 Obtain access to a press tool, moulding die or a similar item of tooling. Sketch the design and list the materials used for the major components. They analyse the material requirements and specify suitable materials together with any heat treatment or surface finish processes that are required.

11 Electrical properties of materials

The topic areas covered in this chapter are:

- Atomic structure.
- Electrical conduction.
- Conductors and superconductivity.
- Dielectrics.
- Semiconductors, diodes and transistors.
- Magnetic materials and magnetic properties.

11.1 Atomic structure

The structure of the atom was considered in Section 2.2 of *Engineering Materials*, Volume 1. As a reminder, Fig. 11.1 shows the simplified structure of a copper atom. It can be seen that it consists of a positively charged nucleus surrounded by orbiting electrons. In reality, experiments performed by Ernest Rutherford in 1909 showed that the atom consists mainly of empty space. The nucleus of an atom is very small in size, about 10^5 times smaller than the overall size. In other words, the size of an atom is around 100,000 times greater than its nucleus. However, the nucleus is by far the heaviest part of the atom, carrying most of the mass. The size of the electron is too small to be measurable and its mass is negligible compared with either the proton or the neutron. Not all the electrons surrounding the nucleus are at the same energy level and it is useful to arrange them into 'shells'. In an *atom*, the number of positively charged protons in the nucleus is electrically balanced by an equal number of negatively charged electrons. The neutrons in the nucleus have the same mass as the protons but carry no electrical charge. (Protons and neutrons are themselves made from more basic particles called *quarks*, that are electrically charged. Protons contain two 'up' quarks and one 'down' quark, whilst neutrons are made up from one 'up' quark and two 'down' quarks. Electrons are believed to be truly elementary particles.)

If the number of electrons in the atom changes, clearly the atom becomes charged – for example, in losing one electron a copper atom gains +1 unit of electrical charge. Atoms having an overall electrical charge are known as *ions*. Diagrams such as Fig. 11.1 are known as *Bohr models*.

The electrons exist in 'shells' with fixed energy. These energies are said to be *quantised*.

Fig. 11.1 *Bohr model of a copper atom*

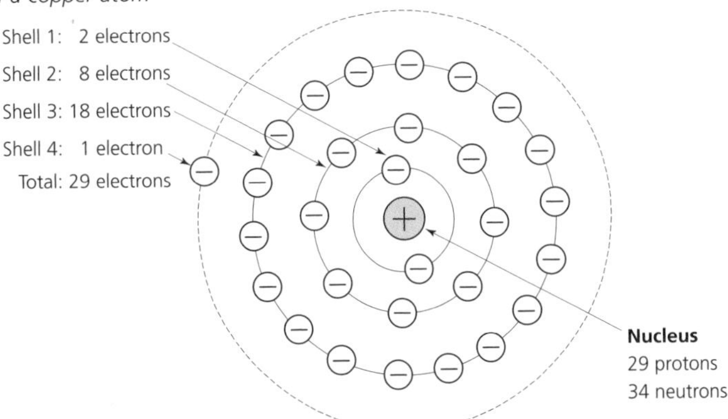

Reference to Fig. 11.1 shows that the first *quantum shell* contains only two electrons, the second shell contains a maximum of eight electrons, the third shell contains a maximum of 18 electrons and, although the fourth shell can have a maximum of 32 electrons, there is only one electron in the copper atom. Note that no matter what the maximum number of electrons may be in a particular quantum shell, when that shell is the outermost or *valency shell* it can never hold more than eight electrons.

The Bohr model concept oversimplifies the nature of the atom as it implies that all electrons in the shell are equal and follow strictly defined orbits in a single plane, whereas they do not. The electron is no longer considered to be a clearly defined particle following a precisely defined orbit, but more as a diffuse 'mist' of electrical charge. It is impossible to ascertain the exact position of the electron because any known measuring technique disturbs the wave characteristics of the electron. Thus only probabilities of electron position can be predicted by calculation. Consider the simplest atom, hydrogen, where one electron orbits around a nucleus of one proton. Figure 11.2 shows the probability of the actual electron positional range. The greatest probability is that the electron will be located at a radius 0.53 Å from the nucleus. (Note that the Ångström unit (Å) is a useful but non-SI unit. $1\ \text{Å} = 0.1\ \text{nm} = 10^{-10}\ \text{m}$.) Hence in Fig. 11.2 the position of the electron is shown not as a clearly defined path but as a shaded zone of probable positions with the shading darkest where the probability is greatest. The size of a hydrogen atom is about 1 Å or 10^{-10} m.

The Bohr model also breaks down in showing all the electrons in a given shell as lying in the same plane and having the same energy levels. *Pauli's exclusion principle* tells us that there are definite rules governing the energy levels and probable positions of the electrons. Further, there cannot be more than two interacting electrons with the same orbital quantum number, and even these are not identical since they have inverted magnetic behaviour (opposite 'spins').

Reference to Fig. 11.1 shows that the first shell has only two electrons and, therefore, complies with Pauli's exclusion principle. However, the subsequent quantum shells have many more electrons and, to comply with Pauli's exclusion principle, each quantum shell has to be divided into *subshells* with the electrons orbiting three dimensionally so that there are no more than two electrons in any one orbit. The electron notation for part of the first three periods of the periodic table is shown in Table 11.1

Fig. 11.2 *Possible position of a single hydrogen electron*

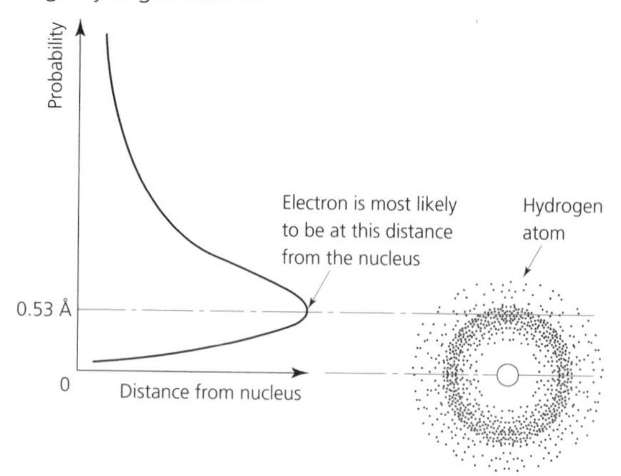

Table 11.1 *Electron notation for common elements*

(From the first three periods of the periodic table)

Element	*Atomic number Z*	*Electron configuration*						
		Shell K (n = 1)	*Shell L (n = 2)*		*Shell M (n = 3)*			
		1s	*2s*	*2p*	*3s*	*3p*	*3d*	*4s*
H	1	1						
He	2	2						
Li	3	2	1					
Be	4	2	1					
B	5	2	2	1				
C	6	2	2	2				
N	7	2	2	3				
O	8	2	2	4				
F	9	2	2	5				
Ne	10	2	2	6				
Na	11	2	2	6	1			
Mg	12	2	2	6	2			
Al	13	2	2	6	2	1		
Si	14	2	2	6	2	2		
P	15	2	2	6	2	3		
S	16	2	2	6	2	4		
Cl	17	2	2	6	2	5		
Ar	18	2	2	6	2	6		
K	19	2	2	6	2	6		1
Ca	20	2	2	6	2	6		2
Sc	21	2	2	6	2	6	1	2
Ti	22	2	2	6	2	6	2	2
V	23	2	2	6	2	6	3	2
Cr	24	2	2	6	2	6	5	1
Mn	25	2	2	6	2	6	5	2
Fe	26	2	2	6	2	6	6	2
Co	27	2	2	6	2	6	7	2
Ni	28	2	2	6	2	6	8	2

Note: Sub-shell 4*s* commences to fill up before sub-shell 3*d* is full.

Each electron in any atom has its own set of quantum numbers denoted by the symbols: n, l, m_l and m_s. These symbols are defined in Table 11.2. The maximum number of electrons in any one shell is $2n^2$. The method of calculating the possible combinations of quantum numbers for any given shell is shown in Example 11.1

Table 11.2 ***Quantum number symbols***

Symbol	*Description*
n	This is the *principal quantum number* for any given shell and it can have an integer value of 1, 2, 3, etc.
l	This is the *angular momentum quantum number* and its integer value ranges from 0 to $n - 1$
m_l	This is the *magnetic quantum number* and its integer value can only be 0, $\pm 1, \pm 2, \ldots \pm l$ (e.g. for $l = 1$, m_l can be $-1, 0, +1$; for $l = 2$, m_l can be $-2, -1, 0, +1, +2$)
m_s	This is the *spin quantum number* and it can only have values of $+\frac{1}{2}$ or $-\frac{1}{2}$

EXAMPLE 11.1

Using the data given in Tables 11.1 and 11.2, determine the maximum number of electrons for any given shell, together with the possible combinations of quantum numbers for the K shell and for the L shell.

Solution

(a) K shell

From Table 11.1, $n = 1$. Therefore, using the expression $2n^2$, the maximum number of electrons that can be present in the shell is $2 \times 1^2 = 2$.

Since $l = n - 1$, $l = 1 - 1 = 0$, and the possible combinations of quantum numbers will be:

l	m_l	m_s		
0	0	$+\frac{1}{2}$	subshell 1s (2 electrons)	shell K 2 electrons
0	0	$-\frac{1}{2}$		

Thus the K shell can only contain a maximum of 2 electrons with opposing spins. This satisfies the requirements of hydrogen and helium and also the requirements of the Pauli exclusion principle.

(b) L shell

From Table 11.1, $n = 2$. Therefore, using the expression $2n^2$, the maximum number of electrons that can be present in the shell is $2 \times 2^2 = 8$.

Since $l = n - 1$, $l = 2 - 1 = 1$, and l can be 0 or 1. The possible combinations of quantum numbers will be:

l	m_l	m_s		
0	0	$+\frac{1}{2}$	subshell 2s (2 electrons)	shell L 8 electrons
0	0	$-\frac{1}{2}$		
1	0	$+\frac{1}{2}$	subshell 2p (6 electrons)	
1	0	$-\frac{1}{2}$		
1	−1	$+\frac{1}{2}$		
1	−1	$-\frac{1}{2}$		
1	+1	$+\frac{1}{2}$		
1	+1	$-\frac{1}{2}$		

Thus the L shell can contain a maximum of 8 electrons (4 pairs of opposing spins – one pair per orbit).This satisfies the requirements of Table 11.1 and the requirements of the Pauli exclusion principle.

SELF-ASSESSMENT TASK 11.1

1. Using Table 11.2, list all the possible combinations of quantum numbers for aluminium ($Z = 13$) and iron ($Z = 26$).

11.2 Energy levels

In any single atom the electrons can occupy a number of discrete energy levels that become closer together the more remote they are from the nucleus. Electrical conductivity is concerned only with the electrons in the outermost (valency) shell. A copper atom has 29 electrons arranged in four main shells (seven subshells).

The subshells 1s, 2s, 2p, 3s, 3p and 3d are filled but the valency subshell 4s contains only one electron and could hold another electron according to Pauli's exclusion principle. The electrons in each subshell occupy specific energy levels in which the s-level is lower than the p-level which, in turn, is lower than the d-level and so on. Further, shell K has a lower energy level than shell L and so on. Thus when an electron moves up from one energy level to a higher energy level it has to be given a quantum (or 'packet') of energy. When it moves down to a lower energy level it gives up a quantum of energy in the form of a *photon*. (For example, this could be seen as light.)

When single atoms bond together to form a solid mass, they influence each other and the electrons are forced to seek alternative energy levels so that no more than two electrons occupy the same quantum state, in accordance with Pauli's exclusion principle. Within any crystal there will be a vast number of atoms and it is not possible to consider individual energy *levels*. Instead, it is necessary to consider densely filled energy *bands*. The concept of energy levels and energy bands is shown diagrammatically in Fig. 11.3.

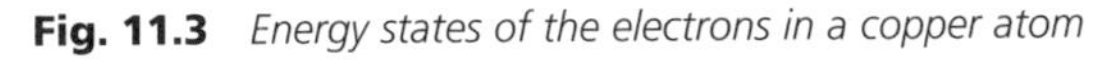

Fig. 11.3 *Energy states of the electrons in a copper atom*

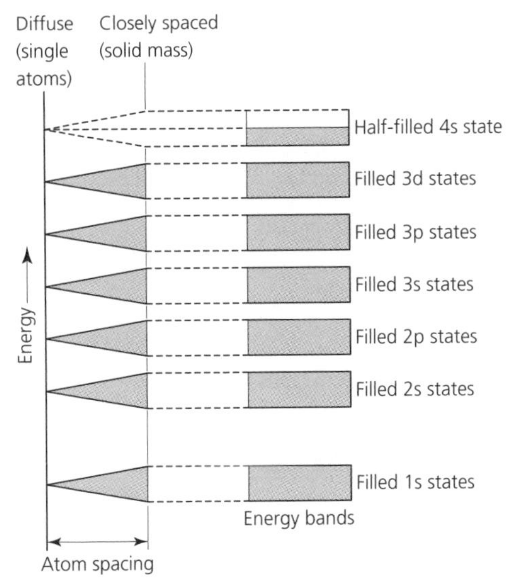

Since copper atoms have only one valency electron (only half the 4s band is filled), the energy required to raise a valence electron to an empty state is negligible. Therefore the valency electrons for copper can move freely within the crystal and conduct electricity (see Section 11.3). Similarly, all metals with an odd number of electrons in their valency bands have good conductivity. Although bivalent metals such as beryllium, calcium and magnesium have their valency bands completely filled, these are only 's' bands and some electrons transfer easily to the next energy bands which are 'p' bands and this endows such metals with good electrical conductivity. On the other hand, the transition metals with the 3d band incompletely filled show relatively high resistivity because the electrons in the 3d band overlap with the electrons in the 4s band causing scattering.

In materials, such as the diamond allotrope of carbon, the valency band is filled by the four outer shell electrons and a wide energy gap exists between the valency band and the next possible energy band. In fact, at 20 °C, about 8.5×10^{-19} J of energy is required by each electron to move from the 'valency band' to the 'conduction band'. This is a relatively large amount of energy and accounts for diamond having such a high resistance to the flow of an electric current that it can be classified as an insulator. Graphite, which is also an allotrope of carbon, has a much lower resistivity than diamond and is classified as a high resistance conductor.

Some materials, known as *semiconductors*, also have energy gaps between their completely filled valency bands and the next available energy or conduction bands. However, as can be seen from Fig. 11.4, the energy required to promote an electron from the valency band to the conduction band for semiconductor materials is very much less than for diamond.

Fig. 11.4 *Energy bands in conductors and insulators: (a) good conductor; (b) relatively poor conductor; (c) very poor conductor–insulator; (d) semiconductor*

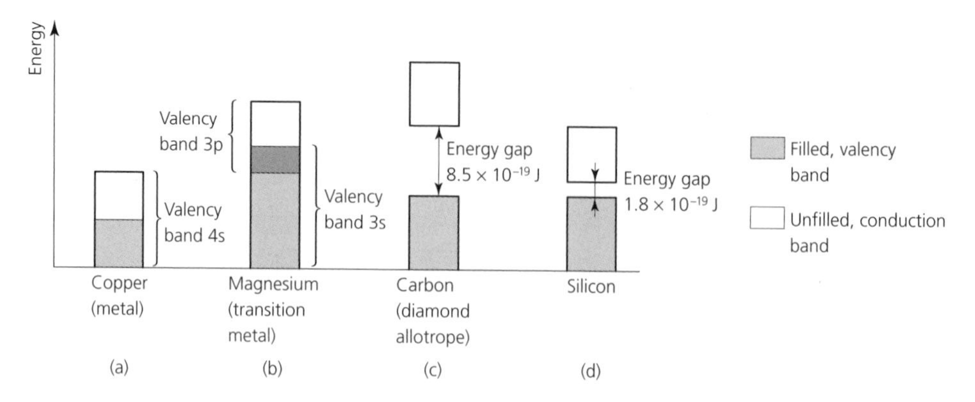

11.3 Electrical conduction

It has already been shown that the high electrical conductivity of metals such as copper is due to the ease with which the outer, valency shell electrons can escape into the conduction band for these materials. Normally there is a random migration of negatively charged electrons amongst the positive ions in the metallic crystals that make up the material. (A positive ion is an atom that has lost one or more electrons and has a residual positive charge.) However, when such a material forms part of an electrical circuit the random migration of electrons becomes a directed flow, as shown in Fig. 11.5. Since like charges attract and unlike charges repel, the (negative) electrons will flow away from the negative pole of the cell and towards the positive pole of the cell. Hence the *electron current* flows from negative to positive whilst the traditional concept of *conventional current* flow is from positive to negative.

Fig. 11.5 *An electric current as a flow of free electrons in metals: (a) random movement of electrons in a conductor; (b) directed movement of electrons in a conductor which forms part of an electric circuit*

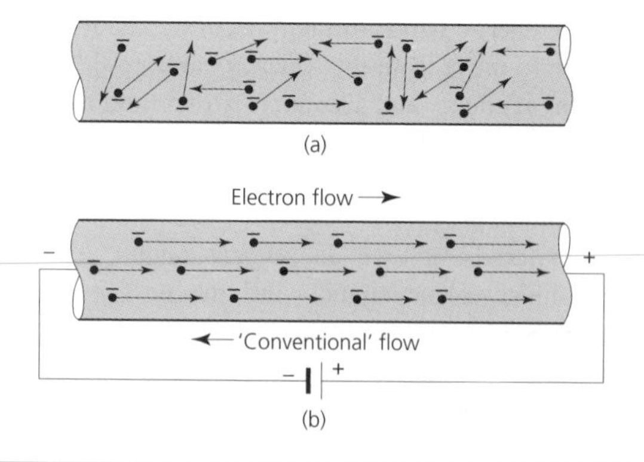

The negative charge of one electron is too small for practical purposes and a much larger unit is used, called the *coulomb*.

$$1 \text{ coulomb (C)} = \text{the charge on } 6.3 \times 10^{18} \text{ electrons}$$

An electric current of 1 ampere (A) is said to flow in a circuit when a charge of one coulomb passes any point in that circuit in one second. An electric current can also flow in non-metallic materials by the drift of positive and negative ions. This occurs, for example, in electrolytes – see Section 4.2, corrosion cells.

SELF-ASSESSMENT TASK 11.2

1. The electron charge on a single electron is 1.6×10^{-19} C. Show that 6.3×10^{18} or 6.3 million million million electrons are required for just 1 C of charge.
2. The capacity of secondary cells (accumulators) is stated in ampere-hours (A h). Calculate the charge in coulombs equal to a charge of 25 A h.

11.4 Conductor materials

It has already been shown that metals are good conductors of electricity because of the relative ease (low energy requirement) with which their valency electrons can be moved from the valency energy band to the conduction band. It follows, therefore, that anything interfering with the free movement of the valency electrons results in increased energy being required to make the transition and lowers the conductivity of the material. Since an electric current is a directed flow of electrons amongst the positive ions in a metallic crystal, anything that impedes this flow will also lower the conductivity of the material.

The *conductance* (G) of a conductor, measured in *siemens* (S), is the reciprocal of the *resistance* (R) of that conductor measured in *ohms* (Ω).

$$G = 1/R$$

Similarly, the *conductivity* of a material (symbol σ) is the reciprocal of the resistivity of that material (symbol ρ).

$$\sigma = 1/\rho$$

The resistivity of a material is defined as *the resistance between any two opposite faces of a unit cube of that material*. Resistivity is measured in ohm-metres (Ω m). Table 11.3 lists the resistivity for a number of typical engineering materials at 0 °C. Care must be exercised in the use of such a Table as the resistivity of a material is affected by such factors as temperature, impurities and distortion of the crystal lattice by hot or cold working. Example 11.2 shows how the resistance of a conductor may be calculated.

Table 11.3 *Approximate specific resistances of common metals*

Material	*Resistivity (ρ) at 0 °C (Ωm)*
Polythene	$\sim 10^{16}$
Germanium } intrinsic Silicon	$\sim 10^{-3}$
Aluminium	3×10^{-8}
Brass	7×10^{-8}
Carbon (graphite)	4400×10^{-8} to 8600×10^{-8}
Constantan or Eureka	49×10^{-8}
Copper	2×10^{-8}
German silver	21×10^{-8}
Iron	9×10^{-8}
Manganin	42×10^{-8}
Mercury	94×10^{-8}
Nickel	12×10^{-8}
Tin	13×10^{-8}
Tungsten	5×10^{-8}
Zinc	6×10^{-8}

EXAMPLE 11.2

Calculate the resistance of a copper conductor 4 m long and having a cross-sectional area of 1 mm^2. Assume the temperature is 0 °C.

Solution

The formula for calculating resistance is:

$$R = \frac{\rho l}{A}$$

where R = resistance (Ω)
ρ = resistivity (Ω m)
l = length of conductor (m)
A = cross-sectional area (m^2)

From Table 11.3, $\rho = 1.59 \times 10^{-8}$ Ω m at 0 °C. Therefore:

$$R = \frac{1.59 \times 10^{-8} \times 4}{1 \times 10^{-6}}$$

$$= 6.36 \times 10^{-2}\ \Omega$$

$$= \mathbf{0.0636\ \Omega}$$

11.4.1 *Temperature*

Any increase in temperature of a conductor produces greater thermal agitation of the metallic ions as they vibrate about their mean positions. Therefore the chance of collision between the electrons and ions becomes greater and the mean free path is effectively reduced. This restricts the flow of electrons and reduces the conductivity of the material. The effect of temperature on the resistivity of some typical conductor materials is shown in Fig. 11.6. It should be noted that the temperature effect on graphitic carbon is the opposite of the temperature effect on the metals. The decrease in resistivity with temperature rise that occurs within graphitic carbon is a phenomenon it shares with all other non-metals. *The temperature coefficient of resistance of a material is the change in resistance of 1 ohm at 0 °C per degree of temperature rise (symbol* α_0*).*

Table 11.4 list the temperature coefficients for a number of typical engineering materials. Note the negative coefficient for graphitic carbon that has been referred to previously. The reference temperature is rarely 0 °C and the method of calculating the effect of temperature change is shown in Example 11.3.

Fig. 11.6 *The effect of temperature on resistivity (source: Higgins)*

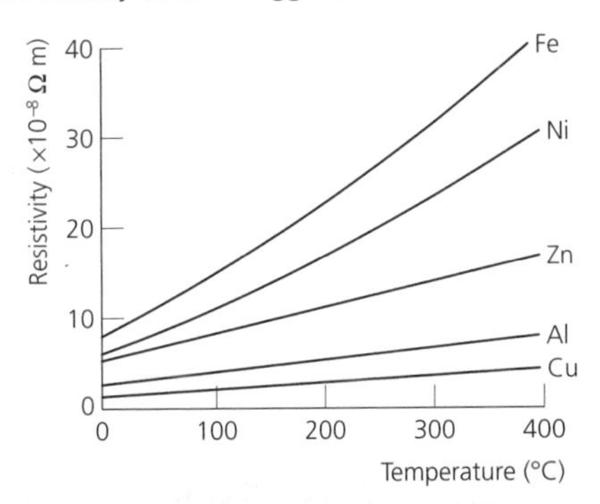

Table 11.4 Temperature coefficients of resistance of common metals

Material	*Temperature coefficient* (α_0) *at 0 °C* (Ω/Ω)
Aluminium	38×10^{-4}
Brass	10×10^{-4}
Carbon (graphite)	-5×10^{-4}
Constantan or Eureka	$+0.1 \times 10^{-4}$ to -0.4×10^{-4}
Copper	43×10^{-4}
German silver	27×10^{-4}
Iron	63×10^{-4}
Manganin	0.3×10^{-4}
Mercury	9.8×10^{-4}
Nickel	62×10^{-4}
Tin	44×10^{-4}
Tungsten	51×10^{-4}
Zinc	37×10^{-4}

EXAMPLE 11.3

The resistance of an aluminium conductor is exactly 5 Ω at 10 °C. Calculate its resistance at 25 °C.

Solution

$$\frac{R_1}{R_2} = \frac{1 + \alpha_0\,\theta_1}{1 + \alpha_0\,\theta_2}$$

where R_1 = initial resistance (Ω)
R_2 = final resistance (Ω)
α_0 = temperature coefficient of resistance Ω/Ω at 0 °C/°C
θ_1 = initial temperature
θ_2 = final temperature

From Table 11.4, $\alpha_0 = 38 \times 10^{-4}$ Ω/Ω at 0 °C/°C. Therefore:

$$\frac{5}{R_2} = \frac{1 + (38 \times 10^{-4} \times 10)}{1 + (38 \times 10^{-4} \times 25)}$$

$$= \frac{1.038}{1.095}$$

$$R_2 = \frac{5 \times 1.095}{1.038}$$

$$= \mathbf{5.275\ \Omega}$$

SELF-ASSESSMENT TASK 11.3

1. Rearrange the formula in Example 11.2 to make ρ the subject.
2. Hence, by substituting for R, l, and A show that the unit for resistivity is Ω m.
3. A wire made from a certain metal had a length of 10 m and its diameter measured 0.2 mm. The resistance was measured to be 156 Ω across the two ends of the wire. Calculate the resistivity and, using Table 11.3, decide which metal was used to make the wire.
4. Using the data from Table 11.4, rework Example 12.3 for copper wire instead of aluminium.

11.4.2 *Composition*

Alloys, in general, have a higher resistivity than pure metals and this is largely due to the presence of solute atoms in solid solutions no matter whether such solutions are interstitial or substitutional. The solute atoms interfere with the flow of electrons amongst the positive ions of the metallic crystals. Figure 11.7 shows the effect of composition on the electrical conductivity of copper–nickel alloys (also referred to as cupro-nickel alloys).

Fig. 11.7 *Resistivity of copper–nickel alloys; alloys have lower conductivities than pure metals (source: Higgins)*

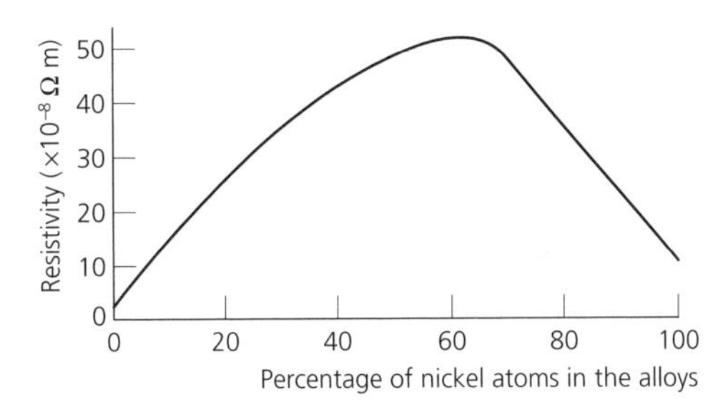

11.4.3 *Impurities*

Even small amounts of impurities can cause large increases in resistivity. Hence the high purity of 'high-conductivity copper'. For example, the presence of traces of phosphorus, silicon, arsenic or even the metal iron can substantially lower the conductivity of copper. It only requires 0.05 per cent phosphorus to be present to reduce the conductivity of copper by some 40 per cent. Figure 11.8 shows the effect of impurities on the electrical conductivity of copper.

Fig. 11. 8 *The effects of impurities on the electrical conductivity of copper: silver and cadmium have little effect and can be used to strengthen overhead conductors (e.g. telephone lines), whilst arsenic and phosphorus must not be present in copper destined for electrical purposes (source: Higgins)*

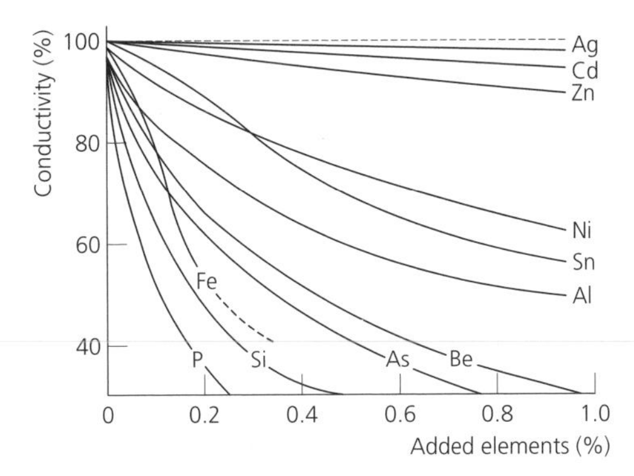

11.4.4 *Structure*

The presence of faults in the crystal lattice, such as dislocations, can also impede the flow of electrons and increase the resistivity of the metal. For example, annealed copper is 11 per cent more conductive than heavily cold-worked copper.

11.4.5 *Superconductivity*

Superconductivity refers to the sudden drop in electrical resistivity of certain metals and compounds at temperatures approaching absolute zero, as shown in Fig. 11.9. Superconductors have zero resistance and zero resistivity. This phenomenon (discovered in 1911) cannot be explained by the normal application of the temperature coefficient of resistance. Quantum physics is necessary to explain the effect. The temperature below which superconductivity occurs is called the *superconducting transition temperature* (T_c). The state of superconductivity for a material continues for as long as the material is maintained below its transition temperature. The state of superconductivity is destroyed by strong magnetic fields, either applied from an external source or caused by allowing an excessively heavy current to flow through the conductor.

Fig. 11. 9 *Superconductivity*

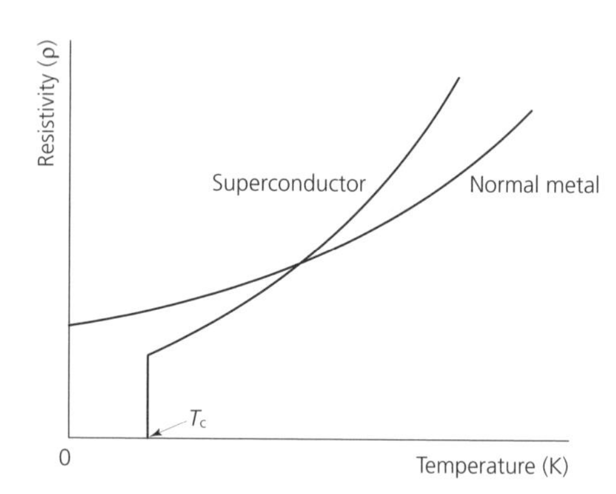

Some materials that are relatively poor conductors at room temperature become superconducting at very low temperatures. Some superconducting materials are listed in Table 11.5 together with their superconducting transition temperatures. Note that although some materials require very low temperatures before they become superconductors, certain ceramic compounds have transition temperatures that can easily be achieved using liquid

Table 11.5 *Transition temperatures of common superconductors*

Material	*Transition temperature (T_c) (K)*
Aluminium	1.2
Lead	7.2
Niobium	9.5
Tantalum	4.5
Technetium	11.2
Vanadium	5.3
Nb_3 Sn	18.1
V_3 Ga	16.8
Mo–Re alloy	10.0
Thallium–barium–calcium copper oxide (ceramic)	122

nitrogen at normal room temperature. Ceramic superconductors were first discovered as recently as 1986.

Because superconductors have no electrical resistance, no heat is generated when a direct current (d.c.) is passed. This means that no energy would be wasted in the transmission of (d.c.) electric power if cables were to be made from a superconducting material. Also superconducting wires can transmit much larger amounts of electricity than conventional wires of the same diameter. Current research is directed towards developing materials that become superconducting at higher and more easily achieved temperatures so that they can be used on a commercial basis.

It was stated earlier that superconductors lose their superconductivity properties in the presence of strong magnetic fields. This is particularly true of the simpler type-1 superconductor materials. However, the more complex type-2 superconductor materials can operate successfully in magnetic fields with flux densities as high as 10–20 tesla. Type-2 superconductors are currently employed in very powerful electromagnets such as those used in magnetic resonance imaging (MRI) equipment in hospitals. Doctors can use MRI scans to examine patients' soft tissues (e.g. the brain) without the need for exploratory surgery.

11.5 Insulating materials (dielectrics)

Devices for isolating electric currents and charges are called *insulators* and are made from materials called *dielectrics*. Dielectric materials must be capable of separating electrical conductors without conducting an electrical charge between them.

An insulating material, or dielectric, is a material in which the valency shell is completely filled and there is a very wide energy gap between its valency band and the next energy band. Thus a very large input of energy is required in order for an electron to cross the gap. A very small number of electrons may acquire sufficient energy to cross the gap if the material is subjected to a very high potential difference. Therefore, there is no such thing as a perfect insulator (non-conductor), only materials with very high values of resistivity that, for all practical purposes, can be used for insulating purposes.

Dielectric materials may be gases, liquids or solids. Gaseous dielectrics are not used commercially, with the exception of air. Air is the dieletric between the bare conductors of the overhead electrical grid system. Liquid dielectrics are used mainly as impregnants for high-voltage paper-insulated cables and capacitors, and as filling and cooling media for transformers and circuit-breakers. Such dielectrics may be petroleum oils or, alternatively, silicon oils and fluorinated hydrocarbons where the operating temperature is sufficiently high to cause the oxidation of petroleum oils. Some solid dielectric materials are listed in Table 11.6 together with typical values for their resistivity and dielectric strengths.

11.5.1 *Dielectric strength*

Dielectric strength is the maximum intensity of electric field (measured in V/mm) that can be placed across an insulating material of unit thickness without breakdown occurring. It should be noted that the breakdown voltage does not increase in direct relationship to an

Table 11.6 ***Properties of common electrical insulators***

Material	*Resistivity at 20 °C (Ω m)*	*Dielectric strength* (V/mm)*
Ceramics		
Soda-lime glass	10^{13}	10 000
Pyrex glass	10^{14}	14 000
Vitreous silica	10^{17}	10 000
Mica	10^{11}	40 000
Steatite porcelain	10^{13}	12 000
Mullite porcelain	10^{11}	12 000
Polymerics		
Natural rubber	—	16 000–24 000
Phenol formaldehyde	10^{10}	12 000
Polybutadiene	—	16 000–24 000
Polyethylene	10^{13} to 10^{16}	20 000
Polystyrene	10^{16}	20 000
Polyvinyl chloride (PVC)	10^{10}	12 000

*Not constant with thickness (see text).

increase in dielectric thickness. This variation is due to the presence of imperfections in the dielectric that may allow local leakage currents to flow, resulting in premature failure. In fact it is common practice to use several thin layers of dielectric material in capacitors, rather than a single thick layer, as it is improbable that all the imperfections would coincide at the same point. Moisture, contamination, elevated service temperatures, ageing and mechanical stress, all tend to decrease the dielectric strength of insulating materials.

11.5.2 *Relative dielectric constant*

This is also called the *relative permittivity* of an insulating material. It is a measure of the displacement or charging effect of a dielectric, and is expressed as a ratio of the capacitance of a capacitor containing the dielectric material to the capacitance of the same capacitor using a vacuum as the dielectric. As a ratio it has no unit.

$$\varepsilon_r = C/C_0$$

where ε_r = the relative dielectric constant
C = the capacitance of a capacitor using the dielectric
C_0 = the capacitance of the same capacitor with a vacuum between the plates instead of the dielectric

Table 11.7 lists some dielectric materials and their relative dielectric constants. In many cases the relative dielectric constant changes with frequency. Figure 11.10 shows two types of capacitor used in electronics. Figure 11.10(a) is used for 'tuning in' a particular radio station. Notice that the dielectric is air. Figure 11.10(b) shows a capacitor suitable for smoothing the rectified alternating current of a power pack.

Table 11.7 ***Relative dielectric constants ε_r***

Material	*Relative dielectric constant (ε_r)*
Air	1.000 59
Barium titanate*	6000
Glass	6
Insulating oil	3
Mica	6
Paper	2.5
Polythene	2.3
Vitreous silica	3.5

*The dielectric film in miniature electrolytic capacitors.

Notes:

(1) The above values are relative to the absolute dielectric constant ε or ε_0, for a vacuum, which is unity.

(2) Because of the similarity of their ε_r values, C_0 is often taken (for practical purposes) as the capacitance using dry air as a dielectric.

Fig. 11.10 *Typical capacitors used in radio receivers*

(a)

(b)

11.5.3. *Temperature*

We have already seen that, in the case of metallic conductors, the resistivity of the conductor material increases when its temperature increases. We have also seen that in the case of graphitic carbon the resistivity of the material falls when its temperature rises. This latter phenomenon is common to all non-metals, and the resistivity of insulating materials becomes less as the temperature rises. For example, glass is a good insulating material at room temperature yet becomes a conductor when it becomes red-hot. Thus metals become poorer conductors as their temperatures rise and non-metals become poorer insulators as their temperatures rise. Further, as has already been mentioned, the dielectric strength of an insulating material becomes less as the temperature rises and this restricts the operating temperature of electrical devices if an insulation breakdown is to be avoided. For this reason, the manufacturers of such components as motors, transformers and capacitors usually state the maximum safe operating temperature.

SELF-ASSESSMENT TASK 11.4

1. Pure water has a high relative constant, where relative permittivity is around 80. State possible reasons why water is not used as a dielectric material.
2. Suggest reasons why a variable tuning capacitor (Fig. 11.11(a)) uses air as a dielectric.
3. Figure 11.11(b) shows an electrolytic capacitor.
   (a) Explain how it is constructed and how the dielectric is formed.
   (b) Explain how this form of dielectric enables such high values of capacitance to be achieved.
   (c) Explain why this type of capacitor has *polarised* connections.
4. Compare and contrast the advantages and limitations of rubber and PVC as insulating materials for electric cables.

11.6 Semiconductors (intrinsic)

Like the diamond allotrope of carbon, silicon and germanium crystallise into structures in which each atom is covalently bonded to four similar atoms. This is shown in Fig. 11.11 for a silicon atom. By sharing their valency atoms in this way each outermost or valency shell

Fig. 11.11 *The silicon atom and its covalent bond: (a) a single silicon atom with 4 electrons in its outer (valency) shell; (b) covalent bonding between adjacent atoms results in the valency shells being 'filled' with 8 electrons – in practice, this occurs three dimensionally*

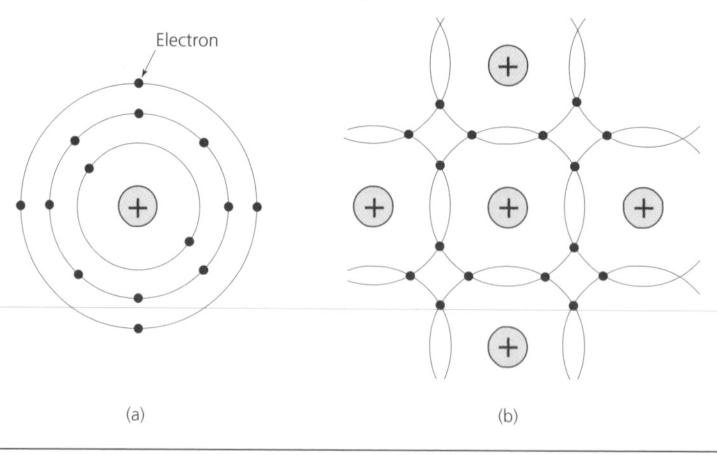

of the atom has, effectively, eight electrons and is filled. Both silicon and germanium differ from diamond in that the energy gaps between their valency bands and their conduction bands are much smaller and it is easier for an electron to move between the bands.

When thermal agitation causes an electron to leave its position in a lattice there is a resulting deficiency in the lattice referred to as a 'hole'. A valency electron from an adjacent atom can then move into the hole leaving yet another hole. This process is repeated in a random manner throughout the material. However, if a potential difference is applied across the material, there is a migration of electrons in one direction and a migration of 'holes' in the opposite direction. Thus, since the migration of the 'holes' is opposite to the migration of electrons, the migration of holes can be considered as the migration of positive charges. In semiconductor parlance, electrons, being negatively charged, are referred to as *n-type charge carriers* and 'holes', being considered to behave as positive charges, are referred to as *p-type charge carriers*. Semiconductor materials that are free from impurities are referred to as being *intrinsic*.

11.6.1 *Thermal conduction*

In intrinsic semiconductor materials the energy required to free electrons from the valency band and raise them through the energy gap into the conduction band is thermal energy, and the electrical conduction resulting from a temperature rise is referred to as *thermal conduction*. Thus, for any intrinsic semiconductor material, the conductivity rises as the temperature rises.

11.6.2. *Photoconduction*

Light as well as heat energy can raise electrons from the valency band of intrinsic silicon to the conduction band, as shown in Fig. 11.12(a). For example, a photon of red light has an energy of 1.9 eV, which is more than sufficient to cause an electron to migrate across the 1.1 eV energy gap of silicon (*Note*: The electron-volt (eV) is a useful unit of energy, where $1\,eV = 1.6 \times 10^{-19}\,J$.)

Fig. 11.12 *Photoconduction luminescence*

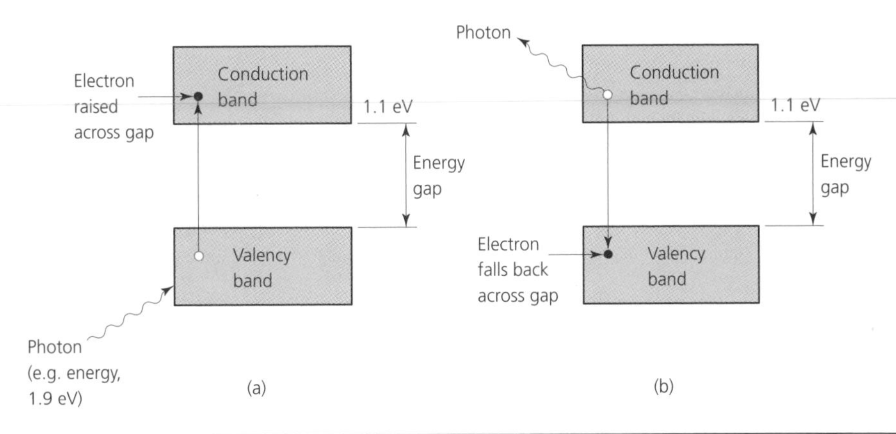

Since all materials are more stable when they reduce their energies, the electrons drop back from the conduction band to refill their holes in the valence band when the external light source is removed. This is referred to as recombination. For many applications intrinsic semiconductors have to be shielded from external light by encapsulation.

11.6.3 *Luminescence*

When recombination takes place as described above, the energy absorbed initially from the light source has to be given up either as heat energy or as luminescence, as shown in Fig. 11.12(b).

11.7 Semiconductors (extrinsic)

Unlike intrinsic semiconductor materials that are of very high purity, extrinsic semiconductor materials are deliberately 'doped' with impurities during manufacture to give them either n-type or p-type characteristics. Since silicon and germanium are Group IV elements (that is, they have four electrons in their valency shells) they are doped with either Group III or Group V elements that have either three or five electrons, respectively, in their valency shells.

Figure 11.13(a) shows the effect of adding a Group V element such as phosphorus to a semiconductor material such as silicon. It can be seen that the phosphorus atom has a 'spare' electron bonded into its valency shell. Very little energy is required for this electron

Fig. 11.13 *Extrinsic semiconductor materials*

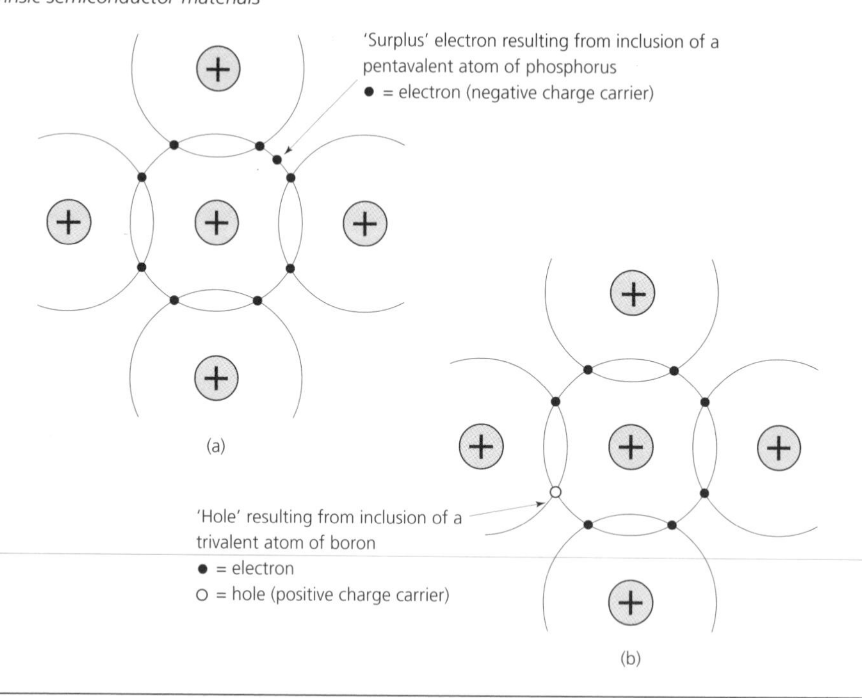

to break free and rise to the conduction band, and the thermal energy available at room temperature is more than sufficient. An extrinsic semiconductor material in which there are free negatively charged electrons is said to be an *n-type material*. The element that provides the free electrons is called a *donor*.

Figure 11.13(b) shows the effect of adding a Group III element such as boron to a semiconductor material such as silicon. It can be seen that as the boron atom has only three electrons in its valency shell it will leave a 'hole' in the bond pattern. As previously explained, 'holes' in semiconductor technology are considered to be positive charge carriers. Therefore an extrinsic semiconductor material in which there are positive charge carriers is said to be a *p-type material*. The element that provides the positive charge carriers ('holes') is called an *acceptor*. Semiconductor materials, and in particular silicon, are used in the electronics industry where their use has revolutionised microelectronics in recent years.

SELF-ASSESSMENT TASK 11.5

1. Assuming photons of energy 0.9 eV were used to bombard the silicon in Fig. 11.13(a), explain, giving reasons, whether the electron would be raised to the conduction band.
2. Pure silicon is to be doped with the trivalent metal, indium. State whether the resulting semiconductor material is n-type or p-type and hence whether electrons or holes are the principal charge carriers.
3. Elements in the periodic Table that are vertically arranged are called groups. Both germanium and silicon are in Group IV. State the information this gives regarding the electronic structure of both germanium and silicon.

11.8 Semiconductor diodes

Let's now look at some applications of extrinsic semiconductor materials as used in electronic components. The most commonly used diode is the *p–n junction diode*. This consists of a piece of semiconductor material that has been doped to give p-type characteristics for half its thickness and n-type characteristics for the other half. Where the two regions meet is called the *junction*, hence the name 'junction diode'. Such a diode is shown in Fig. 11.14. Where the positive and negative charge carriers face each other across the junction, the charge carriers tend to neutralise each other by recombination. p-Type, acceptor carriers, capture electrons and become negatively charged ions, whilst n-Type, donor carriers, lose free electrons and become positively charged ions. Thus the depletion layer develops a negative charge on the 'p' side and a positive charge on the 'n' side of its interface. This results in an electrical potential, called the *barrier potential*, appearing across the depletion layer, which has to be overcome before the diode will conduct.

Fig. 11.14 *Junction diode*

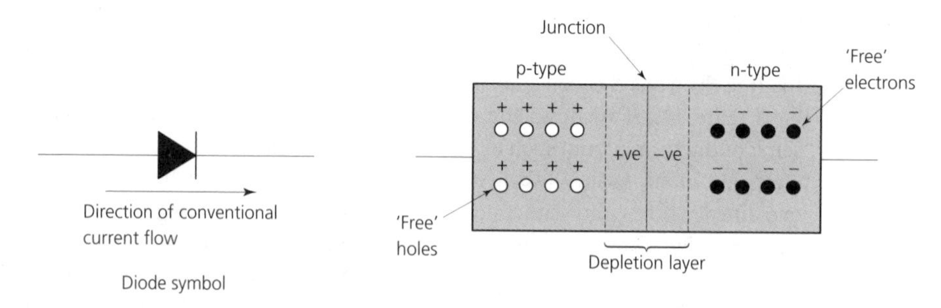

The characteristics for silicon and germanium junction diodes are shown in Fig. 11.15 and their respective barrier potentials can be clearly seen. Although the silicon (Si) diode has a higher barrier potential, its steeper VF/IF curve and its ability to withstand a greater peak inverse voltage (piv) makes it superior to germanium for power handling applications. Further, silicon can sustain higher operating temperatures without destruction than germanium. On the other hand, the lower barrier potential of the germanium (Ge) diode (about 200 mV) makes it more responsive and suitable for small signal radiofrequency applications.

Fig. 11.15 *Diode characteristics (the V_R value can range from a few volts to several hundred volts depending upon the type of diode)*

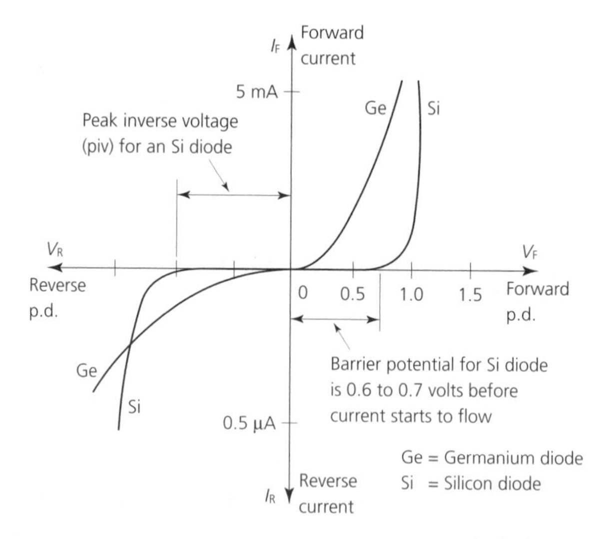

Figure 11.16 shows two ways of connecting such a diode. In Fig. 11.16(a) the lamp will light. This is because the diode has not only been *biased in the forward direction*, but because this bias is sufficient to overcome the barrier potential of the depletion layer. This causes the charge carriers to cross the junction and an electric current to flow. In Fig. 11.16(b) the lamp will not light because *reverse bias* has been applied to the diode and this

has resulted in the depletion layer widening so as to provide an insulating gap in the circuit preventing a current from flowing. Thus a diode can be considered as an electronic switch which will allow current to flow only in one direction. The diodes indicated in Fig. 11.17 are used to convert alternating current (ac) to direct current (dc) in this mains-operated radio set.

Fig. 11.16 *Diode – forward and reverse bias: (a) forward bias – lamp on; (b) reverse bias – lamp off*

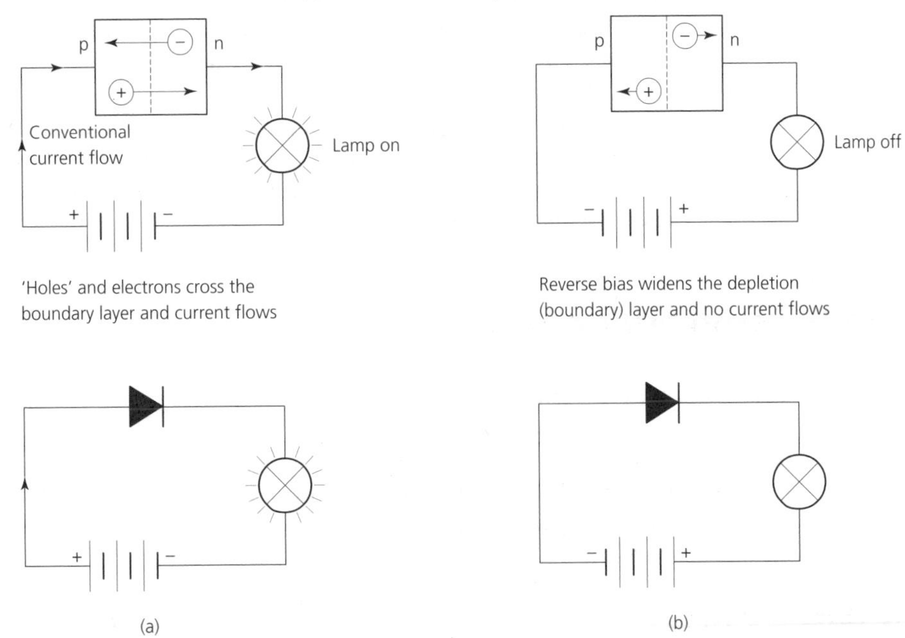

Fig. 11.17 *Diodes used for rectification in a mains-operated radio receiver (reproduced courtesy of the University of Luton)*

Since there is no such thing as a perfect insulator, and semiconductor materials are no exception, a very small current of a few micro-amperes will flow in the reverse direction when the diode is reverse biased. Reference to Fig. 11.15 shows that if the peak inverse voltage reaches a certain critical value, the very small reverse leakage current suddenly increases and there is a heavy flow of current. This is called an *avalanche current* and can be sufficient to destroy the diode. It is caused when the 'insulating' properties of the depletion layer break down because the 'dielectric strength' has been exceeded.

Let's now consider some other types of junction diodes.

- *Light-emitting diode* (L.E.D.) This emits light when biased in the forward direction. It has a longer life and is more reliable than a filament lamp for indicator purposes as well as being physically smaller. Various colours are available.
- *Photodiode* This, like all diodes, does not normally conduct when reversed biased, but will conduct when light falls upon it. This property makes it useful as a sensor for the automatic switching of security lights.
- *Zener diode* This is designed to operate under avalanche current conditions without being destroyed when reverse bias is applied. Under these conditions small changes in potential result in very large changes in current and the Zener diode can be used as a voltage reference source in voltage-stabilising circuits.

11.9 Semiconductor transistors

Transistors were invented in 1984 and were responsible for miniaturising electronics, gradually replacing thermionic valve technology. Nowadays individual transistors have been replaced largely by integrated circuits ('chips') that can contain the equivalent of millions of individual components. Figure 11.18 shows the comparison between a single transistor and an integrated circuit. Nevertheless, it is still important to understand the method of operation of individual transistors, and this will now be described.

Fig. 11.18 *Comparison between a single transistor and an integrated circuit (reproduced courtesy of the University of Luton)*

Figure 11.19(a) shows the circuit symbol and construction of an n–p–n bipolar transistor, whilst Fig. 11.19(b) shows the circuit symbol and construction of a p–n–p transistor. In both cases the transistor consists of layers of n-Type and p-Type semiconductor material. Wires are connected to each layer and these layers are called the *emitter*, *base* and *collector*. The emitter emits (sends) charge carriers through the thin base layer to be collected by the collector layer. In an n–p–n type transistor the emitter sends electrons through the base to

the collector, whilst in a p–n–p type transistor the emitter sends positively charged 'holes' through the base to the collector. In both cases the arrowhead on the emitter symbol shows the direction of *conventional current* flow.

Fig. 11.19 *The bipolar transistor: (a) the n–p–n transistor; (b) the p–n–p transistor*

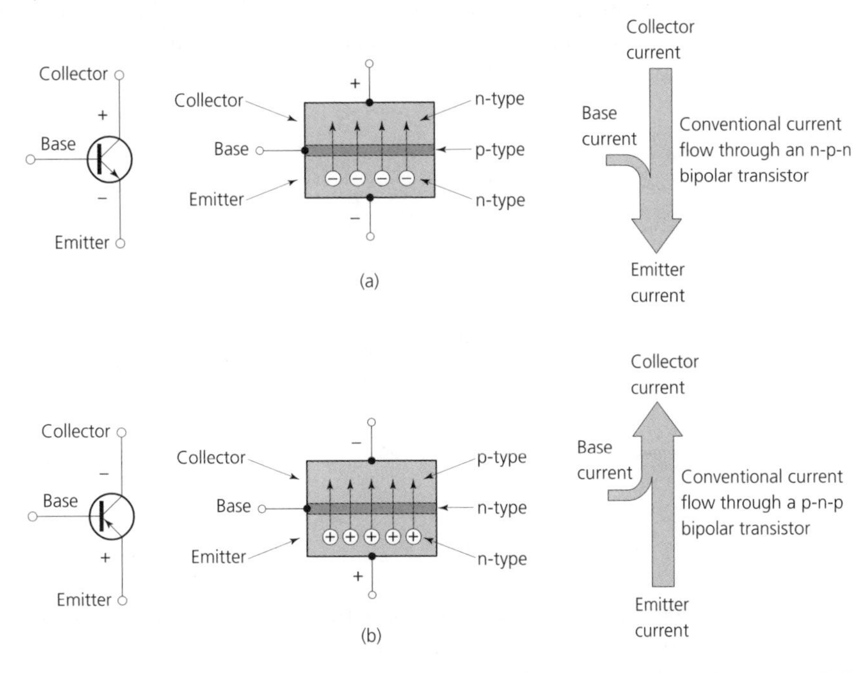

One of the uses of a transistor is as a switching device (electronic relay) with a small current in the base circuit controlling a larger current in the emitter–collector circuit. Figure 11.20 shows such a simple switching circuit. When the switch is closed, as shown in Fig. 11.20(a), the base of the transistor is connected to the 6 V supply via the 10 kΩ current-limiting resistor, the emitter-collector circuit conducts and the lamp lights. When the switch is opened, as shown in Fig. 11.20(b), and base current ceases to flow, the lamp goes out. Since

Fig. 11.20 *The bipolar transistor as a switch: (a) switch open – zero base current – transistor not conducting – lamp off; (b) switch closed – small base current flows – transistor conducts – large collector–emitter current flows – lamp on*

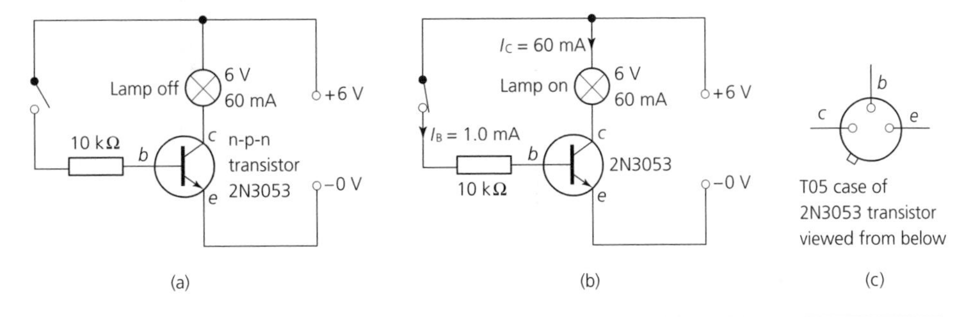

in this circuit a base current of 1 mA is controlling an emitter–collector current of 60 mA, the current gain (amplification) is 60.

$$\text{Current amplification } (\beta) = I_C/I_B$$

The transistor can also be used for amplifying the fluctuating or alternating base currents found in audiofrequency and radiofrequency amplifiers and also in oscillator circuits. A typical common emitter, small signal audiofrequency amplifier using a bipolar junction transistor is shown in Fig. 11.21.

Fig. 11.21 *Common-emitter amplifier*

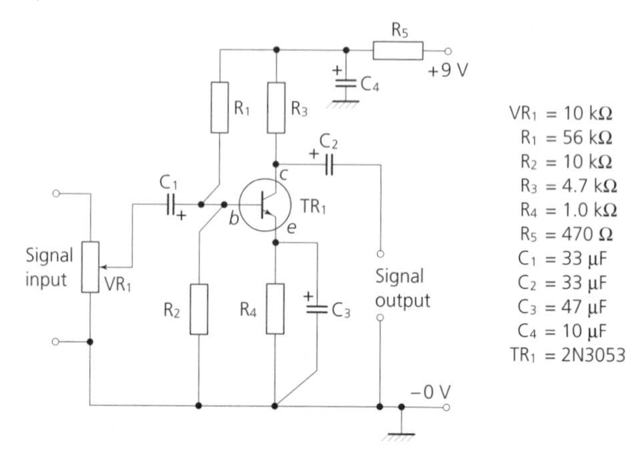

Another widely used device is the field effect transistor (FET). The symbol for a junction unigate field effect transistor (JUGFET) is shown in Fig. 11.22(a) whilst its construction is shown in Fig. 11.22(b). This is only one of many different types of field effect transistors. Figure 11.23 shows a typical FET amplifier circuit. Unlike the bipolar transistor, negligible current flows in the control electrode (gate) of a FET. This results in the input of a FET having a very high impedance, thus it does not load the input source. Since negligible

Fig. 11.22 *Field effect transistor: (a) JUGFET symbol; (b) construction of a JUGFET (there are many other types of field effect transistor (FET) for special applications)*

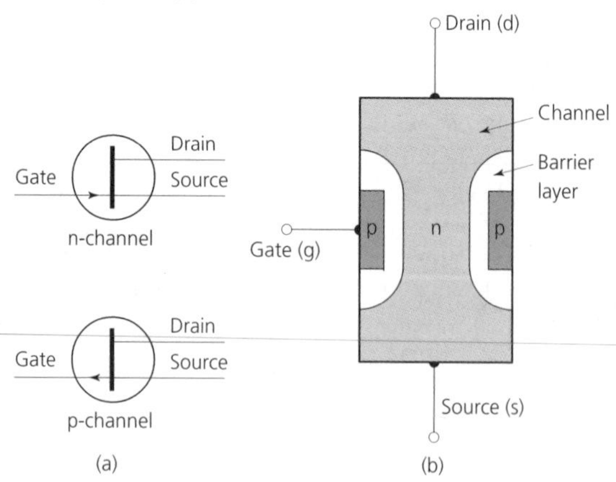

Fig. 11.23 *FET as a small signal amplifier*

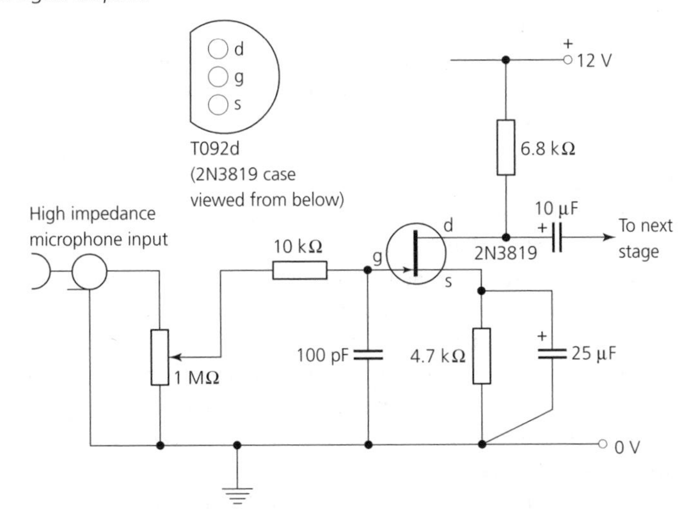

current flows in the gate circuit and the FET only responds to gate potential, it is used as a *voltage-amplifying device*, unlike the bipolar junction transistor which behaves as a *current-amplifying device* with low input and output impedances compared to the FET.

11.10 Magnetic materials

Every electron has a magnetic moment – in other words it behaves like a tiny bar-magnet with a North- and a South-seeking pole. Relatively few elements have sufficient nett magnetism from their complement of electrons to become useful magnetic materials. Iron is the most common of the metallic elements to be associated with magnetism, but three other transition metals – cobalt, nickel and gadolinium – are also strongly magnetic. Such materials are said to be *ferromagnetic*. This distinguishes them from other elements and materials that have weak magnetic properties. Most other metals are *paramagnetic*, that is, they are weakly attracted by strong magnetic fields. Some metals and all the non-metals are *diamagnetic*, that is, they are repelled by strong magnetic fields.

Reference to Pauli's exclusion principle has shown that in a stable atom not more than two electrons can occupy the same energy level and that these two electrons will have opposite directions of spin. It is this 'spinning' of the electrons that produces the magnetic field of an atom. The opposing spins of any two electrons in a given energy level causes their magnetic fields to cancel out. Therefore, in most materials there will be as many electrons spinning in one direction as there are in the opposite direction and the individual magnetic fields will largely cancel out, resulting in paramagnetic or diamagnetic properties.

However, in the ferromagnetic materials, found amongst the 'transition' elements, there are unfilled subvalency shells resulting in unpaired electrons being present. Therefore, overall, in any ferromagnetic atom more electrons will spin in one direction than in the other and there will be a resultant magnetic moment. In metals such as α-iron, nickel and cobalt not only do their magnetic moments have sufficient magnitude, but their atoms are sufficiently closely packed together to produce a powerful magnetic field when the atoms become magnetically aligned.

The effects of atomic spacing on the magnetic properties of a material are critical. If the atoms are too widely spaced, the forces between them are weak and the electron spins can be easily thrown out of alignment by thermal agitation. This results in their individual magnetic fields cancelling each other. If the atoms are too closely spaced, then the inter-atomic bonds are too rigid and alignment of the electron spins cannot take place. The ferromagnetic materials previously mentioned have suitable atomic spacings. However, it should be noted that other 'transition' elements such as chromium, manganese and titanium have atomic spacings only just outside the ideal. In fact if manganese contains interstitial nitrogen atoms, the resulting modification to its atomic spacing results in it becoming ferromagnetic.

Within the crystals of ferromagnetic materials are small regions referred to as magnetic *domains*. These are regions in which the magnetic fields of groups of atoms are aligned in the same direction. When the material is not magnetised, the magnetic fields in the domains are arranged in a random manner so that there is no resultant external field (Fig. 11.24(a)). However, when the material becomes magnetised the fields of the domains become oriented in the same direction and the material has a resultant magnetic field (Fig. 11.24(b)).

Fig. 11.24 *Alignment of domains in a bar magnet: (a) random; (b) aligned*

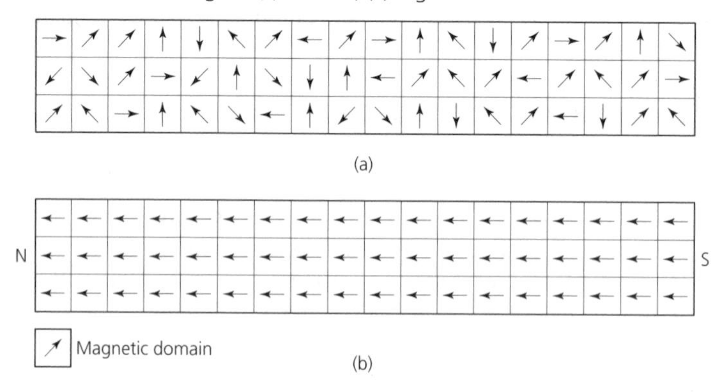

Some ceramic materials can also show magnetic properties. For example, the mineral magnetite (Fe_3O_4), formerly known as 'loadstone', is an example of a naturally occurring magnetic ceramic material. A commercial ceramic magnet material is 'Ferroxdur' ($BaFe_{12}O_{19}$). Such materials are said to be *ferrimagnetic*.

Magnetic materials are classified as *hard* magnetic materials or *soft* magnetic materials. *Hard magnetic materials* are used for 'permanent' magnets since they retain their magnetism when the magnetising field is removed. Such materials are also physically hard and can only be machined by grinding (for example, quench-hardened high-carbon steel). The magnetism of permanent magnets can be destroyed by subjecting them to:

- An alternating magnetic field.
- Temperatures above the Curie point.
- Mechanical vibration (hammering).

All the above provide sufficient agitation to allow randomisation of the domain alignment.

Soft magnetic materials retain their magnetism only as long as they are energised by an external magnetic field. Once the magnetising field is removed, soft magnetic materials immediately lose most of their magnetism. Such materials are used for such applications as transformer cores, motor rotor and stator laminations and electromagnet cores. An example is shown in Fig. 11.25. In this example soft iron laminations are used to construct the field magnet (stator). It is energised by a coil of insulated copper wire wound round the laminations to the left of the figure. The armature (rotor) coils wound round another set of soft iron laminations rotates between the poles of the field magnet. The armature can be seen in the right of the illustration.

Fig. 11.25 *The soft iron laminations used in this vacuum cleaner motor surround the coils and are clearly visible*

11.11 Magnetic properties (hard magnetic materials)

Table 11.8 lists some typical 'hard' magnetic materials together with their magnetic properties. The most important characteristics of a permanent magnet are its:

- *Coercive force* This is the resistance of the material to demagnetisation by electromagnetic techniques.
- *Remanence* This is the intensity of the residual magnetism after the magnetising field has been removed.
- *Energy product value* This is an index of energy required to demagnetise and to reverse the polarity of a permanent magnet and thus a measure of the amount of magnetic energy stored in a magnet after the magnetising field is removed.

Figure 11.26 shows a typical *magnetic hysteresis loop* for a hard magnetic material. The *induced magnetic flux density* B (teslas) is plotted against the *magnetising field* H (ampere-turns per metre). The starting point, O, on the curve indicates zero magnetic field strength. The magnetising field strength H is gradually increased, and corresponding values of B and H are plotted, until magnetic saturation occurs at point P and the curve 'levels off'. The magnetising field strength is then reduced to zero and the corresponding values of B and H

Table 11.8 *Permanent magnet materials*

Compositions and properties

Name	Composition (%)										Magnetic properties		
	C	*Cr*	*W*	*Co*	*Al*	*Ni*	*Cu*	*Nb*	*Ti*	*Fe*	B_{rem} (*T*)	H_c (*A/m*)	$H_{B(max)}$ (J/m^3)
Quenched-hardened high-carbon steel	1.0	—	—	—	—	—	—	—	—	Rem	0.9	4 400	1 560
35% cobalt steel	0.9	6.0	5.0	35.0	—	—	—	—	—	Rem	0.9	20 000	7 800
Alnico				12.0	9.5	17.0	5.0	—	—	Rem	0.73	44 500	13 500
Alcomax III*				24.5	8.0	13.5	3.0	0.6	—	Rem	1.26	51 700	38 000
Hycomax III*				34.0	7.0	15.0	4.0	—	5.0	Rem	0.88	115 400	35 200
Columax**				24.5	8.0	13.5	3.0	0.6	—	Rem	1.35	58 800	52 800
Ferroxdur ($BaFe_{12}O_{19}$)	—	—	—	—	—	—	—	—	—	—	0.4	150 000	20 000

*Anisotropic alloys whose magnetic properties are measured along the preferred axis.
**This alloy derives its very high H_B value from the way it is cooled during casting which orientates its columnar crystals parallel to the preferred axis of magnetisation.
C = carbon Cr = chromium W = tungsten Co = cobalt Al = aluminium Ni = nickel Cu = copper Nb = Niobium Ti = titanium Fe = iron

are once more plotted to give the curve PQ. Thus the residual magnetism is represented by OQ, that is the remanence (B_{rem}).

To determine the coercive force the material must be demagnetised by reversing the magnetising field and gradually increasing its field strength from zero at point Q on the curve until point R is reached. Thus OR represents the field strength (force) to demagnetise the material completely and is called the *coercive force* (H_c).

The strength of the reverse magnetising field is increased until 'negative saturation' is reached at point S, whereupon the magnetising field is again reversed and the curve SUP is plotted to complete the loop. The hysteresis loop represents the amount by which the induced magnetic flux lags behind the magnetising field. For a permanent (hard) magnetic material the loop must be as large as possible. The ultimate requirement of a magnetic material is that the product of B and H (the energy product value BH_{max}) must be as large as possible, as this is the maximum energy that the magnet can provide external to itself. The BH_{max} value occurs when the product of CD and DE on the demagnetisation curve is a maximum.

Fig. 11.26 *Hysteresis loop for a magnetic material*

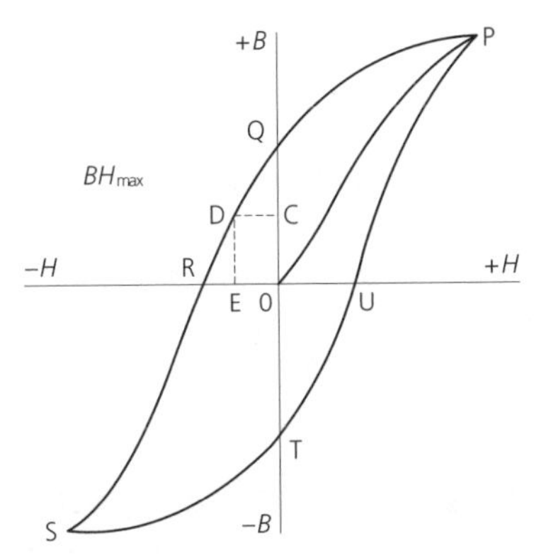

The selection of a permanent magnetic material depends upon whether a high value of remanence is required or whether a high value of coercive force is required. For example, 'Columax' has a high remanence whilst 'Hycomax III' and the ceramic 'Ferroxdur' have high coercive forces.

Some permanent magnetic materials are said to be *anisotropic* – that is, their properties along a preferred axis are much enhanced. To achieve this effect, the magnetic alloy material is raised to a high temperature and then cooled in a powerful magnetic field. The thermal agitation decreases as the temperature falls resulting in groups of atoms becoming aligned along the direction of the external field, so that this alignment becomes 'frozen in' by the time the magnetic material has reached room temperature.

11.12 Magnetic properties (soft magnetic materials)

A major requirement of soft magnetic materials is that they should have a high permeability (μ). This is a measure of the material's ability to 'concentrate the magnetic field' and is defined as the ratio of the flux density (magnetic induction) B to the total magnetic field H.

$$\mu = B/H$$

The permeability value, for a given material, is not constant but alters with the magnetising field, as shown in Fig. 11.27.

Fig. 11.27 *Magnetising curve for a 'soft' magnetic material*

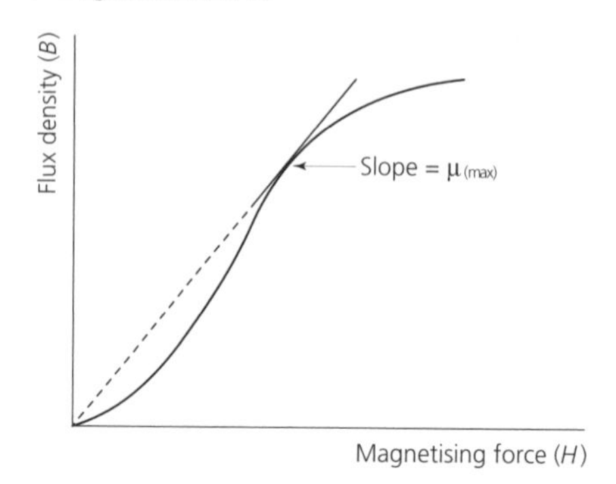

Permeability is measured in henry per metre (H/m). More frequently, the relative permeability (μ_r) is used and, being a ratio, it has no unit. Relative permeability is the permeability of a given material compared with that for a vacuum. For a vacuum, $\mu = 1$, and for air, μ_r is slightly greater than 1. Table 11.9 shows the maximum values of μ_r for various soft magnetic materials. Clearly, high values of μ_r must be used where highly magnetisable materials are needed.

Table 11.9 *Properties of common 'soft' magnetic materials*

Material	*Saturation induction B_s (T)*	*Coercive field H_c (A/m)*	*Relative permeability μ_r (max.)*
α-Iron	2.2	80	5 000
Silicon-ferrite transformer sheet	2.0	40	15 000
Permalloy Ni–Fe	1.6	10	2 000
Superpermalloy Ni–Fe–Mo	0.2	0.2	100 000
Ferrox cube A (Mn Zn) Fe_2O_4	0.4	30	1 200
Ferrox cube B (Ni Zn) Fe_2O_4	0.3	30	700

Since soft magnetic materials often operate under the influence of magnetic fields generated by alternating currents they are magnetised, demagnetised, then remagnetised and demagnetised with reverse polarity, this cycle being repeated many times each second. Such an alternating cycle produces a closed hysteresis loop and the area of the loop represents the energy wasted in overcoming the remanence for the material. This wasted energy causes heating of the magnetic material. Therefore, soft magnetic materials are formulated to have very narrow hysteresis loops and Fig. 11.28 compares typical hysteresis loops for hard and soft magnetic materials.

Fig. 11.28 *Hysteresis loops for 'hard' and 'soft' magnetic materials*

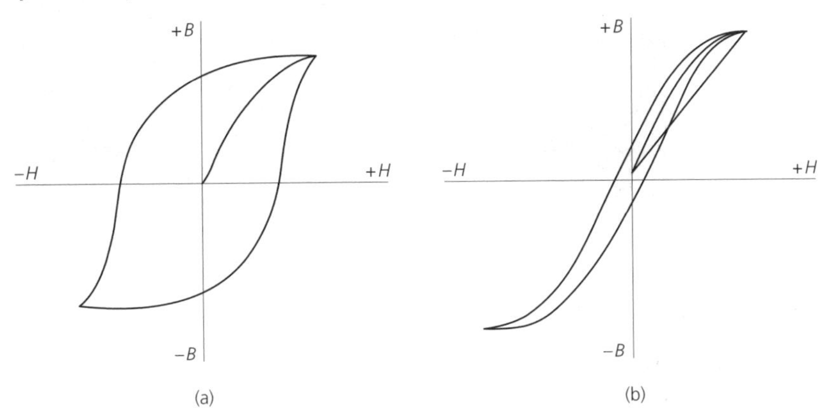

Permeability is not the only property to be considered when selecting a soft magnetic material, otherwise the ceramic magnetic materials (Ferroxcube A and B) would not be used. However, at high frequencies (over 1 MHz) a compromise between permeability and electrical conductivity is required. Metallic materials are unsuitable at such high frequencies because of their high conductivity, which results in excessive eddy-current heating. Eddy-current heating not only represents wasted energy, but very high temperatures are attained. In fact, it is possible to melt metals using high-frequency eddy-current heating.

The silicon–ferrite sheet is used for transformer core laminations and motor and generator rotor and stator stampings. This is a relatively low-cost material and is therefore suitable for such large-scale production. The more expensive 'permalloy' alloys are used for more specialised applications such as screening cans and the cores for audiofrequency coupling and matching transformers in telecommunications equipment.

SELF-ASSESSMENT TASK 11.6

1. Write down several ways of destroying the magnetism of a permanent magnet.
2. Briefly explain the meaning of:
   (a) magnetic domain
   (b) coercive force
   (c) remanence
3. Describe the properties that are necessary for the core of an electromagnet.
4. The ferrite rod used in the aerial of a radio (Fig. 11.29) is employed to concentrate the magnetic component of the incoming radio waves. State the properties that you think are possessed by the ferrite rod.

Fig. 11.29 *The ferrite rod shown forms part of the aerial in a radio receiver*

EXERCISES

11.1 With the aid of diagrams explain what is meant by:
(a) Bohr model
(b) quantum shell
(c) valency shell
(d) Pauli's exclusion principle

11.2 Explain how materials may be classified as conductors, insulators or semiconductors in terms of their 'energy levels', 'valency bands' and 'conduction bands'.

11.3 Explain how an electric current flows through a conductor material and distinguish between conventional current flow and electron current flow.

11.4 (a) Describe the effects of temperature change, composition, impurities and structure on the electrical properties of conductor and insulator materials.
(b) Explain what is meant by superconductivity.

11.5 Discuss the essential differences between intrinsic and extrinsic semiconductor materials.

11.6 Explain how an intrinsic semiconductor material conducts electricity in terms of its n-type charge carriers and its p-type charge carriers.

11.7 Explain how extrinsic semiconductor materials are given their n-type or p-type characteristics by the addition of dopants.

11.8 (a) Explain the basic principle of operation of a junction diode and suggest **two** typical applications of this device, giving reasons for your choice.
(b) Explain the basic principle of operation of a bipolar junction transistor and show how it may be used as:
(i) a direct current amplifier
(ii) an alternating current (AF or RF) amplifier.

11.9 Explain what is meant by the terms:

(a) atomic magnetic moment
(b) magnetic domain
(c) ferromagnetism
(d) ferrimagnetism
(e) paramagnetism
(f) diamagnetism

In the cases of (c) to (f) above, give examples of typical materials with these characteristics.

11.10 With reference to their hysteresis diagrams, discuss the essential differences between 'hard' and 'soft' magnetic materials and, in each instance, suggest **two** typical applications, giving reasons for your choice.

12 Semiconductor materials and manufacturing processes

The topic areas covered in this Chapter are:

- Silicon as a semiconductor material.
- The manufacture and purification of silicon.
- The production of silicon wafers.
- The fabrication of semiconductor devices using planar, MOS, CMOS and bipolare technology.

12.1 Silicon as a semiconductor material

Solid-state semiconductor devices were introduced in Chapter 11. Such devices are manufactured from a variety of high-purity semiconductor materials of which silicon is the most common. Germanium, the material from which the first solid-state devices were manufactured has been superseded by silicon except for certain specialised applications. More recently 'compound' semiconductors such as gallanium arsenide (GaAs) and indium phosphide (InP) have been developed. Again these have specific applications of which the most common are optical devices, such as light-emitting diodes (LEDs) and lasers, and very high-frequency transistors and integrated circuits. Whatever the semiconductor material used, it should be moncrystalline and free from crystallographic defects such as dislocations. It will usually be doped n-type or p-type and have the resistivity required for a particular application.

Since silicon is the most widely used of the semiconductor materials, this Chapter will be restricted to its manufacture and use. Silicon must not be confused with the *silicones* which are inorganic silicon–oxygen structures whose many uses include the production of silicone polymers that are useful heat-resistant, flexible, insulating materials.

Pure elemental silicon is a silvery-grey material with a density of 2300 kg/m^3, a melting point of 1680 K, and a boiling point of 2628 K (at 760 mm Hg). Pure silicon has a resistivity of the order of 2.3×10^3 ohm metre (at 20 °C), and thus lies part way between the values expected for conductors and for insulators. However, the resistivity of commercially hyperpure (*intrinsic*) silicon rarely exceeds 2×10^2 ohm metre, and it will have residual n-type or p-type characteristics. The conductivity of silicon not only increases as its temperature increases, as explained in Section 11.7, but the conductivity is also dependent upon its purity.

Most silicon is supplied to manufacturers of semiconductor devices as an extrinsic semiconductor material; traces of trivalent or pentavalent impurities (dopants) having been added during manufacture to give the material specific p-type or n-type characteristics and levels of resistivity to suit customer requirements. The presence of the dopants increases the conductivity of the material significantly, depending upon the amount and type of dopant present. The level of dopants present is only of the order of parts per million (p.p.m.) and close quality control is essential.

SELF-ASSESSMENT TASK 12.1

1. Discuss the reasons for silicon still being the most widely used semiconductor material.
2. Distinguish carefully between intrinsic and extrinsic semiconductor materials in relation to physical structure and electrical characteristics.

12.2 Purification of silicon

Silicon is the second most abundant element occurring in nature where it is nearly always in association with oxygen (silicon dioxide) and other elements to form mineral silicates as in quartz, sand and clays. It is from these raw materials that commercially pure silicon is nearly always produced by chemical decomposition (reduction).

A two-stage process is nearly always used to produce the high-purity silicon used for the manufacture of semiconductor devices.

1. The production and purification of a volatile silicon compound, usually trichlorosilane ($SiHCl_3$).
2. The decomposition or reduction of trichlorosilane to produce pure polycrystalline silicon.

A simplified diagram of the process used to produce trichlorosilane is shown in Fig. 12.1. In this process hydrogen chloride is combined with silicon in a fluid-bed reactor at approximately 300 °C.

$$Si + 3HCl \longrightarrow SiHCl_3 + H_2$$

Trichlorosilane has the advantage of a low boiling point (31.8 °C) that allows it to be purified by fractional distillation.

Trichlorosilane is the most widely used gaseous compound for the manufacture of hyperpure silicon because of its low price, high volume availability, ease of handling, low toxicity, low flammability, and commercially viable convertibility into silicon by a high-temperature reduction process.

Polycrystalline silicon, of high purity, is produced by *chemical vapour deposition* (CVD). In this process trichlorosilane is reacted with hydrogen gas in the presence of a thin, high-purity silicon rod as shown, diagrammatically, in Fig. 12.2. The rod is heated by the passage of an electric current and silicon is deposited upon it coherently in polycrystalline form. Hydrogen chloride gas is generated as a by-product. Provided that the rod is at the correct temperature and the ratio of trichlorosilane to hydrogen lies within critical limits, the reaction proceeds quite readily. The deposition rate and the evenness of the deposit depend upon a number of parameters, including the reaction vessel size, the temperature of the 'thin rod', and the ratio of the reaction gases. By careful control, polycrystalline rods having a coherent structure and high density can be formed.

Fig. 12.1 *Preparation and refining of trichlorosilane (simplified) (reproduced courtesy of Wacker Chemitronic GmbH)*

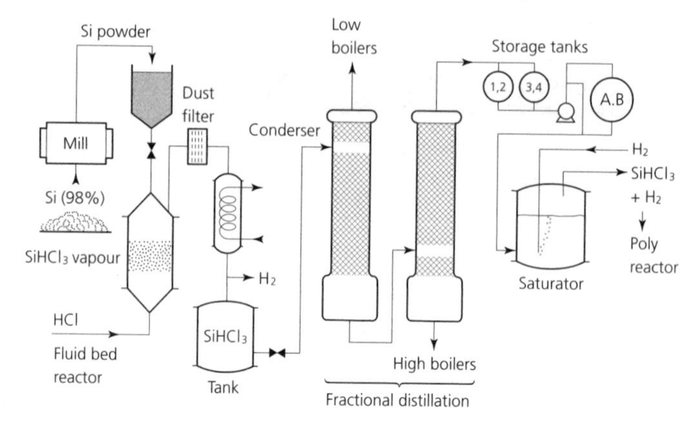

Fig. 12.2 *Production of high-purity polycrystalline silicon by chemical vapour deposition (reproduced courtesy of Wacker Chemitronic GmbH)*

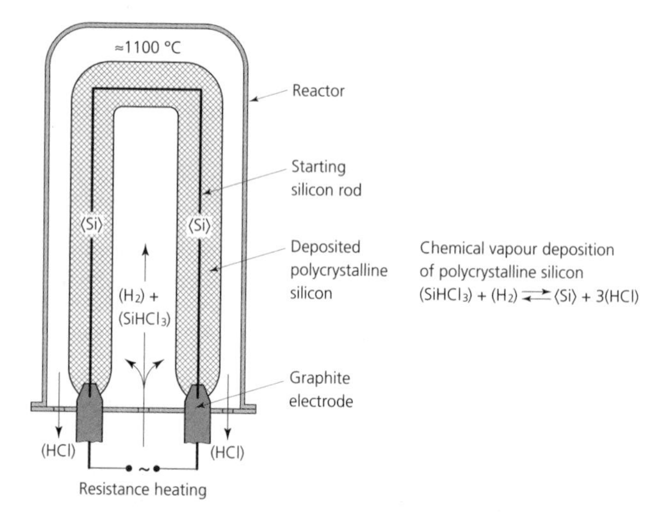

To ensure silicon of the highest purity, care must be taken in the choice of raw materials and in the design and operation of the plant. The following precautions must be taken:

- Only a limited range of materials can be used for the reaction vessel and for those parts of the plant coming into contact with the reaction gases to prevent contamination.
- The trichlorosilane must be purified by careful distillation so that the level of impurities (particularly phosphorus and boron, which are electrically active) are reduced to a fraction of one part per million (p.p.m.).

- The hydrogen gas must be free from traces of oxygen, dried to a dew point of −100 °C and dust free. Hydrocarbons and ammonia must be rigorously excluded.

The point in the process at which the dopants are added depends upon the quantity and type of dopant, and the manufacturing processes used. In practice, the electrical characteristics required by the customer may be given to the polycrystalline silicon by adding the dopants under very close control during the refinement and/or the polycrystallisation process, or it may be added to the crucible prior to crystal pulling. The amount of intrinsic silicon produced is very small compared to the amount of extrinsic p-type or n-type silicon produced.

Before the high-purity polycrystalline silicon can be used in the manufacture of electrical semiconductor devices, it has to be converted into monocrystalline rods of sufficient size that can be sliced into wafers. Such monocrystalline rods must have a dislocation-free atomic structure. The two processes most widely used for growing single crystals of silicon are the *Czochralski* (CZ) process and the *float-zone* (FZ) process. Let's now consider these in detail.

12.3 Czochralski (CZ) process

Worldwide, 80 per cent of all solid-state devices are manufactured from monocrystalline silicon produced by the Czochralski process. Crystals grown by this process are particularly suitable for low-power devices such as large-scale integrated (LSI) circuits and very large-scale integrated (VLSI) circuits that are used in logic and control electronics.

The principle of the Czochralski process is shown in Fig. 12.3. Polycrystalline silicon, produced as described in Section 12.2, is crushed or sawn into small pieces and cleaned. It is loaded into a crucible under clean room conditions and the dopant is added. (Where volatile dopants are used, these are added directly to the melt.) The most common dopants are boron, phosphorus, antimony and arsenic. Where high dopant concentrations are required these can be weighed and handled without difficulty. However, in the case of medium and low dopant concentrations this is not possible on a commercial basis. For example, to grow p-type silicon crystals with a resistivity of 0.1 Ω m, only 0.1 mg of boron is required for a 10 kg melt. To achieve adequate control of such concentrations on a commercial basis, alloys of silicon are made that contain the dopant in a higher concentration. An appropriate weight of such an alloy can then be added to the melt to give the required low dopant concentration. Such alloys are prepared by adding the dopant as a vapour during the growth of polycrystalline silicon by the CVD process. To avoid contamination of the melt, only crucibles of pure quartz are suitable for the CZ process.

The pieces of polycrystalline silicon in the crucible are melted under an inert gas such as argon or under high vacuum conditions. High vacuum conditions can only be used where the melt is small (less than 1 kg); for larger melts crystal pulling can only take place under flowing inert gas. There are two main pressure ranges: atmospheric pressure or 5–50 millibars. The inert gas has to flow permanently downwards through the pulling chamber, as shown in Fig. 12.3, to carry off the reaction products which are evaporated in considerable amounts.

Fig. 12.3 *Czochralski single crystal pulling apparatus: (a) initial stage of process; (b) advanced stage of process; (c) final stage of process (reproduced courtesy of Wacker Chemitronic GmbH)*

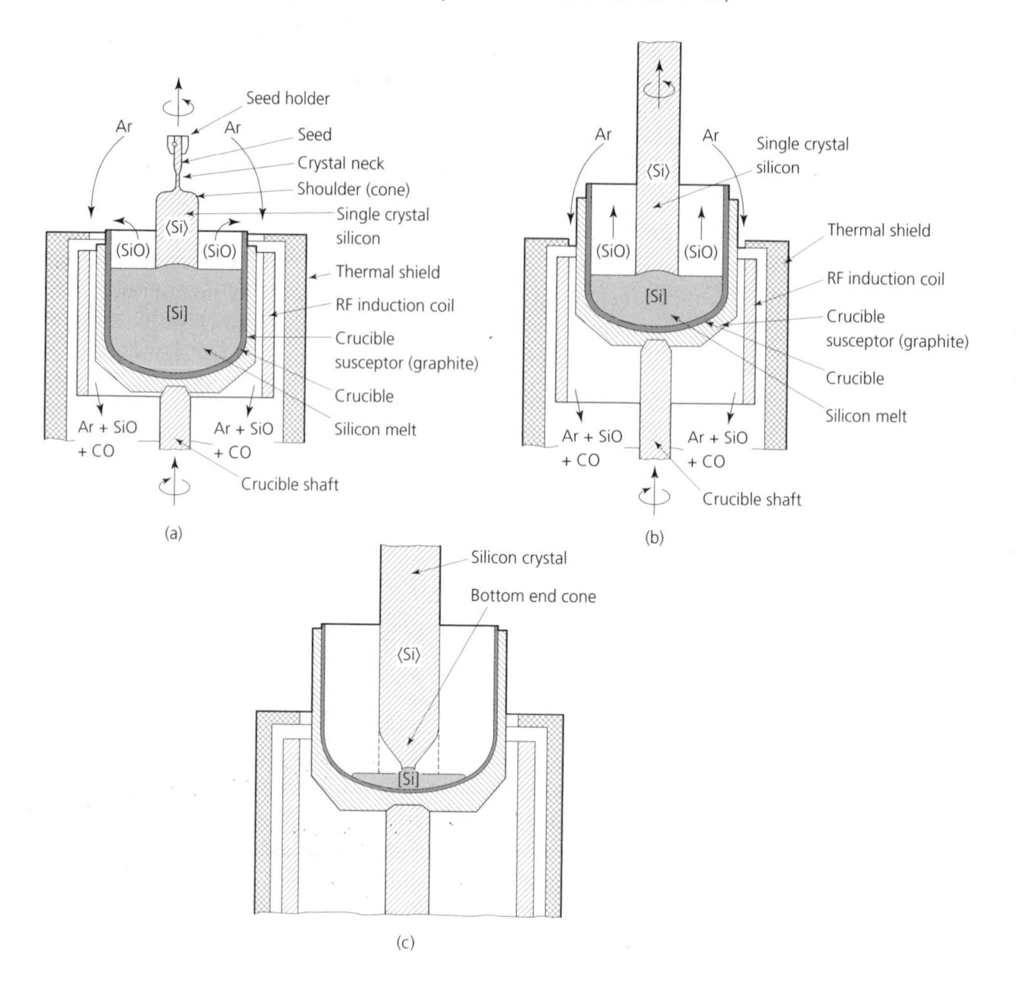

To give close temperature control, modern furnaces use radiofrequency (RF) induction heating. The crucible is surrounded by a graphite *susceptor* in which eddy currents are induced, causing heating of the susceptor. This heat energy is transmitted to the quartz crucible by conduction. Older furnaces used a cylindrical graphite-resistance heating element surrounding the crucible and heating it by radiation.

When the silicon in the crucible is completely molten, the electrical energy supplied to the graphite-resistance heating element is reduced so that during crystal pulling the whole furnace, including the crucible and melt, are kept in thermal equilibrium. To produce a high-purity single crystal of silicon by the CZ process, a tiny seed crystal, having the perfection of the crystal to be grown, is dipped into the quartz crucible of molten pure silicon that is held accurately at the process temperature. The crystal commences to grow and, whilst growth takes place, it is slowly withdrawn from the crucible (*crystal pulling*). At the same time as the crystal is withdrawn, it is rotated so that crystal growth is uniform. The crucible is rotated in the opposite direction. Under correctly controlled conditions the nascent crystal that grows on the seed crystal follows the structural perfection of the seed crystal itself.

Although the seed crystal is dislocation free, the action of dipping the crystal into the melt causes thermal shock and surface tension effects which can cause dislocations to appear which move into the growing crystal, particularly in crystals of large diameter. Because of the high strains and temperatures in the crystals, the dislocations are not confined just to their own glide planes, but can spread to adjacent glide planes by cross-slip, multiplication processes and by climb (see Chapter 8).

To ensure dislocation-free crystals the crystal-pulling technique developed by W. C. Dash has to be used. This technique generates a 'neck' between the seed crystal and the crystal being pulled. This neck reduces the cooling strain to a very low value and, as a result, the remaining strain energy may not be sufficient to move the existing dislocations or generate new ones. In any case, necking will result in any residual dislocation motion being slower than the rate of crystal growth.

The Dash technique has two basic stages:

- The crystal diameter is gradually reduced to about 2–4 mm.
- The growth velocity is raised to a maximum of 6 mm/min, depending upon orientation.

A (111)-orientation crystal does not require a very small crystal neck but it does need a high growth velocity. However, a (100)-orientation crystal requires both a thin neck and a low growth velocity. With a suitable combination of neck diameter and growth velocity the crystal becomes dislocation free after a few centimetres of growth. The dislocation-free state of the grown crystal shows itself in the development of 'ridges' on the crystal surface. The transition region from the seed node to the cylindrical part of the crystal is referred to as the seed-cone and can vary between almost flat to a very pronounced taper. Shortly before the desired crystal diameter is reached, the pulling velocity is raised to the specific velocity at which the crystal grows. Rotation of the seed crystal results in the monocrystalline silicon being almost circular in cross-section.

The pulling velocity is reduced towards the bottom end of the crystal. This is because the heat loss from the walls of the crucible increases as the level of the melt sinks; the heat transference required for crystallisation becomes more difficult, and more time is required to grow a given length of crystal. To complete a crystal free from dislocation an end cone has to be produced so that the crystal diameter is gradually reduced. For this reason the pulling speed is raised and the crystal diameter decreases and, if the diameter becomes small enough, the crystal can be separated from the melt without dislocations forming in the cylindrical part of the crystal from which the wafers are made. Care must be taken so that the increase in pulling velocity is not too great otherwise thermal shock will occur, resulting in plastic deformation (slip) in the lower part of the crystal. Conventionally, the residual melt left in the crucible is discarded together with the crucible since there are many practical difficulties not only in using up the whole of the melt but also in emptying the crucible.

Since several hours of machine time is lost in cooling the furnace, replacing the crucible and charge, reheating the furnace, evacuating the furnace if a vacuum is used or purging and flooding the furnace if an inert atmosphere is used, solitary charging is costly and inefficient. Therefore current practice is moving towards techniques for recharging the hot crucible whilst pulling is in progress so that continuous or semicontinuous pulling can be achieved.

However, there are severe limitations to the number of times the crucible may be recharged. The more important of these limitations are:

- A build up of impurities in the melt.
- Contamination from the crucible itself.
- Lack of mechanical strength in the crystal itself which results in handling problems and limits its physical size.

An important aspect of the CZ process is the possibility for in-situ remelting of faulty crystals.

- The entire crystal can be remelted and growth started again.
- The crystal is only partially remelted until the defect is reached, whereupon the Dash technique is applied to create a neck and a new dislocation-free crystal is grown from the new neck and shoulder.

The possibility of remelting is an important economic advantage of the CZ crystal-pulling process. This is not possible in the float-zone or pedestal-pulling processes. Table 12.1 lists typical data for modern pulling apparatus used in Czochralski silicon growth, whilst Table 12.2 lists the electrical and mechanical characteristics of wafers made from monocrystalline silicon produced by the CZ process.

Table 12.1 ***Czochralski silicon growth***

Typical data for modern crystal pulling apparatus

Crucible:	diameter	from 180 to 350 mm
	height	from 160 to 280 mm
	capacity	from 6 to 50 kg Si
Crystals:	diameter	from 50 to 150 mm
	length	from 500 to 2200 mm
	weight	from 5 to 48 kg Si
Seed shaft travel		up to 2500 mm
Seed shaft rotation		from ≈0 to 50 rev/min (reversible)
Seed shaft speed		from ≈0 to 10 mm/min (slow),
		from 20 to 800 mm/min (fast)
Crucible shaft travel		up to 500 mm
Crucible shaft rotation		from ≈0 to 20 rev/min (reversible)
Crucible shaft speed		from ≈0 to 1 mm/min (slow)
		from ≈0 to 200 mm/min (fast)
Power control		up to 150 kW
Overall height		up to 9 m
Overall weight		up to 9500 kg
Pressure range		10^{-5} to 2 bar
Gas flow (argon)		0.4 to 3 m^3/h
Vacuum pumps (mechanical)		up to 200 m^3/h

Source: Wacker Chemitronic GmbH.

Table 12.2 *Polished Czochralski silicon wafers*

Properties and geometries

Resistivity

Type, dopant	*Range ohm-cm (target)*	*Minimum tolerance (%)*	*Maximum radial variation (%) (6 mm edge exclusion) (111)*	*(100)*
p, Boron	0.005–25	±10	8	8
	>25–60	±20	12	12
n, Phosphorus	0.030–15	±15	20	12
	>15–40	±20	25	15
n, Antimony	<0.020 for diameter up to 125 mm		20	12
	<0.025 for diameters up to 125 mm		22	8
	>0.025–0.050		22	9

Geometry

Diameter (mm)		*76.2 ± 0.3*	*100 ± 0.3*	*125 ± 0.1*	*150 ± 0.1*
Thickness (μm)	Standard	381	525	625	675
	Minimum	290	350	400	500
Thickness tolerance (μm)	Minimum	±10	±10	±10	±10
Global flatness (μm)	Typical	<2.0	<2.0	<2.5	<3.0
Front side reference (TIR)	Maximum	3.0	4.0	5.0	5.5
Local thickness variation (LTV) (μm)	Typical	—	<1.0	<1.0	<1.0
	Maximum	—	2.0	2.0	2.0
Total thickness variation (TTV) (μm)	Typical	<3.0	<5.0	<5.0	<5.0
	Maximum	7.0	8.5	8.5	10.0
Warp (μm)	Typical	<10	<15	<20	<25
(damage-free etched backside)	Maximum	20	25	30	40

Source: Wacker Chemitronic GmbH.

12.4 Float-zone (FZ) process

The principle of the *float-zone* process is shown in Fig. 12.4(a). A rod of polycrystalline silicon is mounted vertically and there is no containing vessel. A narrow zone of silicon is melted using radiofrequency induction heating, and the molten zone is made to traverse the rod vertically by relative movement of the rod and the induction coil. Surface tension and levitation forces from the electric field prevent the molten zone from falling out even in large diameter rods of 100–150 mm diameter. The process may be carried out in vacuum or in an inert atmosphere such as argon. For the highest purity, a vacuum is preferable. If a seed crystal is introduced at one end of the polycrystalline rod and the molten zone is made to traverse the rod from this end, then a single crystal will be produced which reflects the structural perfection of the seed crystal. To ensure complete freedom from dislocation, the Dash technique of introducing a crystal 'neck' is also used in the float-zone process.

In the Czochralski (CZ) process some reaction occurs between the polycrystalline charge and the quartz crucible at the high melting temperatures involved. For the very highest purity and crystallographic integrity, the more expensive float-zone (FZ) process is used. In this process there is no contamination from the crucible since none is used. Further, refinement takes place as well as crystal growth, and any residual impurities are slowly swept along the rod from one end to the other and then discarded.

A variation on the float-zone process is *pedestal pulling* and this is also shown in Fig. 12.4(b). Again there is no crucible to cause contamination and crystals of the highest purity and crystallographic integrity can be produced by this process. The Dash technique of 'necking' is also used when pedestal pulling to ensure freedom from dislocation.

Fig. 12.4 *Principles of crystal pulling: (a) float zone pulling; (b) pedestal pulling (reproduced courtesy of Wacker Chemitronic GmbH)*

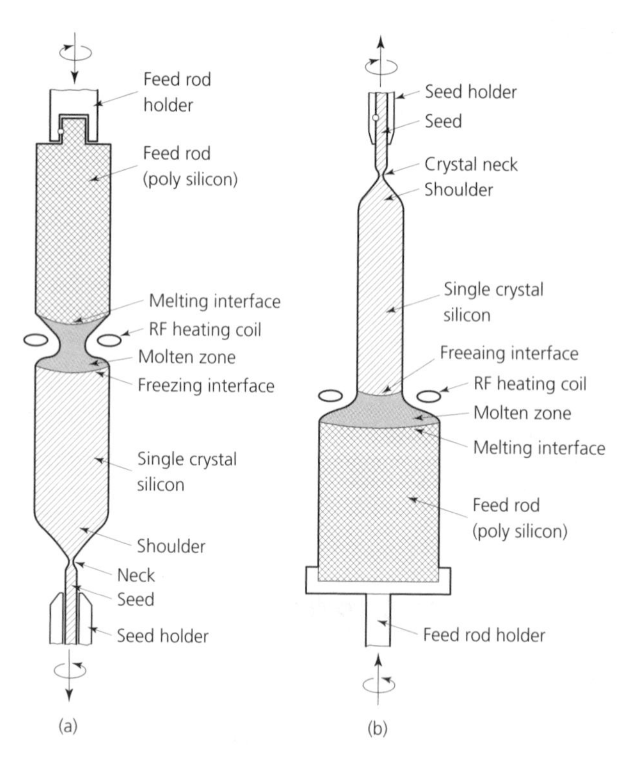

Monocrystalline silicon produced by the float-zone and pedestal-pulling processes are used for high-power and very high-power devices, as well as for detector devices. Table 12.3 compares some of the properties of crystals produced by the CZ and the FZ processes.

Table 12.3 *Typical crystal data (special materials excluded)*

Parameter	*Crucible pulling (CZ)*	*Float-zone (FZ)*
Crystal quality	Dislocation-free	Dislocation-free
Max. diameter	150 mm	150 mm
Resistivity range		
p-type	0.005–50 Ω cm	0.1–3000 Ω cm
n-type	0.005–50 Ω cm	0.1– 800 Ω cm
Dopants	B, P, Sb, As	B, P
Orientations	[111] [110] [100]	[111] [100] [511]
Lifetime	10–50 μs	100–3000 μs
Oxygen content	10^{16}–10^{18} cm^{-3}	below detection limit
Carbon content	10^{17} cm^{-3}	below detection limit

Source: Wacker Chemitronic GmbH.

SELF-ASSESSMENT TASK 12.2

1. List the advantages and disadvantages of the CZ and FZ processes as means of producing large ingots of high purity, monocrystalline, dislocation-free silicon.
2. Explain how the Dash technique helps in the production of dislocation-free silicon. (You should refer to Chapter 8 for further understanding of dislocations.)

12.5 Production of wafers

Modern equipment for silicon crystal production is highly automated, including automatic sizing. However, although variations in ingot size and roundness are very small, the ingot is still not dimensionally accurate enough for modern device-processing equipment. Therefore the monocrystalline ingot is centreless ground to an accuracy of better than ±0.2 mm. Flats are then surface ground along the ingot for identification and location purposes as shown in Fig. 12.5. The identification of crystal orientation by *Miller indices*, for example (100), is described in Chapter 8.

Fig. 12.5 *Wafer characterisation by different flats (reproduced courtesy of Wacker Chemitronic GmbH)*

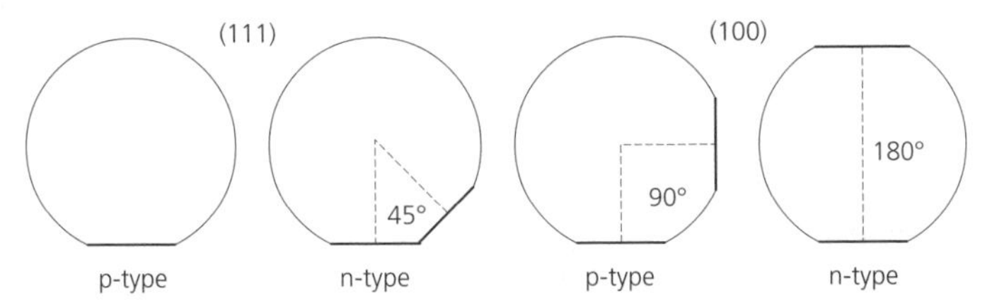

Unfortunately, all grinding processes result in some structural damage to the crystal surface and if this is not rectified slipping of the crystal lattice and stacking faults will develop during subsequent high-temperature treatment. The usual treatment is to chemically etch the ground surfaces to remove the damage layer when all the mechanical cutting processes have been completed.

Next, the ingot is cut into thin wafers ready for device processing. Depending upon the application for which the wafers are required, the ingot is cut perpendicular to the crystal axis or with a well-defined misorientation of several degrees. Off-axis wafers are usually used, after polishing, as the substrates for epitaxial processes used in the manufacture of discrete devices or bipolar-integrated circuits. Cutting techniques are now becoming sufficiently sophisticated that, after cleaning and etching, the wafers can be used without further processing.

However, for the majority of applications, the cut wafers are cleaned and lapped to meet the thickness and flatness tolerances required by some discrete devices and by most modern mask-printing techniques. The majority of wafer diameters used are in the 100–150 mm range with a typical thickness of 0.3–0.7 mm. Since the electrical performance of semiconductor materials is affected by changes in temperature, chips produced from silicon wafers have to be kept as thin as possible in order to allow adequate heat dissipation. For this reason, wafers for low-voltage, heavy-current diodes may only be 0.15 mm thick. The production processes for silicon wafers are summarised in Fig. 12.6. Note that unlike most circular saws, those used for cutting silicon are in the form of an annulus with the teeth on the inside diameter.

Fig. 12.6 *Crystal machining and polishing: flowchart (reproduced courtesy of Wacker Chemitronic GmbH)*

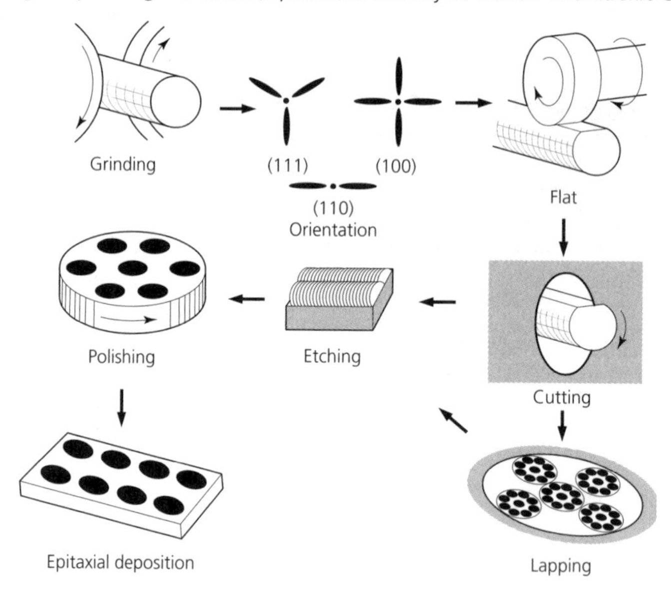

Wafer yield during device processing has been increased considerably by 'edge rounding'. This offers the following advantages.

- Rounded edges prevent the formation of an 'epi-crown' during epitaxial deposition processes. An 'epi-crown' is caused by the build up of epitaxial layers on the sharp edges of untreated wafers at a greater rate than the build up of the epitaxial film on the flat surface of the wafer.

- Similarly, the photoresist film is smoother on an edge-rounded wafer and there is no increase in thickness adjacent to the wafer edge.
- Edge rounding also improves the mechanical properties of the wafers with less tendency to chipping and cracking at the edges during handling. This results in less chance of process-induced crystallographic faults developing.

After edge rounding the wafers are etched in order to remove any mechanical damage. Care has to be taken during etching not to disturb the plane-parallel surfaces of the wafer otherwise it would be difficult to obtain the required flatness during subsequent polishing of the wafer surfaces. Modern photolithographic processing techniques require very high surface quality. To avoid damage of the wafer surface, abrasive polishing is impractical and a combined chemical and mechanical technique is used. After polishing a carefully controlled, multi-step, cleaning process is used so that the wafers are completely free from particles and residues.

12.6 Diode fabrication process

Junction diodes, as described in Section 11.9, are the simplest of the many solid-state devices manufactured using silicon wafers. For power diodes a high surface finish is not usually required, and wafers that have been sawn and etched are suitable. The wafer will be supplied already doped to give the p-type or n-type electrical characteristics required by the device manufacturer. The junction is formed by the diffusion of the appropriate complementary dopant into only one surface of the wafer. That is, diffusing an n-type dopant into a p-type wafer or vice versa. The dopant may be brushed on, sprayed on, spun on or applied in the form of impregnated paper discs. The wafers are stacked back to back and heated in a controlled atmosphere furnace for 30–40 hours until the applied dopant has penetrated to a depth of some 50 μm.

The wafer is then treated chemically to remove any residue and to ensure that it can be metallised ready for electroplating with gold or nickel to provide contact surfaces. Finally, the wafer is scribed and broken into 'chips' or 'dice'. The size will depend upon the required current-handling capacity. For the diodes used to rectify the output from motor vehicle alternators the chip is, typically, 4.75 mm (3/16 in.) square. The individual chips are then encapsulated to provide appropriate protection and heat dissipation, and provided with wire 'tails', solder tags, or terminals for connection to the external circuit.

12.7 Planar fabrication

Devices with a *planar* configuration are manufactured so that the preparation of the various p-type and n-type layers and the metallised contacts are on a flat surface of the chip and not on its sides or ends. A section through a typical bipolar junction transistor is shown in Fig. 12.7. The substrate is made from a relatively thick wafer of n-type silicon that has been doped to give it a low resistivity. The layers that are built upon this substrate are referred to as *epitaxial* layers. These are layers that grow onto the substrate surface as a continuation of the underlying crystal. The growth comes from the gaseous mixtures in which the wafer is heated to give p-type or n-type conducting layers or silicon dioxide insulating layers.

The fabrication of planar devices requires wafers with one surface lapped and polished to a high degree of flatness and surface finish. Figure 12.7 shows only a single transistor. However, since the chip on which it is fabricated is only some 4 or 5 mm square, very many such devices can be made at the same time on wafers whose diameters lie between 100 and 150 mm. High-precision photographic processes are used to produce the masks used during the fabrication of the device. Only one such device is drawn out and the camera takes a succession of photographs of the device on the same negative, moving by an increment equal to one chip spacing between each exposure. Thus the negative is covered in a pattern of chip masks suitable for printing photographically onto the wafer, which is coated with a photosensitive emulsion. The resolution required to reproduce the fine detail and intricacy of modern solid-state devices precludes the use of white light, and ultraviolet light and laser light sources are used for making the exposures. For simplicity, the fabrication of only one device on one chip will be considered.

Fig. 12.7 *Bipolar transistor manufactured using planar technology (reproduced courtesy of Wacker Chemitronic GmbH)*

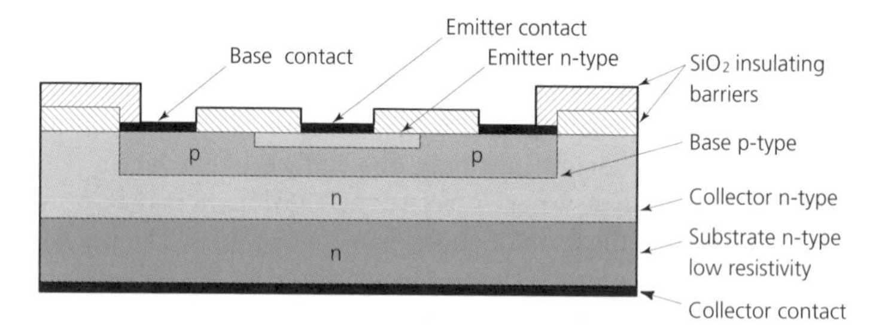

The overlying (epitaxial) layer of n-type silicon is grown on the substrate and it has a higher, but controlled, resistivity. Following a photographic masking procedure, boron is diffused from the surrounding gas into the epitaxial silicon. This changes the n-type material into p-type material in the zone unprotected by the photo-resist mask and forms the base of the transistor. A second masking and diffusion process using phosphorus establishes the n-type emitter layer. Final masking and processing allows the formation of silicon dioxide (SiO_2) insulating barriers. Metallisation and electroplating produces the contact surfaces on those surfaces not protected by the silicon dioxide.

The thickness of the diffused base layer is limited to only 0.5 mm in order to prevent any recombination. The thickness of the layers and the dopant concentrations are controlled by the diffusion time, the furnace temperature and the furnace atmosphere gas composition. However, although the diffusion technique has been developed into a highly reproducible production process, it is difficult to achieve a well-defined wall (abrupt profile) by conventional techniques. The need for an abrupt profile becomes increasingly important with the development of large-scale integration (LSI) where very large numbers of components are built up on a single chip with component separation measured on a molecular scale. Such an abrupt profile can be more easily attained using *ion implantation* than with diffusion techniques. When ion implantation is used, the dopants are bombarded into the semiconductor surface using a high-energy ion beam. The semiconductor material does not have to be heated and this has the added advantage of reducing the possibility of contamination and dislocation. The number and depth of the implanted ions is controlled by the process time and by the electrical potential of the ion beam which is usually about 10,000 eV. Figure 12.8 shows a 100 mm diameter wafer with 640 planar high-power diodes fabricated on it. Before the wafer is cut up into individual chips, the individual devices are checked with automatic testing equipment. Those devices not meeting the required specification are identified with a spot of black ink for subsequent rejection. This particular wafer has an abnormally high reject rate.

Fig. 12.8 *Silicon wafer with planar fabricated diodes (reproduced courtesy of the University of Luton)*

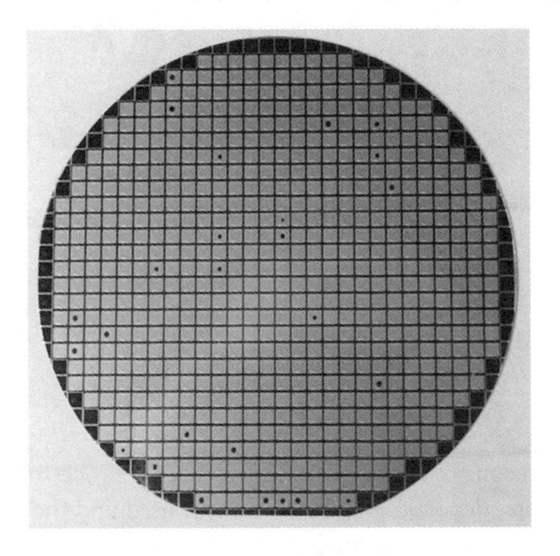

12.8 Metal-oxide-silicon (MOS) technology

The development of metal-oxide-silicon (MOS) devices established the superiority of monocrystalline silicon as a semiconductor material. One of the main reasons for this was the ease with which an ideally isolating silicon dioxide film can be grown on silicon wafer by simple heat treatment in an oxygen-rich atmosphere. Silicon dioxide layers serve many useful functions in device manufacture. For example, such oxide layers are not easily penetrated by dopant elements and thus form useful diffusion masks; further, such films form the gate oxide in MOS devices and this determines the electrical characteristics of MOS transistors. Silicon-dioxide films are also used during the processing and finishing of solid-state devices to protect p–n junctions (passivation) and, in multi-level integrated circuits, such films serve as electrical insulators for the various metal or polysilicon layers.

The field effect transistor (FET) concept only became a reality with the perfection of MOS technology. MOS devices are essentially planar surface devices and the drain and source regions, even in large MOS-integrated circuits, have a thickness equal to or less than 1 mm. Therefore, several insulation and interconnection layers can be situated directly on the wafer surface. Figure 12.9 shows, schematically, a cross-section through a typical MOS-RAM device. The main high-temperature processes that govern the performance of such a device can be outlined as follows:

- Gate oxidation.
- Field oxidation.
- Drain and source implantation.
- Annealing of implantation damage and drive in diffusion.
- Chemical vapour deposition of polysilicon, nitrides and (phosphorus) oxides.

Fig. 12.9 *Cross-section of storage cell of a typical MOS-RAM (reproduced courtesy of Wacker Chemitronic GmbH)*

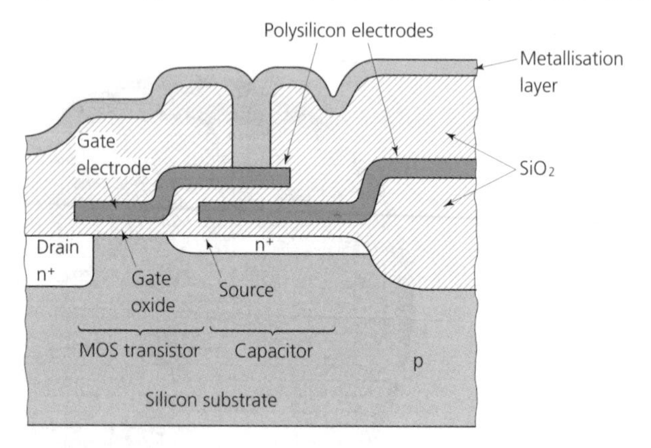

The chemical vapour deposition (CVD) processes may take place at relatively low temperature, depending upon the system used and the nature of the layer to be deposited. Some annealing of ion-implantation damage may be coupled with subsequent processes. To obtain a high accuracy of alignment, modern devices are processed by the so-called self-alignment technique, which requires that the field and gate oxidation have to take place during the first part of the process. To keep the gate oxide, which is the most sensitive oxide, free from impurities, it is covered by a nitride layer and the field oxide is formed at temperatures below 1100 °C. Since the drain and source are rather shallow, they are realised by brief and/or low-temperature diffusion at relatively high push–pull rates, or by ion implantation to obtain well-defined diffusion walls.

Another way of realising a densely packed device, e.g. a memory, starts with the deposition of a phosphorus-doped oxide which, after a masking step, serves as a diffusion source. Such an approach does not allow a subsequent long heat treatment at relatively high temperatures as needed, for example, for the field-oxide formation, since the diffused regions would be washed out. Here other ways have to be found to insulate the different diffused islands. One way is to pre-bias a polysilicon layer electrically.

The p-channel MOS technology, which was first used to manufacture MOS devices, has largely lost its importance. However, medium-scale consumer devices are still produced by this technology. The gate oxides are thick and aluminium layers are used for the interconnections. Since the dimensions need not be controlled with a high precision, the drain and source diffusion can be deeper and the process temperatures higher, in order to obtain a higher throughput.

12.9 Complementary metal-oxide-silicon (CMOS) technology

Complementary metal-oxide-silicon (CMOS) technology combines p-channel and n-channel transistors on one chip. CMOS devices are more economical in power consumption and are superior in reliability to normal MOS devices and, therefore, have gained increasing attention. However, integrating two field-effect transistors on one chip needs more process steps. A distinct feature is the so-called p-well into which the n-channel FET is built. In order to obtain the correct surface concentration of dopants, a long 'drive-in' has to be performed at relatively high temperatures.

12.10 Bipolar technology

Standard bipolar technologies include a process step not used in MOS technologies: the epitaxial deposition of a monocrystalline layer of silicon on the polished surface of the wafer. The electrically active parts of the devices are confined to this epi-layer; thus stringent demands are made on its crystalline perfection.

Figure 12.10 shows a section through a basic element of a standard bipolar integrated circuit. The main process steps to realise such a device structure are:

- Buried layer diffusion.
- Deposition of epitaxial layer.
- Isolation diffusion.
- Collector diffusion.
- Base diffusion.
- Emitter diffusion.

A thermal oxidation plus a photolithography step is needed before every diffusion in order to establish the diffusion masks.

The isolation diffusion has now been replaced largely by an oxidation step, and the silicon dioxide produced is used to isolate the single transistors dielectrically as shown in Fig. 12.11. Since silicon dioxide has a higher specific volume than silicon, part of the silicon in the isolation well has to be removed before oxidation. Oxide isolation techniques are also widely used in MOS and CMOS devices. Figure 12.12 shows part of an integrated circuit manufactured using bipolar technology. The diagonal dark line is a human hair (diameter approximately 100 microns) for comparison. The other 'thick' lines are the very fine wires used to connect from the silicon surface to the external pins on the device. The actual size of the silicon is approximately 2 mm by 1.5 mm.

Fig. 12.10 *Bipolar device with diffused isolation wells (reproduced courtesy of Wacker Chemitronic GmbH)*

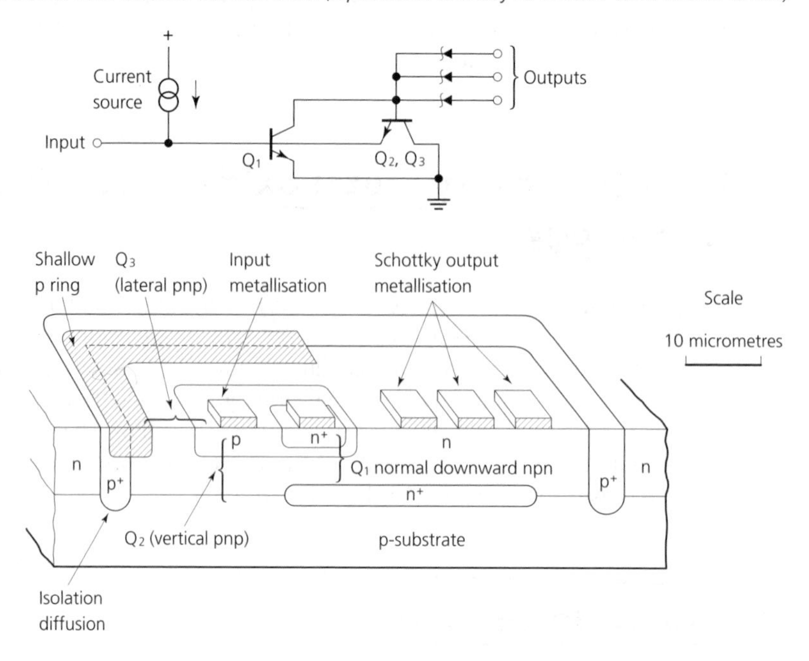

Fig. 12.11 *Bipolar structure with oxide isolation (reproduced courtesy of Wacker Chemitronic GmbH)*

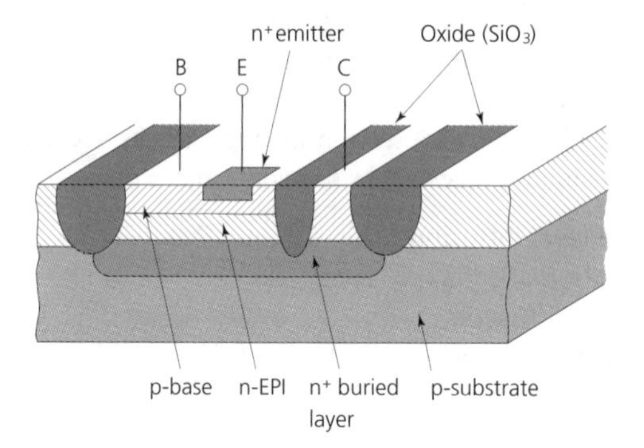

With particular respect to the performance of CZ-silicon wafers, the main differences between the device technologies can be described as follows:

- MOS devices are predominantly surface devices with shallow junctions; they are extremely sensitive to surface inhomogeneities and impurities.
- CMOS devices, although basically still surface devices, extend much more into the substrate because of the relatively deep p-well. The formation of the p-well adds a long-lasting, high-temperature process step to normal MOS technology.
- Bipolar devices have a more pronounced three-dimensional character. The electrically active parts are built into the epilayer. Compared to MOS technology, bipolar technology uses deeper junctions and high-temperature processes have to be

used. This, in turn, requires greater perfection and integrity in the crystalline structure of the wafer and the epilayer if process-induced discontinuities and other defects are to be avoided.

Fig. 12.12 *Part of a bipolar integrated circuit with a human hair for comparison (reproduced courtesy of the University of Luton)*

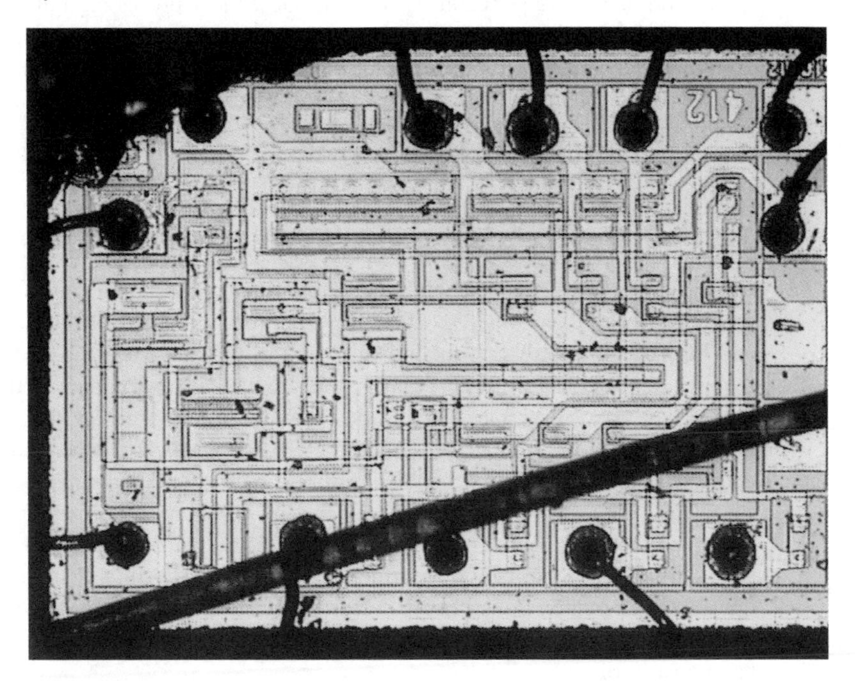

12.11 Postscript

Although the basic processes for the production of silicon wafers has not fundamentally changed since the first edition of this book, there have been a number of developments as a result of the changing requirements of the devices made from the wafers. In general these relate to the size and number of individual devices that can be produced from one wafer and the speed of operation of those devices.

12.11.1 *Wafer size*

There has been a major push to increase the diameter of the wafers so that more devices (or the same number of larger, more complicated devices) can be made from one slice thus reducing the processing costs. Although this requires more careful control of the crystal growing and refining processes, it has not required any fundamental change in the processes themselves. Most manufacturers now produce 200 mm diameter wafers as standard and there has been a major effort to increase this to 300 mm diameter. It can be seen from Table 12.2 that, as wafer diameter increases, so does the thickness (and, incidentally, other parameters relating to the 'flatness'). This is to ensure that the wafer is sufficiently strong to avoid being damaged in the subsequent processing. Of course this results in a less efficient use of silicon, ultimately tending to increase the cost of a finished device.

In addition, the equipment used in the subsequent processing must be redesigned to handle the larger slices. Having recently purchased new equipment to upgrade from 150 to 200 mm wafers, manufacturers are somewhat reluctant to invest further at this stage. At the time of writing (early 1999) it is by no means clear that 300 mm wafers will become standard.

12.11.2 *Device size, complexity and speed of operation*

Complex integrated circuits contain many thousands or even millions of transistors of the types described in Sections 12.8–12.10. In order for such circuits to operate at high speeds and be of small overall size, the individual transistors must become smaller. In turn, this requires that the 'feature size' – that is, the size of the smallest part of the device – must reduce. The classic example of this is the computer processor integrated circuit (chip). The first processor chip, produced by Intel in 1971 (the 4004 chip), operated at 108 kHz and contained 2300 transistors fabricated with 10 micron feature size (note the size bar on Fig. 12.9). The first chip in common use in '286' machines (the Intel 80286), appeared in 1981, operated at 6 MHz and contained 134,000 transistors based on 1.5 micron feature size. Currently processors are operating at up to 450 MHz (Intel Pentium II) and contain upwards of 25 million transistors with feature size now aiming at 0.13 micron. The increase in complexity and the quality of control of the processes required to produce these devices has increased dramatically.

SELF-ASSESSMENT TASK 12.3

1. Describe the technology used to produce:
   (a) computer memory chips
   (b) computer processor chips
   (c) single transistors and diodes
2. As feature size becomes smaller, describe the improvements that must be made in the processes involved in making semiconductor devices.

The production of monocrystalline silicon and the production of solid-state devices from monocrystalline silicon wafers is extremely complex and the basic principles discussed in this Chapter can only be but a very brief introduction to this subject. The author is indebted to Wacker-Chemitronic GmbH, Postfach 1140, D-8263 Burghausen, Germany, for most of the data upon which this Chapter is based.

EXERCISES

12.1 Discuss the relative merits of germanium and silicon as semiconductor materials.

12.2 Describe how high-purity polycrystalline silicon is produced.

12.3 Describe how polycrystalline silicon is converted into monocrystalline silicon by the Czochralski (CZ) process.

12.4 Describe how polycrystalline silicon is converted into monocrystalline silicon by the float-zone (FZ) process.

12.5 Explain why monocrystalline silicon, free from impurities and of high structural integrity, is required for the manufacture of solid-state electronic devices.

12.6 Explain how the Dash technique is used to prevent dislocations occurring during the production of monocrystalline silicon.

12.7 Explain the stages during the manufacture of wafers dopants at which may be added, and how they may be added. Name the dopants in general use and describe the electrical characteristics they impose upon the silicon.

12.8 With the aid of diagrams, outline the production of 'wafers' from monocrystalline silicon and indicate the dimensional and electrical characteristics that may be expected from such wafers when the monocrystalline material has been produced by the CZ process.

12.9 With the aid of diagrams, explain:

(a) how junction diodes are produced
(b) what is meant by an epitaxial layer
(c) how a bipolar junction transistor is produced by planar fabrication processes.

12.10 With the aid of diagrams, explain the basic principles of the manufacture of field effect transistors (FET) and simple integrated circuits, using metal-oxide-semiconductor (MOS) technology.

Index